32

Springer Series in Chemical Physics

Edited by Robert Gomer

Springer Series in Chemical Physics

Editors: V. I. Goldanskii R. Gomer F. P. Schäfer J. P. Toennies

Volume 40 **High Resolution Spectroscopy of Transient Molecules**
Editor: E. Hirota

Volume 41 **High Resolution Spectral Atlas of Nitrogen Dioxide 559–597 nm**
By K. Uehara and H. Sasada

Volumes 1–39 are listed on the back inside cover

V. F. Kiselev O. V. Krylov

Adsorption Processes on Semiconductor and Dielectric Surfaces I

With 70 Figures

Springer-Verlag
Berlin Heidelberg New York Tokyo

Professor Dr. Vsevolod F. Kiselev

Department of Physics, Moscow State University
Moscow SU-119899, USSR

Professor Dr. Oleg V. Krylov

Institute of Chemical Physics, Academy of Sciences of the USSR, Kosygin Street 4,
Moscow SU-117334, USSR

Adapted from the original Russian book entitled:
Adsorbtsionniye protesessi na poverkhnosti poluprovodnikov i dielektrikov
© Nauka, Moskau 1978

ISBN-13: 978-3-642-82053-3 e-ISBN-13: 978-3-642-82051-9
DOI: 10.1007/978-3-642-82051-9

Library of Congress Cataloging in Publication Data. Kiselev, V. F. (Vsevolod Fedorovich). Adsorption processes on semiconductor and dielectric surfaces I. (Springer series in chemical physics ; v. 32). Translation of: Adsorbtsionnye protsessy na poverkhnosti poluprovodnikov i dielektrikov. Includes index. 1. Semiconductors–Surfaces. 2. Dielectrics–Surfaces. 3. Adsorption. I. Krylov, O. V. (Oleg Valentinovich) II. Title. III. Series. QC611.6.S9K5713 1984 537.6'22 85-2709

Softcover reprint of the hardcover 1st edition 1985

Bookbinding: J. Schäffer OHG, 6718 Grünstadt
2153/3130-543210

Preface

This book deals with various physical and chemical phenomena associated with the interaction of a solid surface in a gaseous environment. The authors have gone through a vast body of experimental material on the structure and properties of dielectric and semiconductor surfaces from the point of view of adsorption and catalysis. They have attempted to look into mechanisms of these processes and to outline the ways of controlling them, as long as this seemed possible. A great deal of attention is paid to considering the nature of active surface sites responsible for chemisorption, catalytic conversion of adsorbed molecules, and certain electronic surface phenomena. All the problems concern physicists working in the fields of microelectronics, optoelectronics, thin-film electronics, as well as chemists doing research in adsorption, catalysis, and combustion. The wide scope of surface phenomena included in this study is dealt with from a firmly established standpoint of solid-state physics and the theory of chemical structure and reactivity.

The roots of this monograph go back to our earlier book published with Nauka, Moscow, in 1978. The present edition has, however, been revised substantially and is extended to cover more grounds and, in particular, recent results. We prepared the manuscript in our native language and Mr. A.S. Dobroslavski was extremely helpful in the translation.

For fruitful discussions the authors are grateful to G.F. Golovanova, Yu.A. Zarifyants, S.N. Kozlov, Z.L. Krylova, O.V. Nikitina, L.Ya. Margolis, P.K. Kashkarov, Yu.N. Rufov, K.N. Spiridonov, B.R. Shub. We appreciate the assistance of Ms. N. Culviner in copy-editing the book manuscript and of Mrs. I. Blass in typing the master manuscript. Our thanks are also due to Dr. H. Lotsch of Springer-Verlag for his generous help in all stages of preparing the English version of the book.

Moscow, January 1985 *V.F. Kiselev · O.V. Krylov*

Contents

1. Introduction

The nature and mechanism of molecular and atomic processes on solid-state surfaces are of great interest to researchers in diverse branches of modern science and technology. These processes play an important role in interphase phenomena such as adsorption, heterogeneous catalysis, combustion of solid fuels, friction, and crystal growth. Adsorption has a strong influence on surface electronic processes, which determine the way in which thin-film semiconductors, integrated circuits, and other microelectronic devices work. The development of all of these branches of surface chemistry and physics depends primarily on the success of research on solid-state surfaces and the mechanisms of absorption and catalysis at the molecular level.

Current developments in chemisorption and catalysis are marked by a departure from attempts to construct global (mainly phenomenological) and speculative theories, and by the beginning of extensive investigations on the real solid-state surface, the nature of chemisorptive and intermediate complexes, and the basic stages of chemisorption and catalysis. In other words, the deductive approach is being replaced by an inductive one. This does not, of course, mean that theory is lagging behind experiment—on the contrary: the lack of detailed knowledge about the mechanism of catalysis has had a positive influence on attempts to develop a universal theory of this phenomenon. The theories developed in the twenties, forties, and even fifties of this century were by all means progressive for their time, since they singled out, from the general turmoil of phenomena and interconnections, those which were considered most important. Then, however, scientists were confronted with the problem of explaining the elementary stages of catalysis and the structure of surface complexes; not even the rate of catalysis and absorption could be measured exactly.

By the fifties it was possible to distinguish between purely chemical and diffusional stages in catalysis and absorption, and to measure the surfaces of powders and fine-grained crystals; convenient and exact methods were available for obtaining the rates of stationary-state adsorption and catalysis.

And then the "correlation boom" began. Every conceivable property of solid bodies and constituent atoms was correlated with an exactly measured rate of catalysis. This resulted in certain ideas which were helpful in the search for catalysts, and also threw light upon some aspects of catalyst action [1.1].

The seventies gave rise to a new stage of development characterized by rapid progress in physical methods for investigating solid-state surfaces and the structure and energy levels of surface complexes. While in the fifties the principal techniques were trivial conductivity and work function measurements, today research workers rely heavily on spectroscopic methods. All of the leading laboratories now employ IR and UV spectroscopy, EPR, and NMR, which were previously the privilege of outstanding research centers. New techniques for obtaining information about surface structures, adsorption complexes, and their energetics were developed, such as low-energy electron diffraction, X-ray and UV photoelectronic spectroscopy, Auger spectroscopy, and secondary ion mass spectrometry (SIMS). Respect for the usefulness of correlations diminished, and was replaced by a growing interest in the elementary process of chemisorption and catalysis.

New investigative techniques have recently provided valuable information on surface electronics, chemisorption, and catalysis. These data have made it possible to revise many established concepts and to reconsider and reinterpret classical experiments. All this has laid a new foundation for the further development of the theory of chemisorption and catalysis [1.2].

It should be noted in this connection that developments in the theory of chemisorption and catalysis are quite similar to those in the general theory of chemical compounds and their reactivity. Researchers in the field of catalysis cannot be blamed for not yet being able to predict the catalytic activity of a substance, just as their colleagues cannot be reproached for their inability to calculate in advance the reactivity of a complex molecule. Naturally, the theory of catalysis and chemisorption must be based on the latest achievements of quantum chemistry and solid-state science.

The present monograph gives a general presentation of a vast body of information obtained by chemists and physicists about the structure and properties of semiconductor and dielectric surfaces, and the mechanisms of adsorption and catalysis. The authors sought to develop a unified approach in their treatment of the possible nature of adsorptive and catalytic centers on a surface, the factors that determine their activity, and the role played by these centers in various processes of adsorption. An attempt at such an approach to the real surface of semiconductors was made earlier [1.3]. The rapid development of experimental techniques in recent years and new methods of investigating surface phases have yielded additional information on the nature of many surface complexes and thrown new light on the elementary processes of chemisorption and catalysis. The present book also deals with the role of excited molecular states in adsorption and catalysis, a concept developed quite recently.

The authors have not attempted to give an exhaustive review of all aspects of adsorption and catalysis. The book focuses on the nature of local interaction between the simplest atoms and molecules and active surface centers, and on the disturbances caused by these interactions in both the adsorbed molecules and the solid.

The collective effects connected with electrical charging of solid-state surfaces in the process of adsorption, and with changes in the energetic spectrum of surface electrons, have intentionally been omitted from this treatment. Electronic phenomena surfaces and their role in adsorption and catalysis constitute an important branch of surface science which will be dealt with in [1.4]. The present monograph also does not include a discussion of the chemisorptive and catalytic properties of transition metals, which together with their oxides play an important role in many catalytic and biochemical processes. This topic will be covered in [1.5].

In order to provide a complete and thorough analysis of the phenomena discussed, the authors have restricted themselves to the most typical semiconductors and dielectrics. The specific chemical mechanisms are touched on only briefly, and the formal kinetics of catalytic processes are not discussed. Quantum-chemical calculations of chemisorptive interactions are treated only superficially, because other monographs and reviews are devoted to these questions.

The book starts out with a review of the current state of the theory of adsorption (Chap.2). We analyze the main assumptions, which form the basis of calculations for the energy of adsorptive interactions and the thermodynamic description of the adsorptive phase. We also discuss the problem of adsorbent inertness in the models of physical and chemical adsorption, and supply pertinent information on the main features of catalysis on semiconductor and insulator surfaces.

In Chapter 3 we draw generalizations from experimental data on the structure and the properties of surfaces of most typical semiconductors and dielectrics. Using atomically clean surfaces of monatomic semiconductors (germanium and silicon) as examples, we discuss their surface structure (superlattice), hybridization and the dynamics of surface atoms. This is followed by a comprehensive account of real surfaces on these semiconductors, and of the composition and structure of the superficial oxide layers. We include many relevant details on the structure of hydrate cover of real surfaces of semiconductors and their oxides, on the nature and structure of electron-acceptor and electron-donor sites, responsible for many catalytic and electronic qualities of the surface.

Chapter 4 deals with the elementary mechanisms of adsorption of comparatively simple molecules on real surfaces of semiconductors and their oxides. Much attention is paid to the coordinative adsorptive bondings and the reactivity of coordinated molecules. We also discuss the possible nature of proton-donor surface sites, and ways of controlling their activity.

Finally, Chapter 5 deals with the migration of energy released in the process of adsorption and catalysis. We consider the probability of appearance of vibration-excited and electron-excited complexes in adsorption, and their experimental manifestations. We also analyze whatever role may be played by excited states in the reconstruction of the surface in the course of a catalytic reaction.

2. Principles of Adsorption and Heterogeneous Catalysis

Adsorption is the attraction of molecules, atoms, or ions of a substance to a solid surface. It is traditionally referred to as either physical or chemical. In physical adsorption the adsorbed molecules retain their individuality, and the forces involved are similar to Van der Waals forces in nonideal gases. In chemical adsorption (or chemisorption) the adsorbed molecule forms a chemical connection with the solid surface through exchange interaction forces, which to some extent include ionic interaction [2.1].

2.1 Physical and Chemical Adsorption

The theoretical approaches to physical and chemical adsorption are fundamentally different. Theories of physical adsorption view the adsorbate and the adsorbent as two independent systems, and treat the influence of the adsorbed molecule on the solid adsorbent lattice as a small perturbation. In chemical adsorption the adsorbate and the adsorbent are looked upon as a single quantum-mechanical system. Two different potential curves correspond to the two types of interaction. According to [2.2], Curve 1 in Fig.2.1, with a minimum at point A, describes Van der Waals physical adsorption, while Curve 2, with a deeper minimum at B, describes chemical adsorption. The radius of equilibrium, which corresponds to the minimal potential energy, is always

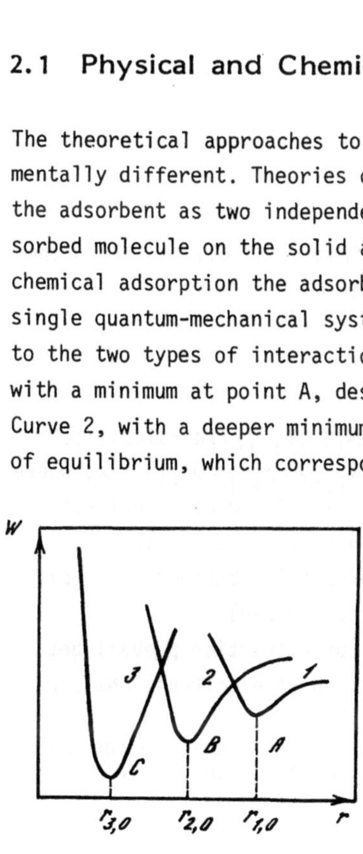

Fig.2.1. Potential curves of adsorptive interaction

greater in physical adsorption ($r_{1,0}$) than in chemisorption ($r_{2,0}$). The intersection of Curves 1 and 2 defines a potential barrier whose height is often interpreted as the activation energy. Sometimes the state of chemisorbed particles is characterized by two potential curves [2.3], for example with Curve 2 in Fig.2.1 corresponding to nondissociative, and Curve 3 to dissociative chemisorption [2.2-4]. The height of the activation barrier between these two states depends on the degree of occupation of the surface, and on the temperature [2.4].

2.1.1 The Energy of Physical Adsorption

Calculations of the energy of physical adsorption are based on formal application of the theory of Van der Waals forces. The energy of adsorptive interaction, W_{ads}, is as a rule presented as a sum of paired interactions $w_{i,j}$ of the ith gas molecule with the jth lattice atom, i.e.,

$$W_{ads} = \sum_i \sum_j w_{i,j}(r_{i,j}) \quad . \tag{2.1}$$

The potential $w_{i,j}$ is taken either in the form of the Lennard-Jones potential

$$w_{i,j} = -\frac{C}{r_{i,j}^m} + \frac{A}{r_{i,j}^n} \quad , \tag{2.2}$$

where m is usually set equal to 6,[1] and n to 12 ("potential 6,12"), or in the form of the Buckingham potential ("potential 6, exp"):

$$w_{i,j} = -\frac{C}{r_{i,j}^m} + A\exp(-\alpha r_{i,j}) \quad . \tag{2.3}$$

The repulsive potential constants $A\,r_{i,j}^{-n}$ or $A\exp(-\alpha r_{i,j})$ are calculated either from the minimum of paired interactions $(dw_{i,j}/dr_{i,j}=0)_{r_{i,j}=0}$, where the equilibrium radius is determined as the sum of Van der Waals (or ionic) radii of interacting particles, or after summation from the condition $(dw_{ads}/dr=0)_{r=r_0}$. The attractive potential constant is obtained from well-known formulae of the theory of Van der Waals forces [2.6-8].

For nonpolar molecules, when dispersive forces of attraction prevail between the fluctuative dipole resulting from the motion of electrons, one can

---

[1] For large molecules, when the photon wavelength corresponding to the activation energy of a molecule becomes smaller than the characteristic dimensions of the molecule (ℓ and $r_{i,j}$), m = 7 [2.5].

calculate the dipole-dipole interaction constant C_1 either via the London formula:

$$C_1 = (3/2) \frac{I_A I_B}{I_A + I_B} \alpha_A \alpha_B \quad , \tag{2.4}$$

where I_A and I_B are the ionization potentials, and α_A, α_B are the polarizabilities of adsorbed atom or molecule A and lattice atom B; or via the Kirkwood-Mueller formula:

$$C_1 = 6\, mc^2 \frac{\alpha_A\, \alpha_B}{\alpha_A / \lambda_A + \alpha_B / \lambda_B} \quad , \tag{2.5}$$

where λ_A and λ_B are the diamagnetic susceptibilities. In general the energy of dispersive interaction W_{disp} is the expansion over multipole-multipole interactions:

$$W_{disp} = -\frac{C_1}{r^6} - \frac{C_2}{r^8} - \frac{C_3}{r^{10}} - \dots \quad , \tag{2.6}$$

where C_2 is the constant of dipole-quadrupole interaction, C_3 is the constant of quadrupole-quadrupole interaction, etc. In adsorption the contribution of C_2 can be estimated to be 10-20% and that of C_3 to be 1-3% of the contribution of dipole-dipole interaction [2.9]. In adsorption of a nonpolar molecule on the surface of the ionic lattice, one must also take into account the polarization component of interaction, $\alpha_A \varepsilon / 2$, ε being the strength of the mean field over the lattice (of the order of $10^5 \mathrm{cgs} \cdot \mathrm{cm}^{-2}$ [2.10]).

The interaction of polar molecules with the ionic lattice is calculated on the basis of well-known formulae of electrostatics. The energy of dipole-ionic interaction is $W_{dip} = \mu_A \varepsilon \cos\theta$, θ being the angle between the direction of field and the dipole axis. The theory of electrostatic adsorption was pioneered by ILYIN [2.6].

It must be noted that all calculations of the constants of dispersive interaction, C_1, C_2, etc., are based on the assumption that the interacting particles are far enough away from one another ($r_{i,j} \gg \ell$) so that their wave functions do not overlap either for the ground state or for excited states of the molecules. The same restrictions naturally hold for the electrostatic component of intermolecular interaction. In calculating the interaction potential of large molecules at distances of this order, we assume additivity of the dispersive and electrostatic forces that act between the constituent atoms and the surface.

There is a large body of literature on the calculation of the energy of adsorptive interactions [2.3a,6,11-14]. The applicability of the theory of

Van der Waals interactions (which was developed primarily for nonideal gases) to the condensed adsorptive phase is not questioned by the overwhelming majority of researchers; in most cases astonishingly good agreement (within 10%) has been found between theoretical values and measured adsorption heat, both in ionic and covalent crystals and in graphite bodies. This was taken by the authors of [2.13,15] as proof of the correctness of all of the assumptions made in the theory and encouraged them to undertake the task of consistently calculating the physical and chemical characteristics of a large class of adsorption systems on the basis of known properties of adsorbate molecules and solids.

This optimism, however, is not justified by a closer analysis of the modern state of the theory of Van der Waals forces [2.8,16-21]. Indeed, since the general expression for adsorption potential includes a number of ad hoc constants, agreement between theory and experiment is not a reliable criterion for judging the theory's applicability. GUGGENHEIM [2.22] is quite right in noting that five constants are needed to describe the potential curve near the minimum A (Fig.2.1), whereas the Lennard-Jones potential requires three. The vicinity of the minimum of potential ($r_{1,0}$ in Fig.2.1) is terra incognita, in which the values for the repulsive (w_r) and attractive (w_a) branches cannot be precisely calculated. The minimum is obtained via extrapolation of the two branches to $r_{1,0}$. However, the right-hand branch (w_a) is calculated under the assumption that the wave functions of interacting atoms do not overlap, while the left-hand branch results essentially from overlapping of the wave functions. One is obliged to agree with VOLKENSTEIN [2.23] that there is little point in looking for the minimum of a curve constructed under incompatible assumptions.

At small distances $r_{i,j}$, such as are encountered in adsorption, one cannot neglect effects caused by the distortion of orbitals, as noted by HELLMAN as early as 1937 [2.24]. A straightforward proof of this was obtained in calculations of the electron terms of interacting hydrogen atoms [2.25-27]. It turned out that at distances of 5 atomic units the energy of Van der Waals interaction is comparable to the exchange energy. At $r < 5$ the exchange forces prevail, and only at $r > 10$-20 does the potential curve lie close to the Van der Waals one. The same conclusion was reached in calculations of He-He interaction [2.28]. These effects are most prominent in large atoms and molecules [2.16], especially molecules with conjugated bonds [2.17].

Attempts have been made to investigate the contribution made to the intermolecular potential by second-order exchange effects resulting from a slight overlapping of orthogonal wave functions of interacting particles [2.29]. In [2.30] this contribution was calculated in a third-order approximation of perturbation theory, resulting in a potential of the type $(r^{-6}, r^{-9}, r^{-12})$. However, quantitative estimates are hardly possible yet, as the interaction on a surface is quite complicated.

Equations (2.4,5), which are valid for small spherical molecules, do not hold for dispersive interaction of large molecules. Furthermore, free-atom values, which are known to differ from those for the solid state, are employed for α and λ, and the anisotropy of α and λ for adsorbed organic molecules and for the adsorbent lattice is not taken into account. At small distances, when $r_{i,j} \sim \ell$ (ℓ being the size of the molecule), the parameters in the expressions for dispersive, electrostatic, and repulsive forces, owing to distortion of the orbitals of interacting particles, become themselves functions of distances [2.16,17,31]. At distances of this order the dispersive forces lose their additivity [2.16,17]. The repulsive forces are always nonadditive. For the sake of simplicity, KICHARA [2.32] approximated the adsorbed molecule by a rigid structure with dimensions derived from molecular constants of the substance. All the assumptions regarding α and λ remain the same for this model as well.

Even more assumptions are made in calculations of adsorption potential on a metallic surface, where the electron gas of the metal takes part in the interaction. The pioneering semiclassical calculations by LENNARD-JONES [2.2], MARGENAU and POLLARD [2.33], which dealt with the interaction of a fluctuating London dipole with its mirror image in metal, yielded a potential of the type $W_{ads} = Cr^{-3}$. The same relationship is obtained in calculations based on the free-electron model [2.34]. PROSEN and SACKS [2.35] investigated degeneration of the electron gas and obtained $W_{ads} \propto r^{-2}$. Within the framework of a one-electron model, however, it would be impossible to obtain a quantitative estimate of the energy of interaction between the molecules and the metal. A review of different models for calculating W_{ads} in this approximation can be found in [2.36].

Another approach to calculating the energy of adsorption was adopted by MAVROYANNIS [2.37] and MCLACHLAN [2.38] on the basis of LIFSHITZ's macroscopic theory [2.39], which describes the interaction of two semi-infinite homogeneous bodies. While in the London theory (2.4) the static values of polarizability α represent the material parameters, in the LIFSHITZ theory, α is a function of the complex component of the electromagnetic field frequency, which is always present in every medium and beyond its limits. Extrapolation of the interaction of two molecules from this theory yielded r^{-6}, and extrapolation of the interaction between a molecule and metal gave r^{-3}. Like the London theory, these approximations of intermolecular potential can hardly be applied to adsorption: at small distances a solid cannot be considered a continuum.

Finally, it is necessary to mention one more significant assumption made in calculations of adsorption potential, an assumption related to the idealization of a solid surface. In most calculations it is taken for granted that the interatomic distances are the same on the surface as inside the solid; i.e., values for d are taken from X-ray measurements. However, as will be shown later, the structure of both dielectric and metallic surfaces differs basically from that of the bulk crystal. This leads to total confusion in calculations of the total energy of interaction, W_{ads}.

From the preceding discussion it follows that the existing theory of intermolecular interactions can at best give only order-of-magnitude estimates for the values of physical adsorption energies. The fact that the theory yields good agreement with experiment despite incorrect basic assumptions can be explained only by its highly empirical nature. The constants in (2.2,3) are taken from experiment, and the convenient form of the potential function is postulated in advance. The form adopted for the potential curve allows the constants employed to vary freely; in (2.2), $m = 4 - 14$ and $n = 8 - \infty$ [2.12,22]. The formulae for W_{ads} are insensitive to the nature of the solid surface. For example, it was found that the values of W_{ads} for the adsorption of argon on surfaces as different as those of metals (Cu, Pt), ionic crystals (KCl), and covalent crystals (graphite, boron nitride, Al_2O_3) differ by only $0.4 - 1.2\ kJ \cdot mol^{-1}$ [2.12,13], which is almost within the accuracy of the measurements of adsorption heat used for reference. Because the theory is highly empirical, it seems improbable that further improvements in the calculation technique with the aid of computers [2.13,15], or consideration of higher terms in (2.6) (which diverges at small distances [2.40]), will lead to any progress in understanding the nature of physical adsorption.

2.1.2 Interactions in Chemical Adsorption

Chemical adsorption is accompanied by the formation of chemical compounds on the solid surface, but this does not necessarily mean that dissociation of molecules takes place. The adsorbed molecule, although it forms a chemical bond, can retain its individuality to some extent. The nature of chemisorptive bonds is practically the same as that in "bulk" compounds, though the specific surface properties can influence the bond and the electron distribution in interacting atoms.

Modern quantum chemistry, on the basis of an analysis of the entire range of chemical compounds, considers classical valence bonds, bonds with multicentered delocalized orbitals, and coordination bonds to be extreme cases

[2.41-45]. A classical valence bond is similar to the two-electron bond in an H_2 molecule. The principal method of describing valence bonds is the molecular orbital (MO) method. It is based on a one-electron approximation (independent motion of an electron in the averaged field created by the rest of the electrons and nuclei) and presents the wave function of a many-electron system as a linear combination of one-electron atomic functions (MO LCAO). This is the well-known Rutaan method in the Hartree-Fock self-consistent field approximation. In real models, however, the number of centers is great, and one has to calculate out a large number of matrix elements, so a compromise method is usually employed which involves simplifying assumptions and approximations. A detailed account of modern semi-empirical methods of MO calculations can be found in [2.45-48].

Recently a new nonempirical method of calculating valence has been receiving wider attention: the X_α method of self-consistent field and scattered waves (SCF-X_α-SW), proposed by Slater and Johnson. In contrast to the MO LCAO method, here a self-consistent total solution is sought from approximate solutions obtained separately for the neighborhoods of nuclei and external space, under the assumption of a certain average distribution of electrons in each of these domains. One great advantage of this method is a drastic reduction by nearly two orders of magnitude in the computer time needed. This technique is described in [2.49]. The SCF-X_α-SW method has been successfully used to calculate simple and complex molecules and to solve chemisorption and catalysis problems.

Another important class of bonds are delocalized multicenter bonds in conjugated systems, e.g., in benzene molecules. These bonds are created, by localized σ bonds resulting from sp^2-hybridized carbon orbitals, and by p_z orbitals normal to the benzene ring plane. These orbitals form a united conjugated system of molecular π orbitals. The delocalization of electrons leads to a violation of the additivity of polarizability, dipole moment and other properties along the direction of the bond. In calculations by the MO method it is usually assumed that the MO σ bonds are localized and presentable as two-center orbitals. Far greater difficulties are encountered in calculations of delocalized orbitals and correlation effects [2.46].

The delocalization of an electron is possible not only within a molecule, but also between two molecules, and the electron need not necessarily be an unpaired one. This phenomenon was analyzed by MULLIKEN and PERSON [2.47]. If an electron from an upper occupied molecular orbital of donor molecule A moves partially or completely to a lower orbital of acceptor molecule B, then a so-called donor-acceptor bond is formed, as in the complexes NH_4^+ or

$AlCl_3^- \cdot NH_3^+$, owing to the creation of a bond involving a common pair of electrons from nitrogen in one case, and from the free sp orbital of Al in the other. The term "donor-acceptor" bond alludes to the origin of the electrons participating in the formation of the bond [2.45]. In the complexes thus formed (e.g., NH_4^+), the bonds around the central atom are indistinguishable, so the complexes can be viewed as compounds with valence bonds within the framework of the MO LCAO method. Depending upon the distance between atoms A and B and the "pullover" of the electron cloud towards B, all intermediate types of interaction are possible, from weak ion-dipole to partially covalent and ionic interaction. In the excited states of interacting molecules, a complete transfer of the electron from A to B is possible, which can be illustrated by the following scheme:

$$A + B \rightarrow \underset{\substack{\text{partial}\\ \text{transfer}\\ \text{of charge}}}{(A^{+\delta}\, B^{-\delta})} \rightarrow \underset{\substack{\text{complex}\\ \text{with}\\ \text{transfer}\\ \text{of charge}\\ \text{(CTC)}}}{(A^+ B^-)} \rightarrow \underset{\substack{\text{isolated}\\ \text{ions}}}{(A^+)(B^-)} \quad . \qquad (2.7)$$

Finally, a very common type of bond is the coordination bond. It involves inner (d and f) orbitals, by virtue of which the number of molecules or ligand atoms coordinated around the central atom can exceed the number of normal valence bonds. Delocalization of the bonding electrons involves the whole complex and is three-dimensional. Coordination bonds cannot be described rigorously by a simple combination of localized one-electron atomic orbitals. Because of the outstanding role which coordination bonds play in chemisorption and catalysis, a detailed discussion of them is deferred until Chap.4.

To turn from a consideration of chemical bonds in molecules to an investigation of chemisorptive interaction is extremely difficult. Even with an ideal lattice, the hybridization of the surface atoms' orbitals differs from that of the atoms in the bulk. The interaction of an adsorbed molecule with a surface atom causes partial rehybridization of neighboring chemical bonds and disturbs the solid-state electronic subsystem. Certain approximations must be made to allow for these changes. We shall restrict ourselves here to a discussion of the major stages in the development of our views on chemisorptive bonds and the models employed to calculate them.

The first estimates of chemisorption energy were based on an analogy between the structure of surface complexes and the corresponding bulk compounds. A widely employed method of calculating covalent complexes, especially for metals was the ELEY formula [2.50] for the energy of dissociative one-point adsorption of two-atom homonuclear molecules AA on a metal surface:

$$W_{AM} = \tfrac{1}{2}(D_{M-M} - D_{AA}) - 23.06(x_M - x_A)^2 \quad , \tag{2.8}$$

where D_{M-M} and D_{AA} are the bond dissociation energies for metal-metal and two-atom molecules, respectively, and x_M and x_A are the Pauling electronegativities. Then the differential adsorption heat is

$$q_d = 2W_{AM} - D_{AA} = D_{M-M} + 46.12(x_M - x_A) \quad . \tag{2.9}$$

In an ionic bond, when the charge transfer complex ($M^{\mp}A^{\pm}$) is formed on the metal surface [see (2.7)],

$$q_d = [-Je + \varphi e + (e^2 / 4r_0)]N \tag{2.10a}$$

for $M^- A^+$, and

$$q_d = [-\varphi e + A_0 e + (e^2 / 4r_0)]N \tag{2.10b}$$

for $M^+ A^-$, where J and A_0 are the ionization potential and the energy of affinity of the adsorbed molecule for the electron, φ is the work function of metal, and r_0 is the radius of the adsorbed ion. Similar formulae were later employed [2.51] to calculate adsorption complexes on semiconductor surfaces. The highly qualitative nature of these estimates is self-evident.

The constants in (2.8-10) which characterize the adsorbed molecule (A_0, J, x_A, r_0) and the solid (φ, D_{M-M}) are usually chosen quite arbitrarily [2.3]. The problem of chemisorption is a quantum-mechanical one. In general, the adsorbed molecule and the solid surface ought to be considered a single quantum-mechanical system in which the changes in the electronic state of the adsorbed molecule during the process of chemisorption and the corresponding changes in the solid must necessarily be taken into consideration [2.23].

The quantum-mechanical approach to the chemisorptive bond is being developed in two basic directions. One of these seeks a universal description of the system comprised by the adsorbed molecule and the solid surface in zone approximation. The other focuses on local peculiarities of the chemisorptive bond, and treats the solid as a cluster of comparatively few atoms.

TSENTER and VOLKENSTEIN [2.52,53] were the first to consider the quantum-mechanical problem of interaction between a univalent atom and an ionic lattice within the framework of the molecular orbital method. Despite rather essential a priori assumptions, they succeeded in demonstrating the formation of local levels in the crystal's energy spectrum in chemisorption. In [2.53] the possible existence of two forms of chemisorption was proved: neutral and "charged." In neutral chemisorption the molecule is bound to the surface due to partial drawing of its electron into the lattice. The local level of the adsorbed particle remains unoccupied. In charged chemisorption a free lattice electron becomes involved in the chemical bond. This electron is localized on the level of the adsorbed particle, which strengthens the chemisorptive bond. The surface then becomes charged with respect to the bulk substance. Similar conclusions about the localization of bonds in chemisorption were reached by GRIMLEY [2.54] and KOUTECKY [2.55], who performed calculations by the MO method for somewhat different models and under different assumptions. HORIUTI and TOYA [2.56] investigated the adsorption of hydrogen atoms on metal surfaces; and BONCH-BRUEVICH [2.57] applied the effective-mass method, borrowed from crystal physics, to surfaces, and calculated the lower energy levels of adsorbed atoms with one and two valence electrons.

Independently of these works, chemisorption had also been treated as the result of electron transitions in a system consisting of adsorbate and solid surface [2.58-64]. These works represent what is generally known as the boundary layer theory. In contrast to the VOLKENSTEIN electron theory of chemisorption [2.23], it considers only the charged form of chemisorption. A detailed account of various aspects of the electron theory can be found in [2.23,65-67].

In all of the quantum-mechanical calculations [2.2,23,57-64] in the electron theory of chemisorption, the crystal surface considered is ideal, i.e., devoid of intrinsic surface states. An adsorbed particle on such a surface is treated as a structural surface defect, creating local perturbation in the crystal which leads to the emergence of a surface electron state. Even in this approximation, however, the theory is far from being able to calculate local electron states on the basis of existing data about the electronic properties of adsorbed particles and the solid state, even for quite simple systems. The task becomes far more complicated when one considers a real finite crystal, because of its wide spectrum of intrinsic surface states. The existing theoretical calculations are based on a one-electron approximation, though there have been attempts at a many-electron approach to the chemisorptive bond [2.68]. What with the lack of data on the energy spectrum of surfaces, however, these approaches are of little help when it comes to calculating real systems. This probably explains the shift in electron theory towards analyses of the statistics of electron transitions on surfaces with prepostulated local levels.

The first detailed investigation of the energy spectrum of surface eigenstates (the so-called Tamm and Shockley levels) of the interface between a finite crystal and a vacuum via the MO method was performed by KOUTECKY and TOMASHEK [2.69,70]. Today, all the existing theoretical methods of calculating electron states in crystals are being applied to the study of surface levels, including the symmetry of surface states and the role of electron correlation on ideal faces of ionic and covalent crystals with widely varying structures [2.71,72]. Considerable success has been achieved in quantum-mechanical calculations of chemisorptive interactions on metal surfaces. The main problem here is connected with the correlation of electrons. Valuable models accounting for correlation effects have been proposed, among them the Newns-Anderson model, the Schriffer-Paulson-Gomer model and the Kohn-Smith-Ying model [2.49, 73-75]. These investigations focus on the classification of surface eigenstates, and treat adsorption states as a special case. Despite the success of theoretical and computational methods, as yet only very general statements can be made about the localization of bonds on crystal surfaces and the pattern of the energy spectrum of surface states. Even for strongly idealized surfaces, theory does not give good estimates of the principal parameters of surface states and the chemisorptive bond energy—and these very questions are crucial to chemisorption and catalysis. For that reason, quantum-chemical calculations of local chemical interactions between adsorbed molecules and solid-state surfaces have recently been receiving much attention.

In the first calculations of local interactions, the active adsorption center had been modeled as a single lattice atom [2.76]. The Hamiltonian of this two-particle system, which entered the Schrödinger equation for the extreme cases of purely ionic or purely covalent bonds, was given by expressions similar to the right-hand sides of (2.8-10). Calculations of chemisorptive bonds for one-point and two-point adsorption were further improved by a more correct definition of the atomic orbitals of interacting atoms and of the system's Hamiltonian [2.3b,49]. Naturally, approximating a solid by a single adsorption center gives only a very rough model. Chemical interaction induces changes in the electron state not only of the adsorption center, but also of surrounding atoms. Therefore, the next step in the development of quantum-chemical calculations of surface complexes was the transition to cluster models.

Of course, the substitution of a semi-infinite crystal body by a cluster of relatively few atoms also results in a rather clumsy physical model; the allowed bands are displaced, and new local levels emerge due to fringe effects, which are strongest in covalent crystals [2.72]. Solid-state band theory does predict, however, that the general pattern of band structure will remain the same after transition to a finite crystal, and, therefore, exact calculations for a large enough cluster ought to agree with calculations for a band model [2.77-83]. It is evident that the optimal size for a cluster employed in theoretical calculations will depend upon the concrete parameter to be estimated.

Calculations in cluster approximation are particularly convenient for estimating chemisorption energy, changes in the effective charges of surface atoms, and the order of chemisorptive bonds. In these cases the inaccuracy of the calculated value decreases rapidly with increasing cluster size. For instance, the energy of ethylene adsorption on a chain of N metal atoms tends to a limit at $N \sim 15$ [2.81]. Larger clusters are needed, of course, to obtain estimates of energy levels in chemisorbed particles which are tied to the band structure of the solid.

The MO LCAO methods have by now been used to calculate many bulk and surface properties of metal and semiconductor clusters. Such calculations have usually been performed via the expanded Hueckel method (EHM), the Mulliken-Wolfsberg-Helmholtz method (MWH) and two methods of calculating molecular integrals in the approximation of neglect (complete neglect) differential overlapping of orbitals (NDO and CNDO).

In recent years the SCF-X_α-SW method has been very successfully used to calculate surface properties of clusters [2.82,83]. The calculated values for surface levels are in good agreement with experimental values obtained from photoelectronic spectroscopy.

The importance of donor-acceptor bonds in chemisorption and catalysis was recognized quite early by TERENIN [2.84], SOKOLOV [2.85] and others [2.86-88]. NAGAYEV [2.89,90] analyzed (within the framework of a many-electron approximation) the possibilities of formation of two-electron and three-electron bonds between an H_2 molecule and a partially ionic crystal, and investigated the influence of bonding ionicity on adsorptive activity. All of these investigations avoided the question of whether free charged particles in the lattice take part in bond formations. As was mentioned above, the term "donor-acceptor bond" refers only to the origin of electrons participating in the formation of the bond. They can come not only from s and p orbitals, but from d and f orbitals as well [2.45,91].

As early as 1935 ROGINSKY [2.92] observed high chemisorptive activity in transition elements. BEECK [2.93] suggested that the high chemisorptive activity of transition metals might be due to unoccupied d orbitals in their atoms. Since d bands are rather narrow, many properties of transition metals can be described by the approximation of localized d electrons [2.94,95]. A large number of chemisorptive bonds between simple molecules and clusters of transition metals and their oxides have been calculated in this way [2.75,96].

Finally, let us turn to weak chemisorptive interactions, i.e., to the hydrogen bond, which plays an important role in adsorption on a number of real surfaces. SOKOLOV [2.97,98] showed the electrostatic approach to the hydrogen bond to be incapable of explaining its spectroscopic manifestations, and was the first to propose a quantum-mechanical explanation. Based on a consideration of superpositions of three wave functions of interacting atoms A - N ... B, represented in the form of antisymmetrized sums of AO products for each electron, Sokolov obtained a general expression for the energy of the system via the interaction integrals of atoms, A, N and B. The energy of the system depends strongly on the polarity of A - N bonds. At increased polarity the donor-acceptor interaction becomes more significant. Because of the presence of undivided electrons of B in the bond, its stability depends strongly on the configuration of interacting atoms, a fact that is commonly overlooked in considerations of various schemes of adsorptive interactions involving hydrogen bonds. In interaction with aromatic compounds, the formation of the hydrogen bond can involve π electrons of a conjugated system.

The quantum-mechanical approach to weak interactions on surfaces was further developed by KOUTECKI and co-workers [2.99]. Using the MO LCAO method, they estimated the energy of interaction of benzene, naphtalene and ethylene with a hydroxyl group, which they considered a dipole. The calculations were carried out for distances great enough to allow exchange effects to be neglected, and yielded energy values which were correct in order of magnitude. It turned out that the π electrons of a conjugated system contribute significantly to the energy of interaction, while $\pi - \sigma$ interactions are small in the adopted model. The calculations proved that the bonding of aromatic and unsaturated molecules with the surface's hydroxyl groups is the result of chemical bonds.

The most important, though empirical, method of describing molecular interactions on a surface is at present the molecular orbital (MO) method. Despite its numerous simplifying assumptions, this method, when applied to chemisorption, provides us at least with qualitative data about the real chemical bonds formed on surfaces and about the nature of surface states.

2.1.3 The Thermodynamics of Adsorption

There are two approaches to describing surface phases: the Gibbs method and a method based on a surface layer of finite thickness. The GIBBS method [2.100] requires no special assumptions regarding the properties of interfacial phases, and is applicable to interfaces with arbitrary curvature. The basic equation for the change in internal energy E of a two-phase (α and β) system including an interfacial surface of arbitrary curvature (Fig.2.2) can be written in the following general form [2.101]:

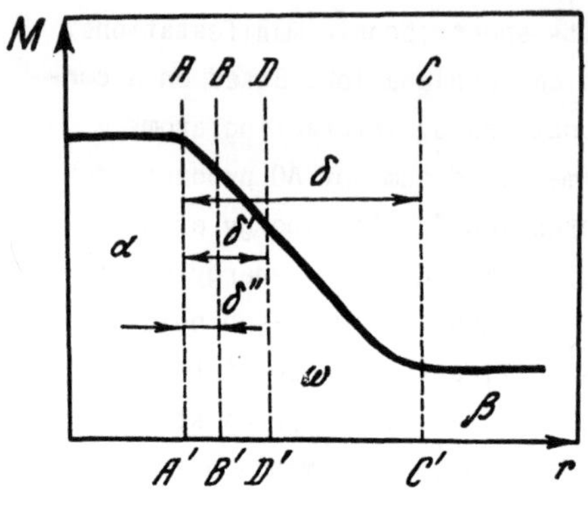

Fig.2.2. Diagram of the surface phase

$$dE = Td\eta - p^{\alpha}dV^{\alpha} - p^{\beta}dV^{\beta} + \gamma ds + s\left(\frac{d\gamma}{dr}\right)dr + \sum_i \mu_i dn_i \quad , \qquad (2.11)$$

where η is the entropy of the whole system, p and V are the pressure and volume of phases α and β, γ is the surface tension, r is the radius of the interface curvature, s is its area, μ_i is the chemical potential of the ith component, and n_i is the number of moles of the ith component. It is extremely difficult to determine the boundaries between an inhomogeneous interfacial phase and homogeneous phases α and β. The Gibbs method deals with this obstacle by introducing a conceptual interface surface DD' the phases α and β are purely homogeneous, and all the extensive properties of the real interface, e.g., the excess number of moles of the ith component per unit volume, $\bar{n}_i$, the excess inner energy $\bar{E}$, and entropy $\bar{\eta}$

$$\bar{n}_i = n_i - (n_i^{\alpha} + n_i^{\beta}) \quad ,$$
$$\bar{E} = E - (E^{\alpha}V^{\alpha} + E^{\beta}V^{\beta}) \quad , \qquad (2.12)$$
$$\bar{\eta} = \eta - (\eta^{\alpha}V^{\alpha} + \eta^{\beta}V^{\beta}) \quad ,$$

belong to this surface. If the surface tension is also ascribed to this surface, then for excess quantities, (2.11) will take the form

$$d\bar{E} = Td\bar{\eta} + \gamma ds + \sum_i \mu_i d\bar{n}_i \quad , \qquad (2.13)$$

which is valid for both flat and curved interfaces.

Integrating (2.12) over extensity factors and differentiating it over intensity gives the well-known Gibbs equation of adsorption

$$d\gamma = -\frac{\bar{\eta}}{s}dT - \sum_i \Gamma_i d\mu_i \quad , \qquad (2.14)$$

where $\Gamma_i = \bar{n}_i / s$ is the Gibbs excess adsorption related to the unit of interface areas. For a rarefied one-component gas phase β (which can be considered an ideal gas) at $T = \text{const}$, the Gibbs equation takes the form

$$d\gamma = -\Gamma d\mu = -RT \ln p \quad . \tag{2.15}$$

For all its universality the Gibbs method has two important drawbacks:

1) Replacing a real interface with a conceptual surface DD' completely eliminates the thickness and physical properties of the boundary layer.
2) The excess quantities in the Gibbs equations depend on the position of the boundary surface DD'. In order to employ these two-dimensional equations one has to determine the exact position of the surface DD', which always involves introducing dubious assumptions [2.101-104].

In many cases, especially those dealing with physical adsorption, the conceptual interface is identified with the surface of the condensed phase, with the very important assumption that this interface is inert. This assumption obliterates all the advantages of the Gibbs general approach and renders it totally unfit to describe such important surface phenomena as chemisorption, catalysis, adsorption on a swelling adsorbent, etc. This question will be discussed again later.

The method developed in [2.101,105-108] is physically more plausible. It is based on the concept of a real interface with an effective thickness δ which is sandwiched between planes AA' and CC' on the sides of bulk phases (Fig.2.2). All changes in thermodynamic parameters during adsorption involve this layer. For instance, the concentration of the ith component in the layer a_i which is related to a unit of surface area is connected with the Gibbs excess concentration in the following manner:

$$a_i = \tau_i + \delta'(n_i^\alpha - n_i^\beta) + \delta n_i^\beta \quad . \tag{2.16}$$

The changes in internal energy, entropy and other thermodynamic potentials ought, strictly speaking, to be ascribed to both layers CC' -BB' and AA' -BB'. If the adsorbent is considered inert, as is customary in physical adsorption, then the surface DD' coincides with AA' and CC', and $a_i = \Gamma_i + \delta n^\beta$.

The main disadvantage of this method as compared to the Gibbs method is its inadequacy in describing a curved interface with $r < \delta$. This becomes important in investigations of highly dispersed adsorbents. It was RUSANOV [2.101] who gave rigorous proof of the applicability of the basic general equation (2.11) to a curved surface layer.

For flat surfaces ($r \gg \delta$), (2.11) becomes

$$dE^{\omega} = T\,d\eta^{\omega} - p_N dV^{\omega} + \gamma\,ds + \sum_i \mu_i dn_i^{\omega} \quad , \qquad (2.17)$$

where p_N is the normal component of the pressure tensor, which in a homogeneous phase is equal to p. The superscript ω is used to emphasize the fact that all the parameters are attributed to the interface. RUSANOV investigated the applicability of (2.11,17) in describing adsorption films on solid-state surfaces, and derived expressions for the isotherm and the isobar of adsorption equilibrium, for variations in surface tension due to shifts in adsorption equilibrium, and for the composition and entropy of the film.

A knowledge of the principal thermodynamic relations for the adsorption of a pure substance on a nonvolatile solid-state surface is necessary for the discussion which follows. Data on the quantity of substance adsorbed per unit area of adsorbent ($\mu mol \cdot m^{-2}$), and on the differential adsorption heat q_d ($kJ \cdot mol^{-1}$) can be obtained from experiment. Based on these data one can then calculate the changes in surface tension $\Delta\gamma$, internal (ΔE) and free (ΔF) energy, and entropy $\Delta\eta$ which result when moles of substance move from the gaseous phase onto the surface. The adsorbent being inert, all these changes will characterize the properties of the adsorption phase ω. For the reference state of the gaseous phase, one usually takes the saturated vapor pressure when dealing with vapors, and 1 atm when dealing with gases.

Integration of the Gibbs equation yields the expression for two-dimensional pressure :

$$\pi = \gamma - \gamma_0 = - \int a\,d\mu = - RT \int a d \ln p \quad . \qquad (2.18)$$

Let us introduce differential molar quantities $d\Delta E^{\omega}/da$ and $d\Delta F^{\omega}/da$. By virtue of simple thermodynamic relations [2.2,101,106-108] one can obtain the expression for isosteric adsorption heat q_{st}, which is idential to the change in internal energy, ΔE, at constant adsorption; it is similar to the Clausius-Clapeyron formula:

$$q_{st} = - \frac{d\Delta E^{\omega}}{da} = RT^2 \left(\frac{d \ln p}{dT} \right)_a \quad , \qquad (2.19)$$

One can also obtain the expression for differential quantities F and η:

$$\frac{d\Delta F^{\omega}}{da} = - RT \ln \frac{p}{p_s} \quad , \qquad (2.20)$$

$$\frac{d\Delta\eta^{\omega}}{da} = \frac{\dfrac{d\Delta E^{\omega}}{da} - \dfrac{d\Delta F^{\omega}}{da}}{T} \quad . \qquad (2.21)$$

Thus from two adsorption isotherms measured at different temperatures, we can calculate all the main differential molar thermodynamic parameters of the adsorbed phase.

Of especial interest when comparing the thermodynamic properties of interface phases with the results of model calculations carried out by the methods of statistical mechanics (Sect.2.1.4) are the integral quantities E^{ω}_{int}, F^{ω}_{int}, and η^{ω}_{int}, as well as the mean molar quantities E^{ω}_{int}/a, F^{ω}_{int}/a, and η^{ω}_{int}/a, which are obtained by integrating (2.19-21) over adsorption, e.g.:

$$E^{\omega}_{int} = \int_0^a RT^2\left(\frac{d\ \ln p}{dT}\right)_a da \quad . \tag{2.22}$$

Naturally, all these relations hold only for a strictly reversible adsorption cycle. This should be borne in mind in thermodynamic treatments of chemisorption phenomena, which in many cases are partially or completely irreversible.

Calorimetric analysis of adsorption heat under isothermal conditions yields the differential adsorption heat q_d, which at low pressure is smaller than q_{st} by the value of RT. The integral heat of adsorption from the vapor phase can be obtained from calorimetric measurements of the heat of wetting of the adsorbent by the relevant liquid:

$$q_{wet} = \int_0^{a_S} q_d da - La_S + \varepsilon s' \quad , \tag{2.23}$$

where a_S is the adsorption corresponding to the formation of a dense adsorptive layer when the adsorption heat is about equal to the condensation heat L; $\varepsilon = \gamma + T(d\gamma/dT)$ is the surface energy of the liquid; and s' is the area of the layer. For porous adsorbents, $s' \ll s$. For nonporous samples, $s' \sim s$, which allowed JURA and HARKINS [2.109] to propose a method of determining specific surface area s by measuring the difference between the wetting heat for a vacuumized sample q_{wet} and that for a sample cured in saturated vapor, q^S_{wet}; $s = (q_{wet} - q^S_{wet})/\varepsilon$.

The question of the thickness of nonhomogeneous interphase δ is very complicated in the method based on a finite layer. Generally speaking, thickness δ in Fig.2.2 ought to be determined for each physical property, since it can be different for each [2.101]. As can be seen from the figure, the interface also includes the subsurface layer of the solid with thickness δ''. All variations in thermodynamic potentials during adsorption that are measured experimentally (2.18-21) involve, in the general case, changes not only in the parameters of adsorbed molecules going from phase β to phase ω, but also in the parameters of the layer AA' - CC'. For adsorption of a pure substance on a nonvolatile adsorbent, various methods have been proposed for separating a two-component system of adsorbent and adsorbate into subsystems, e.g., condensed phase (solid body plus the adsorbed film BB') and gaseous phase β [2.101], or condensed phase (solid body plus surface layer CC') and the adsorbed layer CC' - BB', which is at equilibrium with phase β [2.101]. (The properties of the film-vapor boundary are discussed in [2.110].)

In all cases the thermodynamic functions of the interface phase include a term that takes changes in the thermodynamic properties of the subsurface solid layer into account. In cases featuring additivity of thermodynamic potentials in the subsystems of the interface, the change in energy of the layer AA' - CC' can be formally accounted for by adding to the right-hand side of (2.17) terms of the form $\sum_i X_i dx_i$, where X_i are generalized forces, and x_i generalized coordinates. For instance, in deformation of a solid in the process of adsorption [2.102], the Gibbs equation of adsorption (2.15) will include an additional term $X dx$. Additivity is violated in the transition to microporous adsorbents [2.102].

Microporous bodies whose pores are comparable in size to adsorbed molecules play an important role in adsorption and catalysis; zeolites are an example. The most consistent thermodynamic description of adsorption on microporous bodies is given in the theory of voluminous filling of micropores [2.111,112]. This theory is based on a total departure from the layer mechanism of adsorption, since with pores measuring 1-1.5 nm the very concept of surface has little meaning [2.111].

The adsorptive activity of a porous structure is characterized by a certain distribution of adsorption potential ξ. All the difficulties connected with calculating short-range intermolecular interactions are hidden in this potential. The value of this potential, $\xi = RT \ln(p_S / p)$ (p_S being the saturated vapor pressure), can be thermodynamically proved to be the differential work of transition at $T = \text{const}$ of one mole of substance from the liquid (reference) state into the adsorbed state. Assuming temperature invariance and Gaussian distribution of the adsorption space volume V^ω over adsorption potentials, DUBININ [2.112] obtained a well-known equation for the adsorption isotherm which gives a good description of adsorption on microporous adsorbents:

$$n^\omega = \frac{V_0}{V} \exp\left(- \frac{B'T^2}{\beta^2} \ln \frac{p_S}{p}\right) \quad , \tag{2.24}$$

where V_0 is the total volume of micropores and B' and β are constants characterizing the adsorbent and the adsorbate, respectively. This equation played an important role in the evolvement of the theory of adsorption on porous bodies. Later SERPINSKI and YAKUBOV [2.111b] improved the theory and extended its applicability beyond the region of invariance of ξ. Making use of the well-established experimental fact that adsorption isosteres $f(pT)n_{\omega=\text{const}}$ in coordinates p, T^{-1} are well approximated by straight lines, they proposed elegant methods of calculating the differential heat and entropy of adsorption on the basis of experimental isosteres. The method proved to be quite accurate in the initial stage of filling.

With microporous adsorbents the problem of inertness becomes even more acute. In adsorbents such as zeolites, all or almost all of the adsorbent atoms take part in the interaction with adsorbed molecules. The integral of the Gibbs adsorption equation (2.18) has been shown to represent here not the two-dimensional pressure, but the change in the chemical potential of the microporous solid in the process of adsorption [2.113].

Recently, BERING and SERPINSKY [2.114] obtained very satisfying results on the thermodynamics of adsorption on noninert adsorbents. They analyzed adsorption equilibrium from the viewpoint of the thermodynamics of vacancy solutions, with a vacancy being understood to be an interphase element filled at adsorption by one molecule of adsorbate. Having established a strict analogy between the surface pressure (2.18) and the osmotic pressure of a nonideal binary interface solution (adsorbed molecules plus vacancies), Bering and Serpinsky obtained a number of general thermodynamic relations which are valid for both nonporous and microporous adsorbents. These equations are rigorously applicable to both physical adsorption and chemisorption, as well as to adsorption on swelling adsorbents. For the adsorption of gases and vapors on microporous adsorbents, Bering and Serpinsky also proposed convenient techniques for calculating differential adsorption heat and adsorption entropy on the basis of adsorption measurements.

2.1.4 Statistical-Thermodynamic Description of Adsorption Equilibrium

Statistical mechanics, based on a concrete model of the molecular system and known laws of molecular interactions, allow purely theoretical calculations to be made of the main thermodynamic functions of a system. The starting point for these calculations is the known relationship between the free energy of the system, F, and its statistical sum (or partition function) Ω:

$$F = -kT \ln \Omega \quad . \tag{2.25}$$

For our purpose here we can restrict ourselves to monomolecular adsorption, where the majority of statistical calculations of adsorption equilibrium is based on two important assumptions:

1) both macroscopic and electron properties of the solid surface are assumed not to change during adsorption. The adsorbent is considered an inert body which creates a permanent potential field populated by the adsorbed molecules.

2) The net potential energy of molecules in the surface phase, W, is assumed to be additively composed of the energy of molecules in the field of lattice surface $w(r_i)$ and of paired lateral adsorbate-adsorbate interactions $w(r_{k,\ell})$:

$$W = \sum_i w(r_i) + \sum_{k,\ell} w(r_{k,\ell}) \quad . \tag{2.26}$$

The first assumption immediately restricts the scope of systems that may be considered to relatively weak interactions, e.g., adsorption of inert gases. The second assumption is realized only in highly rarefied adsorption phases. With an increase in occupation, the contribution to W from triple and higher-order interactions can become rather large, amounting to 35% and more of the energy of binary interactions [2.17,19]. The dependence of energy upon distance is also altered. For instance, for three particles A, B and C the dependence is $(r_{AB}r_{AC}r_{BC})^{-3}$ instead of r^{-6} [2.115]. The potential field of the surface strongly influences the value $w(r_{k,\ell})$ [2.116]. The statistical-thermodynamic approach to describing surface phases under the said assumptions has been developed most consistently [2.116-119]. Let us give a brief summary of these calculations.

In its most general form the adsorption phase in material and thermal equilibrium with the environment is described by the canonical Gibbs ensemble. The great statistical sum for a one-component three-dimensional system of N particles is known to be

$$\Xi(T, V, \lambda) = \sum \Omega(T, V, N)\lambda^N \tag{2.27}$$

where $\lambda = \exp(\mu / kT)$ is the absolute activity, and Ω is the statistical sum of the canonical distribution. At high enough temperatures,

$$\Omega = \frac{1}{N!h^{3N}} \int \exp\left[\frac{H(q,p)}{kT}\right] dq^{3N} dp^{3N} \quad , \tag{2.28}$$

where H is the system's Hamiltonian. Integration is carried out in the phase space of coordinates and momenta (q,p).

The classical Hamiltonian of a system is the sum of operators of kinetic and potential energies:

$$H(p,q) = \sum_{i=1}^{N} \left[\frac{p_i^2}{2m} + W(q_i)\right] \quad . \tag{2.29}$$

The expression for the statistical sum for the simplest case of a monoatomic gas has, after integration in the space of momenta, the form

$$\Omega = \frac{Z}{N!\Lambda^{3N}} \quad , \qquad \text{where} \tag{2.30}$$

$$\Omega = \left(\frac{h^2}{2\pi mkT}\right)^{1/2} \quad , \qquad \text{and where} \quad Z = \int_V e^{-W/kT} dq$$

is the configuration integral.[2] By virtue of (2.26) the configuration integral can be presented in the form

$$Z = \int_V \exp\left\{\left[\sum_i w(r_i) + \sum_{k,\ell} w(r_{k,\ell})\right] / kT\right\} dr_i \ \ldots \ dr_N \quad . \tag{2.31}$$

Let the statistical sum for the gas volume be Ξ^0 before adsorption, and Ξ after adsorption. Then the statistical sum of an adsorbed substance Ξ^ω (the adsorbent is assumed to be inert) will be given by the following expression:

$$\Xi^\omega = \Xi \ / \ \Xi^0 \tag{2.32}$$

Knowing Ξ^ω, one can calculate all thermodynamic parameters of the system, the concentration of particles in the surface phase $\bar{n}^\omega$ (from the thermal equation of adsorption), internal energy $\bar{E}^\omega$, and entropy $\bar{\eta}^\omega$ [2.116,117]. From (2.31) it is clear that these parameters are the Gibbs excess values. A physical model of the surface phase can be more easily constructed by using the terms employed in the theory based on an adsorption layer of finite thickness (Sect.2.1.3). This is done (under Assumptions 1 and 2) by splitting the volume of the adsorption system V into two parts: V^ω (the volume of the finite layer) and V^β (the volume of the gaseous phase), and assuming the nonadsorbed gas to be ideal, i.e., $\lambda / \Lambda^3 = p / kT$. Then the properties of the finite layer can be described by a canonical statistical sum (2.25), with the configuration integral taken over the volume of the layer (in monomolecular adsorption, over the surface), i.e.,

$$\Omega^\omega = \frac{Z^\omega}{n^\omega \lambda^{3n\omega}} \quad .$$

Then the thermodynamic parameters of the finite layer are expressed, for example, as

---

[2] At low temperatures, when the adsorption phase is considered in quantum approximation, exp(-W/kT) is replaced by Slater's sums [2.116,117]. Quantum phenomena have been shown in [2.116] to play a significant role only in adsorption of the lightest gases: H_2, D_2, He, Ne.

$$\Xi^{\omega} = kT^2 \left(\frac{\partial \ln \Omega^{\omega}}{\partial T}\right)_{V^{\omega},\, n^{\omega}} , \tag{2.33a}$$

$$\eta^{\omega} = kT\left(\frac{\partial \ln \Omega^{\omega}}{\partial T}\right)_{V^{\omega},\, n^{\omega}} + k \ln \Omega^{\omega} . \tag{2.33b}$$

Both Ξ^{ω} and Ω^{ω} are exactly computable only for the model of an ideal two-dimensional gas. In real systems of interacting particles the configuration integral Z cannot, in either three-dimensional or two-dimensional phases, be presented as a product of independent integrals and taken analytically [2.117]; one has to resort to approximate methods. In fact, all three major statistical methods — virial expansions, correlation functions, and the lattice method — have been employed to describe surface phases.

The method of virial expansions requires no special assumptions regarding the structure of surface phases, and is valid for both localized and mobile adsorption layers. In gas approximation the surface phase is treated as a very rarefied gas. The integrand in (2.31) can then be represented in the form of an Ursell-Maier group expansion. Then, according to [2.116],

$$n^{\omega} = B\left(\frac{p}{kT}\right) + C\left(\frac{p}{kT}\right)^2 + D\left(\frac{p}{kT}\right)^3 + \dots . \tag{2.34}$$

Computing the coefficients of this series is quite cumbersome, so one usually resorts to calculating the second virial coefficient B, which has the form

$$B = \int_V \{\exp[w(r)/\, kT] - 1\}dr .$$

For the sake of simplicity, the potential of lateral interactions $w(r_{k,\ell})$ in (2.26) is assumed to be equal to that in bulk phases or simply to zero. In the latter case, if the bulk phase is an ideal gas,

$$n^{\omega} = \frac{V}{kT}\frac{\Omega_1}{\Omega_0} , \tag{2.35}$$

where Ω_1 is the statistical sum for a single isolated adsorbed molecule, and Ω_0 is the same for a free molecule. In the model of a two-dimensional ideal gas, this equation is equivalent to the well-known DE BOER equation of adsorption for the Henry region [2.120]:

$$n^{\omega} = b'p = \frac{N_A \tau_0 \exp(q / RT)}{\sqrt{2\pi MRT}} , \tag{2.36}$$

which is easily obtained along purely kinetic lines, without involvement of the canonic ensemble. Here N_A is the Avogadro number, M is the molecular weight of the adsorbate, q is the heat of adsorption, and τ_0 is the oscillation period of an adsorbed molecule normal to the surface ($\tau_0 \sim 10^{-13} - 10^{-14}$s). The de Boer equation is very convenient for evaluating adsorption in the initial stage of occupation.

Equation (2.35) has been widely employed to calculate adsorption equilibrium for inert gases and organic molecules on surfaces of graphite bodies, ionic and molecular crystals, and zeolites [2.3a,13,20,121,122]. In all cases good agreement was observed between the measured values and the theoretical calculations of thermodynamic parameters. This is not surprising, since the potential of interaction (2.1) is semi-empiric, and the virial coefficients are insensitive to the type of potential chosen [to the values of m and n in (2.2)].

Despite the universality of the method of virial expansions, its applicability is restricted to highly diluted systems. Only with small n^ω can one be sure that the contribution from the second, third, and higher-order terms in the expansion will be small compared to the first term. None of the works cited investigated the convergency of virial expansions in surface phases. Some authors employ virial expansions to describe adsorption isotherms up to rather high values of adsorption. For instance, the following isotherm is suggested in [2.15]:

$$\ln p = \ln a + C_1 a + C_2 a^2 + \dots \quad . \qquad (2.37)$$

The constants C_1, C_2, ... are calculated by computer. In this approach, however, the physical sense of these constants as coefficients in a virial expansion is lost.

Another rigorous technique for describing adsorption phases is the method of correlation functions developed by N. Bogolyubov and G. Grin. Molecular distribution functions make it possible not only to calculate the thermodynamic parameters of a system, but also to obtain certain information about the molecular structure of the adsorbed substance. A general investigation of the distribution functions for a liquid in the potential field of a solid surface is presented in [2.116.119,123], where the thermodynamic potentials are expressed via the binary distribution function, and the expression for the great statistical sum are obtained. In an approximation of thermodynamic perturbation theory, STEEL [2.116] employed this technique to derive the Frenkel-Halsey-Hill equation of polymolecular adsorption: $\ln(p/p_S) = \mathrm{const}/\theta^3$, where θ is the degree of occupation. Monomolecular adsorption was investigated by MARTYNOV and co-workers [2.124,125]. Making use of the Bogolyubov chain of equations and expanding the distribution functions with respect to density degree, they arrived at expressions for one- and two-particle distribution functions. In the idealized case of localized adsorption without adsorbate-adsorbate interaction, this approach yielded the Langmuir equation (see below). For the case of a mobile monolayer, certain simplifying assumptions regarding the interaction potential were made, and the FRUMKIN equation, one of the first equations of adsorption isotherms to account for lateral interactions (the constant A), was obtained [2.126]:

$$p = \theta \exp(A\theta) / [k_1(1 - \theta)] \quad . \tag{2.38}$$

Insofar as the Bogolyubov-Grin method is the most rigorous way of accounting for correlations between adsorbed molecules, the region of convergency for expansions of molecular functions is much broader than for virial expansions. There is every reason to suppose that the further development of this method will lead to progress in investigations of the structure and properties of adsorption layers. Naturally, in order to determine the asymptotic behavior of molecular functions in highly inhomogeneous interface phases, a number of additional assumptions will be necessary.

Finally, let us discuss the third group of approximate methods for calculating the statistical sums of interface phases. These are the lattice techniques, which are used widely to describe localized adsorption. The surface is approximated here as an ordered system of adsorption centers with concentration n_c, and the adsorption phase is considered a lattice gas. Its statistical sum can be presented in the form [2.117]:

$$\Omega^\omega(n^\omega, n_c, T) = \Omega_1(T)^{n^\omega} \sum_{i=1}^{G} e^{-W_i/kT} \quad , \tag{2.39}$$

where $\Omega_1(T)$ is the statistical sum for an isolated molecule, characterizing its motion around the localization center and the potential energy of interaction with this center, and W_i is the energy of interaction between all nearest pairs of neighbors n^i_{AA} for a given ith configuration ($W_i = n^i_{AA}w_0$, where w_0 is the energy of interaction within a pair). Summation is carried out over all admissible distributions of n^ω adsorbed molecules over n_c adsorption interface centers:

$$G = \frac{n_c!}{n^\omega!(n_c - n^\omega)!} \quad . \tag{2.40}$$

Knowing Ω^ω, one can theoretically calculate all the thermodynamic parameters of the interface according to (2.33). The problem is now reduced to one of calculating the lattice sums via combinatoric [2.127] or matrix methods [2.3a], instead of evaluating the configuration integral.

In localized adsorption on a homogeneous surface, the lateral interaction being neglected ($W_i = 0$), the Langmuir equation is readily obtained from (2.25,39) [2.128]:

$$\theta = b(T)p / [1 + b(T)p] \quad ; \quad b(T) = b_0 \exp(q / RT) \quad . \qquad (2.41)$$

BERING and SERPINSKY [2.114] have been quite right not to consider the Langmuir equation a typical representative of adsorption equations. It is derived without any special assumptions about the distribution of adsorption centers, and therefore the system will still be described by the Langmuir equation even if the discrete adsorption centers are distributed in three-dimensional volume, and if there is no interaction between the molecules of the medium (such as in the case of homogeneous catalysis – formation of complexes).

The lattice gas model is similar to the Ising model, which is widely employed in the theory of order and disorder applied to problems of ferromagnetics and binary alloys [2.129]. The advantage of the Ising model over other methods of approximating condensed phases is that its two-dimensional version (which is of especial interest in the theory of adsorption) provides the basis for rigorous mathematical treatment of a system, including phase transition. Onsager was the first to carry this out. Later, the problem of how rigid diatomic molecules are distributed on a surface was solved via the matrix method [2.130]. A rigorous formal consideration of the two-dimensional Ising model is especially interesting with regard to phase transitions on surfaces [2.117,130]. To avoid cumbersome computations of lattice sums in evaluating the energy of lateral interactions, one usually employs, for sufficiently diluted surface phases, either the quasi-chemical Bethe-Guggenheim approximation or the Bragg-Williams approximation.

The quasi-chemical approximation deals with the equilibrium between mutually independent occupied or free pairs of adsorption centers. The equilibrium constant is equal to 4 $\exp(w_0 / kT)$, where w_0 is the energy of pair interaction.

The Bragg-Williams approximation considers the average number of pairs of nearest neighbors $\bar{n}_{AA}$, which is the same for all possible configurations. Equation (2.39) becomes:

$$\Omega^{\omega}(n^{\omega}, n_c, T) = \Omega_1(T)^{n^{\omega}} \frac{n_c!}{n^{\omega}!(n_c - n^{\omega})!} \exp(\bar{n}_{AA} \cdot 2w_0 / zkT) \quad , \qquad (2.42)$$

where z is the coordination number of adsorption center. At uniform distribution, $\bar{n}_{AA} = 1/2\, z n^{\omega 2} / n_c$ [2.117]. From (2.42) one can readily obtain the isothermal FOWLER-GUGGENHEIM equation [2.128]:

$$p = \frac{\theta}{k_1(1-\theta)} \exp\left(\frac{2\theta w_0}{kT}\right) \quad , \tag{2.43}$$

which is similar to the Frumkin equation (2.38). If interaction is neglected, it becomes the Langmuir equation (2.41).

The lateral forces of attraction between the adsorbed molecules may lead to the formation of surface associations. DUBININ and SERPINSKY [2.131] considered such a case in the adsorption of water on carbonic surfaces and obtained an adsorption isotherm equation of a very simple type:

$$a = n_c k_1 h / (1 - k_1 h) \quad , \tag{2.44}$$

where n_c is the number of primary adsorption centers, and $h = p / p_S$. This equation is very convenient for evaluating n_c for a number of adsorption systems. A more general equation of the same type was derived in [2.132].

For a uniform surface the lateral forces of interaction are the main cause of changes in the differential adsorption heat with increasing occupation θ. In the Bragg-Williams approximation,

$$q_d = q_d^0 - z\theta w_0 \quad , \tag{2.45}$$

where q_d^0 is the adsorption heat at $\theta = 0$. Detailed calculations of tangential interactions for both dispersion forces and electrostatic forces can be found in [2.133], where repulsion ($w_0 > 0$) was shown to cause a decrease, an attraction ($w_0 < 0$) an increase in adsorption heat. Indeed, a number of authors (e.g., [2.15]) have reported a small increase in q_d during adsorption of nonpolar gases on a uniform surface of graphitized carbon black. The sign of the tangential interactions determines the shape of the adsorption isotherm. On the basis of an analysis of the sign of the virial in the two-dimensional equation, BERING and SERPINSKY [2.134] showed the isotherm for a uniform surface to be convex in the presence of repulsive forces and concave in the presence of attractive forces. It also turned out that the sign of w_0 can be judged by the thermal dependence of two-dimensional surface pressure (2.18)[2.135].

It should be noted that the existence of forces of interaction between the adsorbed molecules may lead, in physical adsorption (usually considered to be nonactivated), to the formation of a kind of activation barrier. According to [2.23], this barrier is due to the superposition of potential curves of the normal and the tangential interactions between molecules. An increase in occupation will lead to an increase in the activation energy; that is, in the presence of repulsive forces the trend of the latter will be antibatic to the differential adsorption heat.

All the calculations of the energy of lateral interactions discussed above are based on the tacit assumption that the energy of pair interactions does not depend upon the degree of occupation θ. In reality, however, as adsorption proceeds, the interaction between the molecule and its neighbors works to alter the distance between the potential wells in which the molecules are situated. This effect has been qualitatively accounted for in adsorption of dipole molecules [2.133]. When the occupation numbers reach a certain level, one must be heedful of triple interactions [2.19], and with increased θ, correlation effects can no longer be neglected. Recently, the role of correlation effects was analyzed for lateral interactions of adsorbed molecules with the aid of Green's functions [2.136].

For the sake of simplicity, we have restricted our consideration thus far to adsorption of monatomic molecules. With diatomic (polyatomic) molecules, the Hamiltonian (2.28) and the expression for the statistical sum (2.27) must account for translational *(tr)*, rotational *(rot)*, and vibrational *(vib)* degrees of freedom. Vibrational and rotational degrees of freedom are usually assumed to be separable from translational ones [2.3a,137], i.e.,

$$\Omega^{\omega} = \Omega^{\omega}_{tr}\, \Omega^{\omega}_{rot}\, \Omega^{\omega}_{vib} \quad , \tag{2.46}$$

where Ω^{ω}_{tr}, Ω^{ω}_{rot}, Ω^{ω}_{vib} are calculated according to the well-known formulae of statistical mechanics for a rigid rotator and a rigid oscillator.[3] This representation of Ω is in itself an important assumption [2.119]. The loss of degrees of freedom by a molecule (including internal degrees of freedom of the atoms' motion) upon transition into the adsorbed state is taken into account via a number of model assumptions concerning the nature and magnitude of potential barriers on the surface.

Depending on the temperature and potential shape of the interface, some degrees of freedom may become frozen. Often the molecules possess, as DE BOR put it [2.120], hopping mobility, and switch from one position of quasiequilibrium to another like the molecules in the Frenkel model of a liquid. Such a case was investigated by HILL [2.118], who calculated Ω^{ω} for an adsorbed diatomic molecule, taking into account the retardation of rotation due to the potential barriers of the surface. The calculation was performed using the Pietzer-Gwen approximation, according to which

---

3 In the general case, (2.46) must also account for changes in the electronic state of the molecule that occur in adsorption Ω^{ω}_{e} [2.138,139].

$$\Omega^{\omega} = \Omega^{\omega}_{cl} \frac{\Omega'_{qu}}{\Omega'_{cl}} \quad ,$$

where Ω_{cl} is given by (2.30), and Ω'_{qu} and Ω'_{cl} are the statistical sums for quantum and classical harmonic oscillators, respectively.

Calculations of the entropy of the surface phase are commonly used to obtain an estimate of the mobility of adsorbed molecules. The entropy can be computed from the statistical sum of the surface phase in accordance with (2.33b). KEMBALL [2.137], EVERETT [2.106], and others [2.138], by employing the sums of states for translational, rotational and vibrational degrees of freedom, and accounting qualitatively for the disappearance of some of these upon adsorption, derived comprehensive expressions for two-dimensional models of mobile and localized layers. They considered both the dissociation of molecules in adsorption and the dimeric association of adsorbed molecules. In the case of localized adsorption they also accounted for the configuration entropy η_c, which is determined by the number of possible arrangements of n^{ω} molecules over n_c adsorption centers:

$$\eta_c = R \ln G = R\left[-\ln \theta - \frac{1-\theta}{\theta} \ln(1-\theta)\right] \quad , \tag{2.47}$$

with the combinatorial term G given by (2.40). From a comparison of computed (η_{theor}) and measured (η_{expt}) values, conclusions were drawn about the mobility of molecules on various surfaces.

Methods using lattice sums have been widely employed to obtain numerical calculations of the thermodynamic parameters of adsorption systems of all kinds. Pioneering calculations of the statistical sum for argon adsorbed on an ionic crystal were performed in the early forties by ORR [2.140], and HILL [2.141] did calculations for the adsorption of nitrogen iodide and benzene on graphite surfaces. All later innovations [2.3a,20,121] were essentially improvements in the computation techniques for finding the lattice sums, and applications of these techniques to more complex organic molecules. Algorithms were developed to calculate the electrostatic field and the lattice sums for the interior of zeolite cavities, and calculations were also made of the thermodynamic parameters of adsorption equilibrium [2.121,142].

The next important step in the statistical-thermodynamic description of surface phases lies in the transition from idealized uniform surfaces to nonuniform ones. Without specifying the causes of such nonuniformity (this will be done later), we shall now sketch the main directions in the development of the statistical theory of nonuniform surfaces.

The simplest of these models represents adsorption on a surface without lateral interactions. Centers of different kinds are assumed to populate the surface. Each kind of center is characterized by its own adsorption heat q_i and degree of occupation θ_i. To describe the distribution of centers of different kinds over the surface s, the distribution function $\rho(q) = ds/dq$ is introduced. For discrete distribution of centers the total occupation is $\theta = \sum_i \theta_i \rho$. For continuous distribution, $\theta = \int \theta\rho(q)dq$. Consequently, the problem of calculating the adsorption isotherm $\theta(p)$ is reduced to finding the distribution function $\rho(q)$. ROGINSKY [2.143] solved the inverse problem of finding $\rho(q)$ from experimental measurements of adsorption isotherm $\theta(p)$. Approximate formulae for calculating $\rho(q)$ have been proposed for various typical shapes of adsorption isotherms.

Let us investigate several of the most common cases. If one adopts an exponential distribution of the heat of adsorption over the adsorption centers: $\rho(q) = q_0 \exp(-q/q_m)$, where q_0 and q_m are constants, then, as shown by ZEL'DOVICH [2.144] and later by HALSEY [2.145], integration yields the Freundlich equation

$$\theta = Cp^{RT/q_m} \quad , \tag{2.48}$$

where C is a constant. It is interesting to note that the configurational entropy η_c for such a surface is always smaller than that for a uniform surface (2.47) [2.144].

TEMKIN [2.146] adopted a linear law of changes in the heat of adsorption $q(\theta) = q_0(1 - \alpha\theta)$, where q_0 is the heat of adsorption at $\theta = 0$, and α is a constant. He obtained the following isotherm equation:

$$\theta = \frac{RT}{q_0\alpha} \ln(A_0 p) \quad . \tag{2.49}$$

The expression for the configurational entropy for this distribution function was obtained in [2.147].

TODES and BONDAREVA [2.148] and SIPS [2.149] employed the Laplace transform to find the function $\rho(q)$ from isotherms of type (2.48), a method which, however, led to an infinite number of centers on a finite surface area. To avoid this, certain conceptual assumptions were made, but these were not always supported by experiment. HORIUTI [2.150] resolved this difficulty with the help of the Sommerfeld transform used in the electron theory of metals. SPARNAAY [2.151] analyzed Langmuir adsorption on nonuniform surfaces and derived the equations of adsorption isotherms and entropy. It turned out that if the difference between the free energies of molecules adsorbed on centers of different kinds is small, this equation becomes the Dubinin-Radushkevich equation, which had been obtained from a totally different approach. The equations of adsorption isotherms, including the Temkin equation (2.49), were obtained for various functions of the distribution of adsorption centers over the magnitude of free adsorption energy $\rho(F)$. HILL [2.141] generalized the lattice methods of finding the statistical sum of an adsorbed substance to include nonuniform surfaces. If the distribution of centers over energy is considered to be a continuous function $\rho(w)$, the isotherm equation according to Hill has the form

$$\theta = \frac{\rho(w)dw}{\{1 / \Omega_1(w, T) \exp[-(\mu + w) / kT]\} + 1} \quad . \tag{2.50}$$

We can see that the problem of describing nonuniform surfaces is reduced to finding the distribution function $\rho(q)$. If this function is known, the adsorption isotherms can be determined on the basis of the formulae cited above. The overwhelming majority of the investigations mentioned here is devoted to finding the function $\rho(q)$ from experimental adsorption isotherms (see the review of these works in [2.14]). The physical meaning of the sought-for function $\rho(q)$, however, is not brought to light.

The reason for the discrepancies between the adsorption isotherm and the Langmuir isotherm (2.41) cannot be found within the framework of a phenomenological discussion. These discrepancies may result from lateral interactions between adsorbed molecules on a uniform surface, or from nonuniformity of the surface in the absence of lateral interactions. One and the same isotherm can be obtained either by choosing $\rho(q)$ with no lateral interactions, or by assuming a uniform surface with an appropriately chosen energy function of lateral interactions $w(r_{i,k})$. VOLKENSTEIN [2.23] established correlations between the function $\rho(q)$ (nonuniform surface) and the law of interaction $w(r_{i,k})$ which lead to the same most common equations of adsorption isotherms. Nonuniform surfaces cannot be described within the framework of purely adsorptive concepts. In order to understand the physical meaning of various adsorption isotherms, one must discover the physical meaning of the distribution function. This problem has not yet been completely solved.

2.1.5 Adsorbent Inertness and the Criteria for Distinguishing Different Forms of Adsorption

It follows from Sects.2.1.1,3,4 that the entire theory of physical adsorption is founded on the assumption that the adsorbent is inert. Adsorptive interaction is assumed to be of a local nature, and to cause negligibly small perturbations in the solid. All changes in thermodynamic parameters observed experimentally are attributed to the adsorbed molecules. Comparisons of these data with the results of statistical model calculations provide a basis for conclusions about the freezing of certain degrees of freedom of molecules and about their translational and rotational mobility. Let us investigate the plausibility of the assumption that the adsorbent is inert. As was convincingly proved in [2.113], a purely thermodynamic treatment of adsorption experiments leaves no possibility of dividing the perturbations so as to attribute them either to the adsorbing surface or to the adsorbed molecules.

Let us take a look at those few experiments in which certain changes in the solid surface could be observed in the course of adsorption. It was clearly shown in [2.152] that low-temperature adsorption of N_2 and O_2 on porous charcoals and glasses leads to fully reversible dilatation of the solid. This phenomenon is attributed by the authors to the change in the free energy of the solid surface. A far more pronounced deformation was observed in reversible adsorption of H_2O, NH_3, SO_2, etc. [2.152](Fig.2.3). Porous glass was observed to dilate at small occupations, to contract at greater occupations, and then to dilate again. In [2.20] dilatation was attributed to the formation of coordination bonds; contraction may be due to the formation of hydrogen bonds [2.152]. Great deformations were discovered dilatometrically in zeolites in adsorption of H_2O, CO_2, and C_6H_6 [2.153]. This is in accordance with X-ray radiographic measurements and with the results of Ir-spectrometry of valence oscillations of Al-O and Si-O bonds in a zeolite. There is strong evidence (NMR, EPR, Mössbauer spectroscopy, conductivity methods) that significant

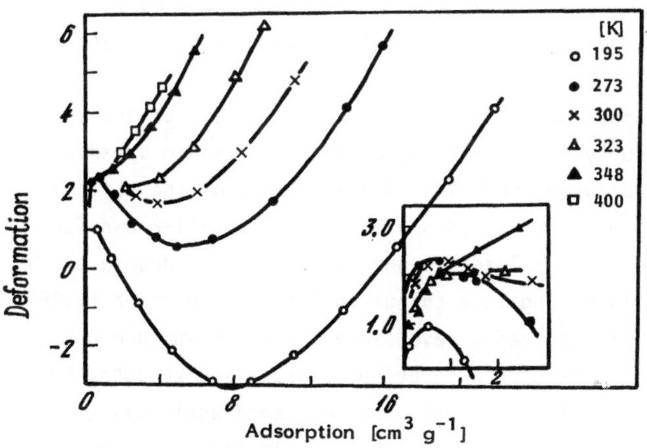

Fig.2.3. Dilatation and compression of porous glass, caused by adsorption of SO_2 at various temperatures [2.152]. The initial sections of the curves are presented in the insert

changes in the localization of cations take place during adsorption. Porous and microporous adsorbents are not inert even in adsorption of noble gases, which is the classic example of physical adsorption. The perturbations in the solid are by no means small: the dilatation of porous glass in adsorption of argon, for example is equivalent to thermal expansion at 550°. This points to large deformations, especially in view of the rigidity of valence bonds. These effects will naturally be much less noticeable in nonporous bodies.

The effects of deformation are most pronounced in swelling adsorbents, such as clayish laminated minerals [2.154]. To a lesser extent they are displayed by graphite bodies, a favorite object of investigations on physical adsorption. For instance, the authors of [2.155-157] observed an appreciable swelling of graphite in low-temperature reversible adsorption of oxygen and water vapor. TYKODY [2.158] has demonstrated that the experimentally measured isosteric heat of adsorption, q_{st}, is not equivalent to the change in enthalpy (ΔH) of adsorbed molecules:

$$\Delta H = q_{st} - \frac{T}{\Gamma} \, n^S \left(\frac{d\Delta V^S / \chi}{dT}\right)_{n^S} , \qquad (2.51)$$

where Γ is the Gibbs adsorption, n^S is the number of moles of the solid, and χ is the coefficient of volume expansion. Part of the energy of adsorption is expended on the change in the volume of the solid, ΔV^S. It is doubtful how valuable attempts are at estimating the energies and mobility of adsorbed molecules by comparing experimental values for the heat and entropy of adsorption with calculated thermodynamic potentials on the basis of statistical models (Sect.2.1.4). In some cases these calculations are performed for zeolites [2.121,159] and swelling adsorbents [2.139], in which noninertness is displayed most vividly.

When dealing with semiconductors and metals, one must take into account changes in the electronic subsystem of the solid which are caused by adsorption. Electronic migration between the bulk and the surface of the solid is by no means the prerogative of chemical adsorption. It is a well-known fact that adsorption even of inert gases leads to the formation of a double electric layer on the metal surface, and to a change in the work function [2.160-162]. FIGUROVSKAYA [2.163] discovered a reversible effect of surface charging in rutile and barium titanate upon adsorption of argon and krypton. Localization of a free carrier on the adsorptive center can drastically change the energy of dispersion interaction between the molecule and this center. ROSENBERG [2.164] noted that with semiconductors the energy of dispersion interaction between a krypton molecule and the surface differs significantly from the experimentally measured heat of adsorption. Further evidence for the interference of reversible adsorption (usually taken to be physical adsorption) with the electronic parameters of a solid surface can be found in [2.20,165]. The effects of adsorption-induced perturbations in the electronic subsystem of a solid are mistakenly ignored in statistical-thermodynamic analyses of physical adsorption.

In a rigorous approach to chemical adsorption, the solid and the adsorbate are treated as a single quantum-mechanical system [2.23]. In such a universal conceptual model, however, it is immensely difficult to discern all the specific features of the complex phenomenon of chemisorption. Therefore, the chemisorption system is usually divided into molecular and electronic subsystems which interact with one another within the framework of simplified models. For example, in [2.136,166,167] the electronic system of a metal and a degenerated semiconductor is represented by degenerated surface electron gas. A very popular model in the physics of metal surfaces with s and p bonds is the "jellium" model, in which the electron-electron interactions of valence electrons and the interactions of electrons with uniform positive ionic background are considered in a self-consistent fashion [2.1,168,169].

The electronic transitions in the adsorbate-adsorbent system lead not only to changes in the bonding energy between adsorbed molecules and the surface due to the involvement of localized charges in the chemisorptive bond, but also to changes in the energy of lateral interactions between the adsorbed molecules. Additional Coulomb forces act between molecules in a charged state, and dipole-dipole interaction occurs between the molecules, which form a neutral bonding with the surface [2.23]. KOUTECKY [2.55] pointed out still another possibility that lateral interactions are due to overlapping of the electronic clouds of interacting atoms spread in the lattice. Using the method of localized states, OSHEROV [2.170] analyzed the effects of the mutual influence of particles chemisorbed on a surface. In [2.171] a theoretical analysis is given of collective effects in a system of noble-gas molecules adsorbed on a metallic surface, and in [2.136] the contribution of lateral interactions to the heat of adsorption is estimated in degenerated surface electron gas approximation. It is evident that the electronic subsystem has a strong influence on the mechanism of interaction within the molecular subsystem.

From the above discussion it follows that the perturbations in the solid during adsorption are not specific to chemisorption — to a greater or lesser extent they are observed in physical adsorption as well. Despite the difference in the theoretical approaches to these two types of adsorption, when it comes real systems, reliable criteria for distinguishing them from one another are urgently needed. As shown elsewhere by KISELEV [2.20,165], the experimental criteria found in the literature are rather ambiguous. Some of these are listed below (see also [2.172]), together with reasons why their applicability is questionable.

1) In physical adsorption the adsorbed molecules retain their individuality. However, this also holds in such typical examples of chemisorptive bonding as coordination bonds and hydrogen bonds.

2) Physical adsorption, as opposed to chemisorption, is reversible. But there are also examples of reversible chemical adsorption.

3) Small values of adsorption heat (4 to 20 $kJ \cdot mol^{-1}$ for simple molecules and 40 to 80 $kJ \cdot mol^{-1}$ for complex molecules) indicate physical adsorption, while larger values (10 to 400 $kJ \cdot mol^{-1}$) are typical of chemisorption. This distinction is also rather arbitrary. If localization of the carrier occurs in physical adsorption, then the energy of interaction includes the energy of interaction between the localized carrier and the adsorbed molecule (as in adsorption of an inert gas upon rutile) [2.163]. At the same time, the heat of chemical adsorption can be quite small if adsorption is accompanied by dissociation of an energy-rich molecule.

4) Activation energy is present in chemisorption but not in physical adsorption. However, there are cases of chemical adsorption without activation energy — for example, adsorption of atoms on a surface with dangling bonds. And conversely, in some cases of physical adsorption, as shown in Sect.2.1.4, an activation barrier can appear as a result of interaction between adsorbed molecules.

Obviously, all these conventional criteria are quite unreliable when dealing with monomolecular adsorption.

Some researchers [2.173] take the transfer of an electron between the adsorbed molecule and the adsorbent to be the main criterion of chemical adsorption. We have already indicated, however, that physical adsorption can also cause electron transfer in the solid ($Ar - TiO_2$). In [2.20] electrical measurements were shown to be incapable in many cases of providing a means for distinguishing chemisorption from physical adsorption. For instance, in experiments with typical chemical adsorption of O_2, a clean atomic graphite surface, high adsorption heat values were observed, while the conductivity and the work function remained unchanged [2.165]. To judge by electric measurements, this is physical adsorption, whereas the adsorption heat points to chemisorption. The same ambiguity characterizes the criteria for charged and reactable forms of chemisorption [2.20].

Attempts have been made to improve the existing classification of adsorptive interactions by introducing the terms "specific" and "nonspecific" interactions. This is of little help, since it remains unclear in the individual case which properties are to be taken as "specific" and what the theoretical criteria are for distinguishing between "specific" interactions and chemical ones. There is probably little point in classifying adsorptive interactions via any specific kind of interactions. It would be more sensible to consider adsorptive interactions in accordance with the existing quantum-chemical classifications of chemical bonds, taking the peculiarities of the interface phases into account.

2.2 Heterogeneous Catalysis

2.2.1 Basic Definitions

Catalysis is defined as the initiation of a chemical reaction or a change in its rate by the action of substances called catalysts. Catalysts take part in the reaction but do not appear in the final products; the catalyst bears no stoichiometric relation to the reaction yield. In positive catalysis the reaction rate is accelerated, while in negative catalysis it is slower; the term "catalysis" usually denotes the former. Negative catalysts are called anticatalysts or inhibitors, though an inhibitor is sometimes defined as a substance which stops chain reactions, that is, breaks a chain. An increase in the rate of reaction under the influence of a final or intermediate product is called autocatalysis.

Some definitions of catalysis include additional information about (i) the chemical immutability of the catalyst in the course of the reaction, (ii) the chemical nature of the intermediate reaction, and (iii) the repeatedness of the reaction [2.174-176]. In most cases these specifics correctly represent the features of catalysis, but sometimes they only can be proved by further investigations.

For instance, the catalyst that is recovered from the reactor differs, as a rule, in chemical composition from the catalyst that had been loaded into it, and changes in the reacting components often lead to alterations not only in the structure of the catalyst, but also in its chemical composition. For this reason the concept of chemical immutability must be understood as being related to the elementary stage of reaction, and consequently, one has to find the relevant elementary stage in the given process.

The chemical nature of the intermediate interaction between the reactants and the catalyst helps draw a line between real catalysis and changes in the reaction rates due to various physical factors. An example of the latter is an increase in the rate of chemical adsorption of gases by liquids which may be achieved with a nozzle made of inert material; it increases the area of contact between phases. However, there are cases that require a more accurate definition of what chemical interaction is. For instance, hydrogen para-ortho-transition (reorientation of nuclear spin an H_2 molecule)

$$p\text{-}H_2(\uparrow\downarrow) \rightarrow o\text{-}H_2(\uparrow\uparrow)$$

at low temperatures proceeds due to reorientation of nuclear spin in a strongly nonuniform field of paramagnetic catalysts (compounds of transition metals: Co_3O_4, Cr_2O_3, etc.). The catalyst here can be a gas: O_2, NO, NO_2. Opinions as to whether this process ought to be considered chemical or physical still differ.

The accelerating action of a solvent on rates of chemical reaction in the liquid phase can be considered both "physical" and "chemical." Usually this is not classified as catalysis on the grounds that the quantity of solvent is much greater than the quantities of dissolved reactants; a catalyst is usually present in quantities which are small compared to the reactants. However, sometimes acceleration is observed in the presence of very small quantities of solvent (e.g., water), making it hard to distinguish between real catalysis and acceleration due to the action of the solvent.

The same uncertainty remains in some cases with regard to repeated participation of the catalyst in the reaction. For example, the catalytic action of iron ions in the decomposition of hydrogen peroxide can be explained, according to some accounts [2.177], by the following set of reactions:

$$Fe^{2+} + H_2O_2 = Fe^{3+} + OH^- + OH^{\cdot} \quad ,$$

$$OH^{\cdot} + H_2O_2 = HO_2^{\cdot} + H_2O \quad ,$$

$$Fe^{3+} + HO_2^{\cdot} = Fe^{2+} + O_2 + H^+ \quad .$$

The dots in the superscripts denote free radicals. The net reaction is the decomposition of hydrogen peroxide:

$$2H_2O_2 = 2H_2O + O_2 \quad .$$

The reaction in this mechanism is genuinely catalytic, since the Fe ions are repeatedly engaged in the intermediate reaction with the reactants. Previously, the Fe ions were merely thought to initiate the chains

$$Fe^{2+} + H_2O_2 = Fe^{3+} + OH^{\cdot} + OH^- \quad ,$$

$$Fe^{3+} + H_2O_2 = Fe^{2+} + HO_2^{\cdot} + H^+$$

and break them in the reaction of Fe ions with free radicals. The chain itself proceeds according to the reactions

$$OH^{\cdot} + H_2O_2 = HO_2^{\cdot} + H_2O \quad ,$$

$$HO_2^{\cdot} + H_2O_2 = H_2O + OH^{\cdot} + O_2 \quad .$$

In other words, the final product, O_2, is yielded without direct involvement of Fe ions. This example shows the difficulty involved in deciding whether the substance that accelerates the process involving radicals is a genuine catalyst, or an initiator that enters just once to start the reaction. In the literature, the initiators of radical reactions are often called catalysts.

Catalysis is considered homogeneous if the catalyst and the reactants are all in the same phase - as a rule, liquid or gaseous. In this case the catalyst is uniformly distributed over the entire volume of the reaction in a dispersed state on the molecular or atomic level. Examples of homogeneous catalysis are: numerous reactions of hydration, hydrolysis, sulfurization, halogenation, etherification and other organic processes accelerated by acids or bases in the liquid phase, and liquid-phase paraffin oxidation in the presence of salts of Co, Mn, Fe, Cu, etc.

When the catalyst and the reactants are in different phases which have a common dividing surface, the terms "heterogeneous" or "contact" catalysis are used [2.178]. The most common case is when a solid catalyst accelerates a reaction between gases or in a solution. Heterogeneous catalysis is vital in ammonia synthesis and oxidation, oxidation of SO_2 into SO_3, hydrogen production by conversion of mixtures of $CO + H_2O$ or $CH_4 + H_2O$, dehydrogenation of paraffins into olefins and further into dienes, the cracking and reforming of oil, and other industrial-scale processes.

Somewhere between these types of catalysis lies microheterogeneous catalysis on colloid particles in a liquid. This type is represented by biochemical processes in living organisms, which are governed by organic catalysts called enzymes.

In what is called heterogeneous-homogeneous catalysis the reaction, having started on the surface of a solid catalyst, continues in the bulk, due to the outflow of active intermediate particles from the surface. Examples of this are hydrogen oxidation on platinum and the oxidation of hydrocarbons on the oxides of transition metals. At low temperatures these reactions occur on the surface, at high temperatures they proceed homogeneously, and at moderate temperatures they follow the heterogeneous-homogeneous pattern. Usually they are exothermal radical chain reactions, which take place in both gaseous and liquid phases.

From a thermodynamic point of view, all catalytic processes are spontaneous ones; in other words, they are accompanied by a decrease in free energy. The catalyst does not induce a shift in the position of equilibrium of the chemical reaction, if one disregards the possibility of the catalyst having an indirect influence on the activity of reactants in solutions or at great pressures in the gaseous phase. Near the point of equilibrium one and the same catalyst accelerates both direct and reverse reactions equally.

Substances are called promoters or activators if, when added to the catalyst, they increase its activity and/or stability. If a promoter is added in relatively large quantities and/or if it is catalytically active as such, then the resulting catalyst is often called a mixed one. The action of mixed catalysts is, as a rule, nonadditive. Often it is impossible to distinguish between promoted catalysts and mixed ones. Usually an admixture is called a promoter if its relative quantity does not exceed the order of hundredths. The majority of industrial catalysts are promoted ones. For instance, the iron catalyst used in ammonia synthesis contains as promoters Al_2O_3, K_2O, and sometimes CaO, totalling several percent. Each of these serves a different purpose: Al_2O_3 is the structure-forming dope, which acts to increase the area of the catalyst and to prevent sintering of Fe crystals, and K_2O is an activator which boosts the effectiveness of catalytic centers of iron, or forms new active centers.

Substances which slow or completely stop the action of a catalyst (usually a solid one) are called catalyst or contact poisons. Examples are oxygen and its compounds (H_2O, CO, etc.) and, in ammonia synthesis, sulfur compounds (H_2S, CS_2). In organic reactions on a solid catalyst the catalytic action is often reduced due to the formation of polymer film on the catalyst surface – so called coke. The catalyst can be regenerated by treating it periodically with oxygen and burning out the coke.

An admixture which is a promoter at some temperatures and concentrations and a poison at others is called a modifier.

In heterogeneous catalysis, catalysts with carriers are often used. A carrier is a substance which by itself exhibits little or no activity, but which boosts the activity of a catalyst deposited upon it. In the simplest case the carrier serves to increase the area of the active catalyst and prevent sintering. For this reason one usually employs as carriers solids with a highly developed surface, both man-made (active carbons, SiO_2, Al_2O_3, aluminosilicates, MgO) and natural (clays, pumice, diatomites, asbestos, etc.). In some cases the carrier ensures the stability of the necessary chemical form of the catalyst.

2.2.2 Phenomenological Description of a Catalytic Reaction

The rate of a catalytic reaction

$$mA + nB + \ldots \rightarrow m'A' + n'B' + \ldots \quad , \tag{2.52}$$

or the change in volume concentration of the reactant c_i per unit of time, is described by the kinetic equation

$$v = kf(c_A, c_B, \ldots, c_{A'}, c_{B'}, \ldots) \quad , \tag{2.53}$$

where k is the reaction rate constant and f is a function of the concentrations of reacting substances and reaction products. In the simplest cases of homogeneous and heterogeneous catalysis, at low concentrations of reactants and provided there is no poisoning of the catalyst by the reaction products, (2.53) takes the form

$$v = kc_A^m c_B^n \quad , \tag{2.54}$$

where m and n are the stoichiometric coefficients in (2.52). The sum of the power indices in the kinetic equation (2.54) is called the order of the reaction. In heterogeneous catalysis the order of the reaction is usually lower than the sum of coefficients in (2.52), and ranges in most cases from 0 to 2.

In many instances of homogeneous catalysis the reaction rate is proportional to the concentration of the catalyst, c_c. In (2.53,54) it can be singled out as a separate factor:

$$k = k_{hom} c_c \quad , \tag{2.55}$$

where k_{hom} is the specific reaction rate constant, reduced to the concentration of the catalyst. Similarly, for heterogeneous catalysis one can write

$$k = k_{het} s \quad , \tag{2.56}$$

where k_{het} is the specific rate constant, reduced to the specific surface area of the catalyst, which has the dimensionally $m^2 \cdot g^{-1}$. From (2.55,56) it follows that the rate of homogeneous catalytic reaction is proportional to the concentration of the catalyst, and that the rate of heterogeneous catalytic reaction is proportional to the surface of the catalyst. Notice, however, that linear dependence is more common for homogeneous catalysis than for heterogeneous catalysis. Equation (2.56) can lose its validity when a heterogeneous catalytic reaction is complicated by diffusion, nonuniformity of surface, etc.

In an earlier period, scientists related the total rate of heterogeneous catalytic reaction (the reaction yield per unit of time) to the unit of volume or mass of the catalyst. In the 1950s, with the development of reliable methods of measuring the surface area of powders, the heterogeneous reaction rate started being reduced to the unit of area. The resulting quantity was called the specific catalytic activity [2.179].

The continued development of experimental techniques has made it possible to consider relating the rate of heterogeneous catalytic reaction to a single active center of the surface (see, e.g., [2.180] on catalysis on ultrapure metals). The employment of ultrahigh-vacuum technology ($< 10^{-10}$) has yielded pure-atomic metallic surfaces and made it possible to find elementary rate constants for the interaction of molecules with surface atoms numbering about $n_c \sim 10^{15}$ per square centimeter. BOUDART [2.181] coined the term "turnover number" (t_n) to denote the number of reactant molecules undergoing the desired transformation on a single catalytic center at a given temperature, pressure and proportion of reactants per unit of time. The rate of the reaction is then

$$v = n_c t_n \quad . \tag{2.57}$$

For many catalytic reactions the turnover number ranges from 10^{-2} to $10^2 s^{-1}$. These values of t_n reflect the conditions of measurement: such rates are usually observed at moderate temperatures and pressures. The upper limit for t_n corresponds to the diffusional retardation limit for reactions taking place at normal pressure. At lower pressures t_n may reach $10^{-4} - 10^{-6} s^{-1}$ [2.182]. For instance, t_n for one and the same reaction of NH_3 oxidation ranged from $5 \cdot 10^4 s^{-1}$ in an industrial-scale diffusional reactor [2.183] to $10^{-6} s^{-1}$ in a reaction on a platinum monocrystal at low pressures [2.184].

The number of active centers on an oxide is harder to evaluate. However, some progress has been made here as well. The reaction rate on a surface has been successfully related [2.185] to the concentration of surface compounds observed spectroscopically, and thus to the concentration of certain surface centers.

The rate constant of a catalytic reaction depends on the temperature in accordance with the Arrhenius equation:

$$k = k_0 \exp(-E_a / RT) \quad , \tag{2.58}$$

where k_0 is the frequency factor and E_a is the energy of activation measured in $kJ \cdot mol^{-1}$. The frequency factors are usually calculated on the basis of the theory of absolute reaction rates, which was first applied to adsorption and catalysis by TEMKIN [2.186] more than 40 years ago. Basically it involves calculating, by methods of statistical mechanics, the concentration of so-called active complexes. An active (activated) complex, or a transition state, is a transient configuration of reacting atoms which corresponds to the peak of the potential barrier surmounted by the reaction (Fig.2.4). The equation for calculating k_0 according to the theory of absolute reaction rates is, in its final form:

$$k_0 = \chi \frac{kT}{h} \frac{\Omega^*}{\Omega_A^m \Omega_B^n \ldots} \quad , \tag{2.59}$$

where χ is the so-called transmission coefficient, which reflects the probability of a particle surmounting the barrier; k is the Boltzmann constant; h is the Planck constant (at normal temperature kT/h is close to $10^{13} s^{-1}$, or to the frequency of molecular oscillations); Ω_A, Ω_B, ... are the statistical sums of reacting molecules, calculated on the basis of the molecular structure (see Sect.2.1.4) and Ω^* is the statistical sum of the active complex, calculated on the basis of the assumed reaction mechanism.

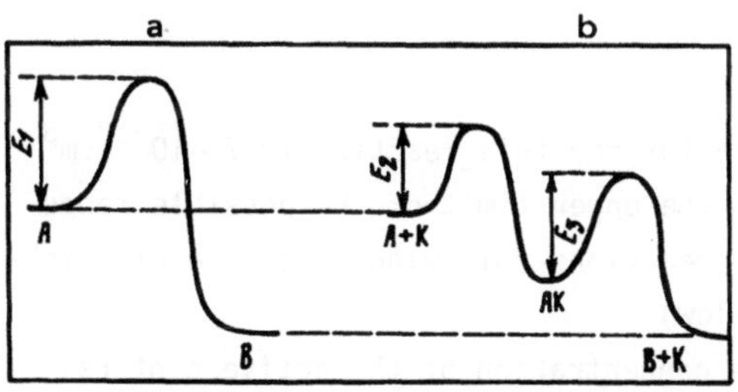

Fig.2.4a,b. Energy diagram of A → B reactions. (a) Without a catalyst. (b) In the presence of catalyst K. E_1, E_2, and E_3 are the corresponding values of the activation energy

The distinctive feature of (2.59), when used for heterogeneous catalytic reactions, is the simplifying assumption that the active complexes can be considered to be static, and devoid of translational and rotational degrees of freedom. At low temperatures, $\Omega^* \sim 1$. The same applies to calculations of the statistical sums of molecules of substances A, B, etc., provided they enter the reaction from the adsorbed state. The application of (2.59) to heterogeneous catalysis is complicated by the necessity of estimating the number of sites (active centers) occupied by the active complex (especially when occupation is far from complete), and accounting for all possible orientations of the active complex.

The application of the theory of absolute reaction rates to heterogeneous catalysis was discussed in [2.3a,174-176,187]. All of these, however, avoid the question of which values of the frequency factor can be taken as normal. The plausible range of variation of k_0 for reaction rate constants in various types of simple heterogeneous catalytic reactions (reduced to a single active center) was calculated in [2.188]. For single-center adsorption of atoms the value of k_0 is of the order of $10^{-11}-10^{-12}\,cm^3\,atom^{-1}s^{-1}$; for dissociative adsorption on two centers $k_0 \sim 10^{-27}-10^{-30}\,cm^5molecule^{-2}s^{-1}$; for bimolecular catalytic reaction in an adsorbed layer $k_0 \sim 10^{-1}-10^{-8}\,cm^2molecule^{-1}s^{-1}$; and for bimolecular reaction following the Eley-Rideal mechanism $k_0 \sim 10^{-10}-10^{-18}\,cm^3molecule^{-1}s^{-1}$.

Comparison of these figures with experimental results is usually complicated by the absence of information on the number of active centers on the catalyst surface. Nevertheless, the calculated data is well confirmed (within 1 or 2 orders of magnitude) by experimental research carried out on metallic surfaces purified in an ultrahigh vacuum. The number of active centers in these experiments is close to the number of surface atoms. For instance, oxidation of CO on platinum proceeds according to the mechanism of interaction between CO and adsorbed oxygen:

$$CO_{(gas)} + Pt \cdot O_2 \rightarrow CO_2 + O \cdot Pt \quad ,$$

$$CO_{(gas)} + Pt \cdot O \rightarrow CO_2 + Pt \quad .$$

According to [2.180], the frequency factor for this reaction is $7 \cdot 10^{-12}\,cm^3\,molecule^{-1}s^{-1}$. This value is close to the upper limit of the possible range of k_0 (ca. $10^{-10}\,cm^3molecule^{-1}s^{-1}$) for reactions following this type of sticking, or Eley-Rideal mechanism (see below).

MAATMAN [2.189] has calculated the concentration of the active centers, n_c, by dividing the experimentally obtained value of k_0 by its value calculated according to the theory of absolute rates. He found n_c to be great for metals ($10^{14}-10^{15}\,cm^{-2}$), and small for semiconductors and dielectrics ($10^8-10^{12}\,cm^{-2}$). One should, however, be aware of the possible inaccuracy of these experiments. MIYAHARA and KAZUSAKA [2.190], having improved calculations of the entropy of an active complex on a surface, revised Maatman's data. According to the latter authors, the turnover number t_n for many reactions featuring removal or addition of hydrogen (hydrogenation, dehydrogenation, $H_2 - D_2$ exchange) is about $10^1 - 1 s^{-1}$, regardless of the type of catalyst. For reactions involving nitrogen atoms, the turnover number t_n is of the order of $10^{-8} - 10^{-9} s^{-1}$. Consequently, in the first case k_0 is greater than in the second. The problem of obtaining reliable values of n_n and t_n from experimental measurements of reaction rate v is far from being completely solved. The latest data on catalysis on so-called stepped (vicinal) surfaces of metallic monocrystals [2.184,191] indicate that the number of active centers n_c on metallic surfaces is also much smaller than $10^{15} cm^{-2}$. The reaction often proceeds on structural defects, dislocations, or edges. After colliding with the surface, the molecules migrate towards the active center, where the reaction occurs. The turnover number, reduced to the number of real active centers, can turn out in this case to be even greater than the number of collisions of interacting particles with the surface.

According to the theory of absolute reaction rates, the energy of activation, E_a, is equal to the mean excess energy of active complexes over the mean energy of initial molecules, i.e., the heat of formation of an active complex. Thus, to be able to calculate the activation energy, one must be able to calculate the height of the potential barrier and, in the general case, the surface of potential energy. Usually, however, the values of E_a are obtained experimentally from the temperature dependence of the reaction rate constant according to (2.58).

Certain authors [2.174] distinguish between the energy of activation and the heat of activation. In their usage, the energy of activation in equal to the minimum energy the molecule must possess for the reaction to proceed, and the quantity E_a in (2.58) is the heat of activation, which exceeds the energy of activation by RT. This difference, however, is within the accuracy of experiment, and therefore the term "heat of activation" is avoided by most researchers.

The energy of activation in a catalytic reaction is usually smaller than that for a noncatalytic one. For example, for noncatalytic decomposition of NH_3 into N_2 and H_2, $E_a \sim 320$ kJ $\cdot$ mol^{-1}, while for the same reaction in the presence of Pt or W, $E_a \sim 160$ kJ $\cdot$ mol^{-1}. For the majority of catalytic reactions E_a is of the order of 10 - 100 kJ $\cdot$ mol^{-1} or even smaller. This reduction of E_a is due to the fact that the catalytic reaction takes a different course, via elementary reactions whose energies of activation E_a^1, E_a^2 ... are each smaller than the energy of activation of a noncatalytic reaction. The scheme in Fig.2.4b corresponds to a so-called stepwise mechanism of catalysis, in which the one-step reaction illustrated in Fig.2.4a,

$$A \rightarrow B \quad ,$$

is replaced by a number of steps:

$$A + K \rightarrow AK \quad ,$$
$$AK \rightarrow B + K \quad ,$$

where K is the catalyst. Each of these steps has a lower activation energy (and displays lower thermal effects), as is shown in the diagram. For a bimolecular reaction

$$A + B \rightarrow C + D$$

the step-by-step mechanism can take place according to the scheme

$$A + K \rightarrow AK \quad ,$$
$$AK + B \rightarrow K + C + D \quad .$$

This pattern is sometimes followed by reactions of catalysis oxidation, e.g., the oxidation of H_2 upon metals M:

$$M + \tfrac{1}{2} O_2 \rightarrow MO \quad ,$$
$$MO + H_2 \rightarrow M + H_2O \quad .$$

Multiple-stage processes should not be confused with one-stage (also called associative or concerted) processes, which follow the scheme

$$A + B + K \rightarrow ABK^* \rightarrow K + C + D \quad .$$

In this case, if ABK* is an activated complex, the catalytic reaction surmounts just one potential barrier, as shown in Fig.2.4a, though with a somewhat lower value of E_a. The lowering of the activation energy in catalytic processes is the main reason for the increased rate of catalytic reactions as compared to noncatalytic ones.

A heterogeneous catalytic reaction is usually a complex reaction which takes place in the monomolecular layer on the surface of the catalyst. Splitting a complex reaction into elementary steps is a very complicated experimental task which scientists have only recently begun to cope with successfully, thanks to the development of novel experimental techniques.

If in a sequence of steps

$$A \overset{1}{\rightleftarrows} A_1 \overset{2}{\rightleftarrows} A_2 \rightleftarrows \ldots A_{\ell-1} \overset{\ell}{\rightleftarrows} A_\ell \rightleftarrows \ldots B$$

it is possible to single out a step ℓ whose rate of direct reaction is significantly slower than the rates of all other reactions in the same direction ($v_\ell \ll v_i$), then this step is called the limiting (rate-controlling) step. The other steps proceed relatively rapidly, and static equilibrium is practically established. It is convenient to introduce the difference in rates between the direct reaction and the reverse reaction: $\Delta v_i = v_i - v_{-i}$. In the overall process $\Delta v_1 = \Delta v_2 = \Delta v_\ell$. For the limiting step $\Delta v_\ell = v_\ell$, and usually

$$k_\ell \ll k_i \quad , \quad E_a^\ell > E_a^i \quad .$$

The overall reaction rate is determined by the rate of the limiting step:

$$v = v_\ell = \Delta v_\ell \quad ,$$

while all the other steps have no influence on the total reaction rate. If equilibrium is not established in the other steps, then the catalytic process will have no limiting step.

Heterogeneous catalysts often are highly porous solids (with a specific surface area of 10 - 100 $m^2 \cdot g^{-1}$ or more), and catalytic reactions take place at high pressures and temperatures. Therefore the overall reaction rate can be limited by the transport of reactants to the surface and the removal of reaction products; these processes then play the role of the lmiting stage. A heterogeneous catalytic reaction can be divided into the following stages: diffusion of components to the catalyst, adsorption, one or more chemical transformations on the surface, and desorption and backflow of the reaction products.

A distinction is made between the kinetic region of catalysis, where the reaction rate is determined directly by chemical transformations on the surface; the region of internal diffusion, where the limiting step is connected with the transport of substance in the pores within the grain of the catalyst; and the region of external diffusion, where the reaction is limited by the transport of reactants from the gaseous or liquid phase toward the external surface of the catalyst. Theory predicts [2.192] that if the reaction rate in the kinetic region is

$$v = kc^n \quad , \tag{2.60}$$

then the reaction rate in the region of internal diffusion is

$$v = s^* \sqrt{D^* k}\, c^{(n+1)/2} \quad , \tag{2.61}$$

where s* is the external surface area of the catalyst, and D* is the effective coefficient of diffusion in a porous medium. From (2.61) it is clear that following an increase in temperature, the observed activation energy is reduced twofold upon transition from the kinetic region to the region of internal diffusion (Fig.2.5). The reaction orders for all components decrease twofold, while the order for the reactant whose diffusion restricts the overall reaction rate takes the mean value between the true order and 1, and equals $(n+1)/2$ [2.192]. Upon transition into the region of external diffusion the temperature coefficient drops yet more abruptly, and the temperature dependence of the reaction rate (and of the apparent reaction rate constant; see Fig.2.5) is determined by the temperature dependence of the coefficient of diffusion, D (in gases and liquids D is usually proportional to $T^{1.5}$ or $T^{1.75}$). The transition from the kinetic region to the region of diffusion can be also induced by an increase in pressure.

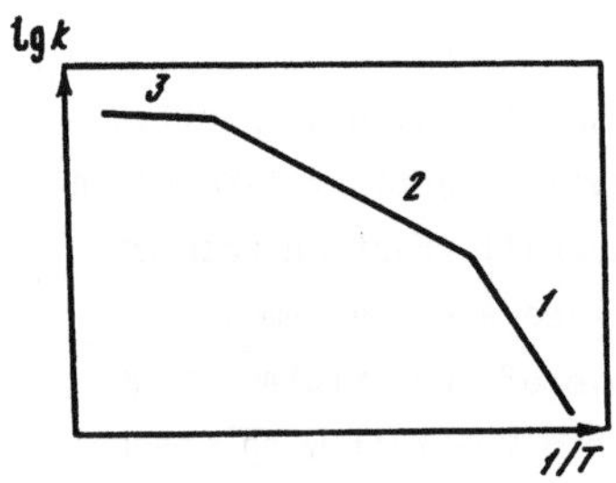

Fig.2.5. Temperature dependence of the apparent rate of catalytic reaction on a porous catalyst. (*1*) Kinetic region. (*2*) Region of internal diffusion. (*3*) Region of external diffusion

Diffusional processes are disadvantageous in step-by-step multistage catalytic reactions, when difficulties involved in the withdrawal of the desired reaction products lead to further transformation of the valuable product. For instance, in ethylene oxidation into C_2H_4O on a porous catalyst, further oxidation yields useless carbon dioxide: $C_2H_4O \rightarrow CO_2$. Consequently, if the desired product is an intermediate product in a chain of catalytic reactions, the employment of porous catalysts or carriers often becomes impractical. Diffusional complications are eliminated by crushing the grains of the catalyst and increasing the agitation of the solution in which catalysis takes place.

It is a generally accepted view that the path of catalysis proceeds via not just any kind of adsorption, but by chemisorption alone (Sect.2.1.2). For catalysis to proceed, it is often necessary that the bonding energy of the surface compound with the catalyst fall within certain limits; it must not be too great or too small. If the bonding between the molecule and the surface is too strong, further reaction may be prevented; if it is too weak (e.g., due to physical adsorption), the molecule is not activated. An ideal "moderate" bonding is the coordination bond, which will be discussed in detail in the chapters to follow.

The potential barrier for a molecule directly hitting the active center is usually low, so adsorption is rapid. There are, however, cases of slow chemisorption, or "activated adsorption." Such chemisorption may play the role of the limiting step. For example, the rate of synthesis of ammonia on iron catalyst is commonly assumed to be limited by activated adsorption of nitrogen on iron [2.193].

Activated adsorption is usually irreversible, and in static conditions will proceed until the surface is completely occupied. Its rate on a uniform surface is proportional to the pressure of the substance being absorbed, p, and to the number of free active centers, $n_c s(1-\theta)$:

$$v_{ads} = -\frac{dn}{dt} = k_a n_c s(1 - \theta)p \quad , \tag{2.62}$$

where k_a is the adsorption rate constant; $(1-\theta)$ is the fraction of the surface that remains unoccupied; s is the surface area of adsorbent or catalyst; and n_c is the number of centers per square centimeter (in adsorption of atoms on a pure metallic surface, $n_c \sim 10^{15} cm^{-2}$). Since the number of collisions of molecules of mass m with a unit of surface area at temperature T is equal to $p(2\pi mkT)^{-1/2}$, where k is the Boltzmann constant, (2.62) can be rewritten in the form

$$v_{ads} = k_S p(2\pi mkT)^{-\frac{1}{2}} n_c(1 - \theta) \quad . \tag{2.63}$$

The adsorption rate constant in activated adsorption is

$$k_S = k_S^0 \exp(-E_a / kT) \quad , \tag{2.64}$$

where E_a is the activation energy of adsorption.

After integrating (2.62) we find the dependence of adsorption a upon time:

$$a = ns\theta = ns[1 - \exp(-k_a t)] \quad . \tag{2.65}$$

The rate of desorption on a uniform surface is proportional to the occupied surface:

$$v_{des} = k_d ns\theta \quad . \tag{2.66}$$

The term $k_S(1-\theta)$ in (2.63) is the so-called sticking coefficient, which is equal to the ratio of the number of adsorbed molecules to the total number of molecules hitting the surface. According to (2.63), it ought to depend upon occupation as $(1-\theta)$, that is, decrease linearly with an increase in occupation. However, meticulous investigations on the adsorption of common gases (H_2, O_2, CO, CO_2) on monocrystals of W, Pt, etc. [2.194] have indicated that the sticking coefficient is independent of θ up to almost-complete occupations (formation of a monolayer). It appears that before the molecule enters the chemisorbed state, in which it is strongly tied to the surface, it wanders across the surface freely in a kind of presorbed (precursor) state, and this wandering takes place over the chemisorbed layer as well. The nature of such a precursor state is not yet clear. Some authors believe the precursor to be a physically adsorbed molecule which takes some time before entering the chemisorbed state. Other researchers think it is an excited molecule (for more detail see Chap.5).

In most cases the surface of adsorbents and catalysts is nonuniform, and its various patches display different catalytic and adsorptive properties. Catalysis and activated adsorption proceed primarily on the most active sections, the so-called active centers. Curvent views on the structure of active centers are discussed in Chaps.3 and 4. Here we shall just mention that active centers are those surface centers on which adsorption and catalysis proceed with minimal activation energy; the constants k in (2.55-58,60-62) consequently take on their maximal values.

The rate of activated adsorption on nonuniform surfaces decreases with increasing surface occupation. A detailed theory of adsorption and catalysis on nonuniform surfaces has been developed by Soviet scientists [2.143,144,146, 193b].

The problem of constructing the kinetic equation of adsorption on the basis of the distribution function of active centers on the surface over the activation energy of adsorption ($\rho(E_a) = ds/dE_a$) is quite similar to the problem of finding the equilibrium isotherm of adsorption from the function $\rho(q) = ds/dq$, which was discussed in Sect.2.1.4. If the distribution of centers is discrete, then the general occupation $\theta = \sum_i \theta_i \rho$. For continuous distribution, $\theta = \int \theta\rho(E_a)dE_a$. ROGINSKY [2.143] solved the inverse problem and found the function $\rho(E_a)$ from experimental data on the kinetics of adsorption. It has been shown, for example, that the rate of adsorption for a uniform distribution of active centers over the activation energies, $\rho(E_a) = \text{const}$ (Fig.2.6a), is described by the Zel'dovich-Roginsky equation:

$$\frac{da}{dt} = bp\,e^{-\alpha a} \quad , \tag{2.67}$$

or in integral form at constant pressure:

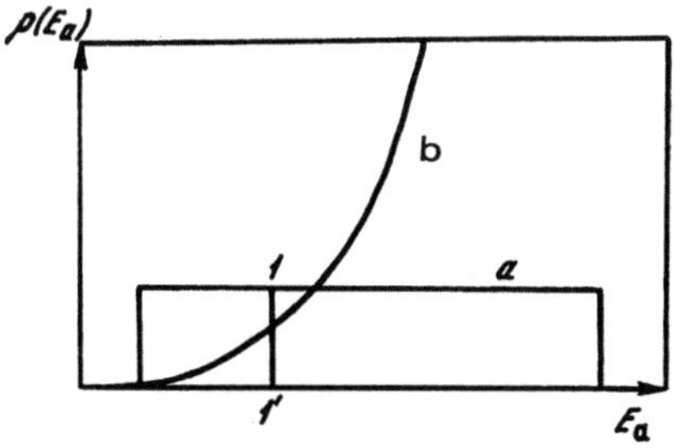

Fig.2.6a,b. Homogeneous (a) and exponential (b) distributions of active surface centers over the activation energy of adsorption. The line extending from 1 to 1' marks the boundary between occupied and free regions

$$a = B \ln t + C \quad , \tag{2.68}$$

where a is the magnitude of adsorption at time t; and α, b, B, C are constants. Equation (2.68) is widely employed to deseribe adsorption on semiconductors and metals. It is often incorrectly called the Elovich equation [2.195]; the pioneering work of ROGINSKY and ZEL'DOVICH [2.196] was published five years earlier than the articles by ELOVICH [2.197-199] which are usually cited. The kinetic equation (2.68) appears to be similar to the Temkin equilibrium isotherm (2.49). These equations can be derived from each other by means of the commonly (though not always justifiably) accepted linear dependence between the activation energy and the heat of adsorption:

$$E_a = E_0 - \alpha q \quad . \tag{2.69}$$

Equation (2.69) is called the linearity relation, or the Brönsted-Polanyi-Temkin relation, and is widely employed in the theory of chemisorption and catalysis. It was first formulated by TEMKIN for heterogeneous catalysis [2.193a] and later modified by Brönsted and Polanyi for homogeneous catalysis.

The rate of adsorption for exponential distribution of centers over activation energies of adsorption, $\rho(E_a) = \beta \exp(\alpha E_a)$ (Fig.2.6b), is described by the Banham-Barth power equation

$$a = B' t^{1/m} - C' \quad , \tag{2.70}$$

where B', C', m are constants. Equation (2.66) gives a good description of many experimental results; it may correspond to the distribution of active centers over the heats of adsorption, which leads to the Freundlich equilibrium adsorption isotherm (2.48).

One cannot tell merely by looking at (2.68,70) whether the divergence of the real kinetics of adsorption from the equation for a uniform surface (2.65) is caused by nonuniformity of the real surface; similar results can be obtained if it is assumed that there is interaction between chemisorbed atoms or molecules (Sect.2.1.4). Various methods have been developed to distinguish between the effects of interaction and the effects of nonuniformity; tracer-atom methods are the most reliable [2.200].

The rate of the catalytic reaction proper on a nonuniform surface can be described by equations which are even more complex. However, in stationary conditions - for instance in a continuous flow of reactants in fixed proportion through a tube filled with a catalyst of nonvarying activity—the catalytic reaction proceeds on areas having approximately the same activity, near the boundary between free and occupied centers on curve $\rho(E_a)$ (Fig.2.6). In this case the rate of catalysis on a nonuniform surface is adequately described by equations for a uniform surface. In some cases the entire surface of the catalyst is uniform, for example in certain simple reactions on the surface of metals purified in an ultrahigh vacuum [2.180]; catalysis apparently takes place on all surface atoms of Pt or Pd, and all the atoms have the same properties.

If the catalytic reaction $A+B+\ldots \rightarrow A'+B'+\ldots$ involves particles adsorbed on the surface, A_{ads}, B_{ads}, etc. (the so-called Langmuir-Hinshelwood mechanism), then the reaction rate is proportional to the occupation of the surface by adsorbed particles. The equation for the reaction rate of a bimolecular reaction $A+B\rightarrow$ (products) has the form:

$$v = ks\theta_A\theta_B \quad , \tag{2.71}$$

where θ_A, θ_B represent the fractions of the surface area occupied by adsorbed molecules A and B. The occupation θ for a uniform surface and reversible adsorption is described by the Langmuir adsorption isotherm (2.41). According to (2.41), at large values of adsorption heat and high pressures, $b(T)p \gg 1$ and $\theta \sim 1$ (saturation region); that is, the entire surface is covered by adsorbed molecules. In this case, in accordance with (2.41), the rate of catalytic reaction does not depend on the pressure of the reactant. At low pressure of low values of adsorption heat, $b(T)p \ll 1$, and the Henry law (2.36) is valid: $\theta = b'p$.

For the rate of a catalytic reaction in the Henry region with respect to the two substances A and B we get the equation

$$v = b_A' b_B' p_A p_B = k' p_A p_B \quad . \tag{2.72}$$

The new reaction rate constant k' includes here the adsorption coefficients b_i as multipliers. The activation energy E_a' is found from the experimental dependence of k' on temperature:

$$E_a' = E_a - q_A - q_B \quad . \tag{2.73}$$

Activation energy E_a' is smaller than E_a, which is measured in the saturation region; the difference is the sum of the adsorption heats q_A and q_B. Sometimes E_a is called the "true," and E_a' the "apparent" energy of activation. If the substances A and B are adsorbed on the same active centers, an additional term $b_B p_B$ appears in the denominator of the Langmuir equation (2.41), creating further complications.

Another common mechanism of catalysis, besides the Langmuir-Hinshelwood mechanism, is the collision (sticking) mechanism, often misnamed the Rideal mechanism. Particle A from the gaseous phase collides with adsorbed particle B, thus forming the reaction product. An example of such a reaction was given above: the interaction of gaseous CO with oxygen adsorbed on platinum. The rate of the reaction

$$A_{gas} + B_{ads} \rightarrow \text{(products)} \qquad \text{is}$$

$$v = ksp_A\theta_B \quad . \tag{2.74}$$

When analyzing the adsorption mechanisms of heterogeneous catalysis (2.71,74), one should bear in mind that overly strong chemisorption is often disadvantageous for catalysis. In their work on the kinetics of CO oxidation on MnO_2, ROGINSKY and ZEL'DOVICH [2.196] showed that strong chemisorption of CO via the formation of a surface carbonate poisons the catalytic reaction. The reaction proceeds through the formation of intermediate $MnO_2(CO)$, which leads either to catalysis or to strong chemical adsorption:

$$MnO_2 + CO \rightarrow MnO_2(CO) \begin{cases} \xrightarrow{+O_2} MnO_2 + CO_2 \\ \xrightarrow{+O_2} MnO \cdot CO_2 \end{cases}$$

The intermediate compound $MnO_2(CO)$ is the very precursor mentioned above. Such a precursor apparently has a certain degree of mobility and may diffuse towards an active center. The latest works on catalysis throw some light on the role of precursors in catalysis.

Many catalytic processes involve a great number of both simultaneous and consecutive reactions. For instance, oxidation of propylene on bismuth or cobalt molybdate yields acetaldehyde (CH_3CHO), formaldehyde (CH_2O), acrolein ($CH_2=CHCHO$), acrylic acid ($CH_2=CH\text{-}COOH$), and products of deep oxidation (CO and CO_2), and the reactions may take place either simultaneously or consecutively:

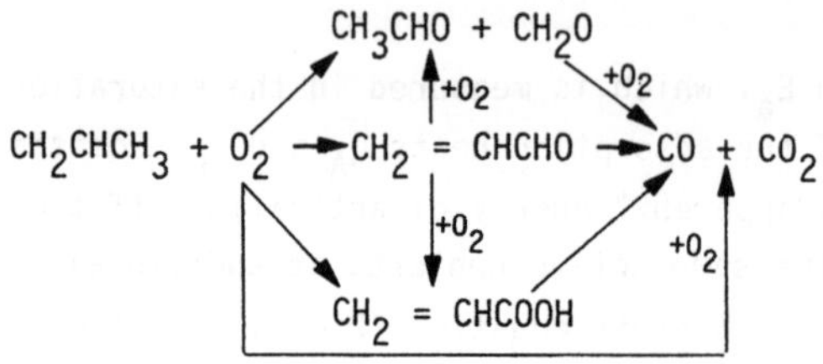

Different catalysts accelerate different paths of this process: $Bi_2O_3 \cdot MoO_3$ shifts the process towards acrolein production, whereas simple oxides of transition metals accelerate the formation of deep oxidation products.

Of the possible paths, a catalyst does not necessarily boost the one which is thermodynamically most efficient, that is, the one which corresponds to the greatest change in the free energy. This ability of a catalyst to accelerate only one of several possible reactions is called selectivity. It is measured by the ratio of the rate of the reaction yield the desired product to the rate of the overall transformation of the initial product: $v_i / \sum_i v_i$. Selectivity is often more important for industrial purposes than high catalytic activity.

Organic catalysts, or enzymes, are especially selective; they usually accelerate just one reaction of one substance from a great number of possible reactions. E. Fischer is often quoted for the metaphor that a ferment is to a substrate what a key is to a lock. The latest investigations have revealed that the high specificity of enzyme action is due to its ability to change its molecular conformation in the presence of the substrate molecule. The enzyme molecule bends in such a way that its active centers K_i are adjusted to the reactable centers of the substrate molecule (Fig.2.7), following a kind of "hand-and-glove" mechanism. There are indications that such an adjustment of active centers in the presence of reactants also takes place in certain instances of heterogeneous catalysis.

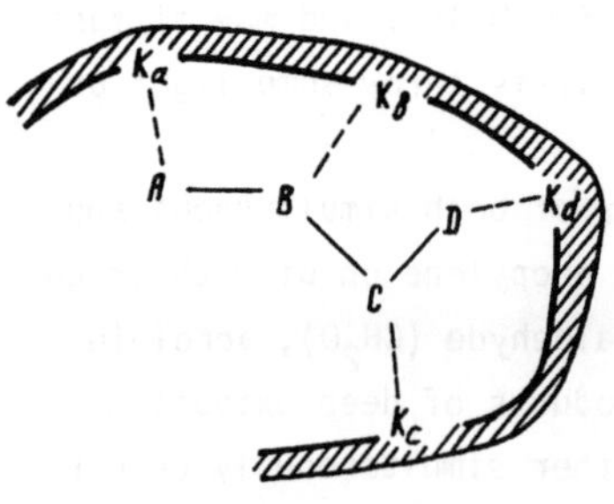

Fig.2.7. Mechanism of the selective action of an enzyme on an ABCD molecule. The active centers of enzyme K adjust themselves to the molecule

To construct the kinetic equation of a complex catalytic reaction, one must divide the overall reaction into possible elementary steps, write the differential equations for the reaction rates of each of these steps, and, having solved this set of equations, compare the solution with the experimentally observed dependence of the rate of transformation upon the concentrations c_i or partial pressures p_i of all participating substances. The differential equations encountered here are often quite complicated, and can be solved only by numerical methods, so that one must be content with approximate solutions.

A widely accepted approximation is the Bodenstein-Semenov method of quasistationary concentrations. If a process takes place via the formation of intermediate active particles of radicals, ions, or complexes, then after a certain time lapse their quasi-stationary concentration is established. In this case the rate of their formation is equal to the rate at which they are used up, or consumed: $v_{form} = v_{cons}$, and the total rate of change in their concentration is zero: $v_{form} - v_{cons} = 0$. Since v_{form} and v_{cons} depend upon concentrations c_i (or pressures p_i) of reactants A, B, ... and intermediate particles M, N, ..., it follows that $v_{form} = kf(c_A, c_B, \ldots, c_M, c_N, \ldots)$ and $v_{cons} = k'f'(c_A, c_B, \ldots, c_M, c_N, \ldots)$. Thus, the set of differential equations is replaced by a set of algebraic equations, which is much easier to deal with. New approaches to the kinetics of complex catalytic reactions are described in [2.201-203].

The agreement between the experimental kinetic equation and the theoretical one derived from certain assumptions about the reaction mechanism is not in itself sufficient proof of the correctness of that mechanism. Further confirmation of a nonkinetic nature, e.g., spectroscopic data on the structure of the active center and the intermediate particles is necessary.

When studying the kinetics of real heterogeneous catalytic reactions, one must take into account the change in the activity of the catalyst in the course of the reaction. Changes in the chemical composition and structure of the catalyst may involve both activation and poisoning, in which case the reaction rate constants and the activation energies will depend on the reaction time or (if the catalytic activity changes quickly) on the composition of the reacting medium. The adjustment of active centers is sometimes a necessary condition for adsorption or catalysis, and it should be accounted for when constructing kinetic equations and analyzing the mechanism of a reaction (Sect.5.3).

ROGINSKY [2.204] was one of the first to point out the possible role of chain mechanisms in catalysis. As early as the 1930s, he analyzed the oxidation mechanism of CO on MnO_2 and surmised that a chain reaction might take place on low-activity preparations of MnO_2, i.e., at low concentrations of active centers. At the same time, the reaction on high-activity specimens of MnO_2 takes a conventional redox course, and oxidation on one active center of the surface is followed by reduction on another.

SEMENOV and VOEVODSKY [2.205] assumed the radical chain mechanism to be the principal mechanism of catalysis. The formation of chains is proved by the fact that radicals form on the surface as a result of interaction between the solid and the reactants; the radicals are then desorbed into the bulk, initiating homogeneous chain reactions (heterogeneous-homogeneous catalysis).

The conjecture that chain mechanisms are common in heterogeneous catalysis was not, however, confirmed by further investigations. Nevertheless, certain experimental results were obtained which prove that chain mechanisms are possible in some cases. For instance, in [2.206] the oxidation of CO on a diluted catalyst 1% V_2O_5 / SiO_2 is illustrated by the scheme:

1) $O^{2-} + CO \rightarrow CO_2 + 2e$
2) $2e + 2V^{5+} \rightarrow 2V^{4+}$

Reduction of catalyst

3) $V^{4+} + O_2 \rightarrow [V^{5+}O^-]$
4) $[V^{5+}O_2^-] + V^{4+} \rightarrow 2[V^{5+}O^-]$
5) $[V^{5+}O^-] + V^{4+} \rightarrow 2V^{5+} + O^{2-}$

Oxidation of catalyst

6) $[V^{5+}O_2^-] + CO \rightarrow CO_2 + V^{5+}O^-$
7) $[V^{5+}O^-] + CO \rightarrow CO_2 + V^{4+}$

Reaction.

If the rates of Reactions (6,7) are much greater than the rates of Reactions (4,5), then Steps (3,4) and (6,7) may start repeating themselves in a cyclic fashion resembling a chain reaction. For a single act of reoxidation of a V^{4+} ion in a diluted catalyst, several hundred elementary reactions of adsorbed oxygen with CO occur. Chain effects similar to those in the mechanism earlier proposed by ROGINSKY [2.204] for the same reaction on MnO_2 are considered possible in [2.206] at a low concentration of active centers. Chain reactions are also probable in deep oxidation of hydrocarbons into CO_2 and H_2O.

Chain reactions on a surface represent a special case of a more general phenomenon: the employment of the energy of adsorption or the energy of the catalytic act to boost the catalytic reaction proper. Another possibility is the formation of excited molecules or the rearrangement of active centers of catalysts (Chap.5).

2.2.3 Specific Features of Catalytic Reaction on the Surface of Metals, Semiconductors, and Dielectrics

The mechanisms by which solid catalysts function are various, and there is no universal theory capable of predicting the catalytic properties of a substance; in this respect the situation differs little from that in noncatalytic chemistry: the theory of noncatalytic kinetics is not able to predict the reactivity of a substance. Both these branches of chemistry employ various analogies and empirical correlations between catalytic activity or reactivity and the physical and chemical properties of a substance. The most fruitful concepts for catalysis have turned out to be those that account for the chemical affinity between the catalyst and the catalytic process.

All the main chemical properties of a substance, as well as many of its physical ones, are ultimately determined by its position on the periodic table, that is, by the structure of the electron shells of atoms and ions. Consequently, comparisons between the catalytic activity of solids and the positions which the elements comprising them occupy on the periodic table have led to certain regularities in the choice of catalysts [2.207].

A classification of catalytic processes according to the mechanism of the catalyst is a helpful frame of reference. According to ROGINSKY [2.208,92], catalytic reactions can be divided into two large classes: redox (electronic) reactions and acid-base (ionic) reactions. The first of these includes processes connected with electron transfer: oxidation, reduction, hydrogenation, dehydrogenation, and decomposition of unstable oxycompounds. Typical catalysts for these reactions are metals and semiconductors, i.e., substances having free or easily excited electrons (or holes). The most common catalysts of this class are the transition metals (Pt, Pd, Ni, Fe, Co) and their compounds: oxides (V_2O_5, MnO_2, MoO_3, Cr_2O_3), spinels (Fe_3O_4, $CuCr_2O_4$) and other complex oxycompounds, sulfides (MoS_2, WS_2, CoS), and certain semiconductors not involving transition elements (Ge, Si, ZnO, ZnS). The second class is represented by reactions of the acid-base type: catalytic cracking, hydration, dehydration, hydrolysis, and many reactions of isomerization, polymerization, and condensation of organic substances. The typical catalysts here are solids with acid properties (amorphous and synthetic aluminosilicates, SiO_2, Al_2O_3, ThO_2) and basic properties (MgO, CaO). Often they are ionic crystal dielectrics or amorphous insulators (gels).

This classification corresponds to the accepted division of organic homogeneous reactions into homolytic and heterolytic reactions according to the mechanism of the elementary act. The elementary act of homolytic reactions involves the splitting of an electron pair to form a chemical bond. In heterolytic reactions the chemical bond is established and destroyed in the elementary act without the splitting of an electronic pair. For instance, homolytic dissociation of a water molecule gives birth to radicals: $H_2O \rightarrow H^{\cdot} + OH^{\cdot}$, while heterolytic dissociation results in a proton and hydroxyl: $H_2O \rightarrow H^+ + OH^-$. Metals and semiconductors can supply electrons for the homolytic breakage of the electron pair and steer the reaction onto the redox course. Solids whose surface atoms can accept (acids) or supply (bases) electron pairs interact with reactants according to a heterolytic (donor-acceptor) mechanism, and boost the acid-base reactions.

This general classification of catalytic processes has proved useful as an orientation aid when choosing catalysts, but in practice one is often confronted with much more complex situations. One and the same reaction can take a redox (homolytic) or an acid-base (heterolytic) course under different conditions and with different catalysts. For example, dehydrogenation of isopropyl alcohol is catalyzed by metals and semiconductors. One could imagine [2.209] that this reaction involves electrons and holes of a metal or semiconductor with heterolytic bond breakage:

$$(CH_3)_2CHOH + e \rightarrow (CH_3)_2COH \ominus (ads) + H \quad ,$$
$$(CH_3)_2COH \ominus (ads) \rightarrow (CH_3)_2CO \ominus (ads) + H \quad ,$$
$$(CH_3)_2CO \ominus (ads) + \oplus \rightarrow (CH_3)_2CO \ (gas) \quad .$$

On the other hand, dehydration of this alcohol is also accelerated by solid bases: CaO, MgO, etc. The removal of hydrogen can then occur in heterolytic fashion, without the involvement of free electrons or holes of the catalyst:

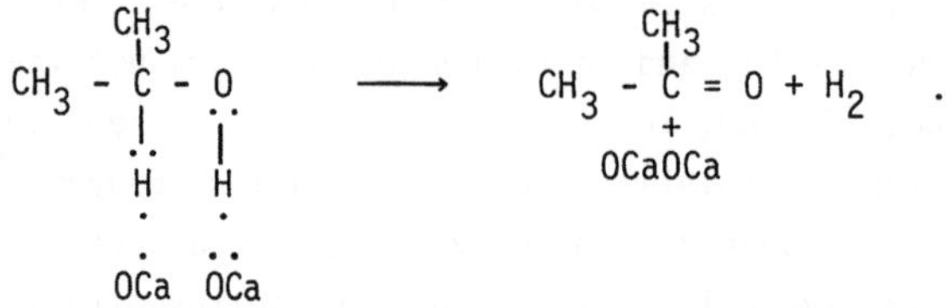

The transfer of an electron from occupied energy bands of the catalyst to the local level of the reactant may be followed by heterolytic dissociation of the molecule, or vice versa—the interaction with the catalyst may be heterolytic, while the rest of the reaction takes a homolytic course.

The oxidation of organic substances is boosted by metals and semiconductors. Partial oxidation of olefins into acids (e.g., oxidation of propylene C_3H_6

into acrylic acid $CH_2=CHCOOH$) is known to be accelerated by substances that are not only electron-type conductors, but also possess weak acidiferous or amphoterous properties, such as the oxides V_2O_5 and MoO_3. The product of the reaction does not form a strong bond with them and is easily desorbed. At the same time the reaction products, being acids, form strong saltlike compounds on basic oxides such as FeO and CdO, and undergo further oxidation into CO_2.

Many oxides of transition metals, such as NiO, Co_3O_4, Fe_2O_3, and Cr_2O_3, and other oxides such as ZnO, PbO, are semiconductors on the one hand and ionic crystals on the other. Their surfaces exhibit both acid and basic properties and consequently they can boost both acid-base (heterolytic) and redox (homolytic) reactions, as well as complex reactions involving both homo- and heterolytic stages.

We see that the difference in the catalytic properties of metals and oxidic semiconductors results from the fact that metals have a much greater concentration of free electrons that can take part in redox reactions; and that the selectivity of oxides differs from that of metals because of the former's ability to influence the rates of both heterolytic and homolytic reactions. Metals are far more active than oxides in redox reactions; they exhibit more universal catalytic activity. The most versatile catalysts are the group-VIII metals, especially Pt and Pd, which accelerate diverse reactions of oxidation, hydrogenation, dehydrogenation, etc., at low, sometimes even room temperatures. In contrast, transition-metal oxides, especially mixed ones, are usually more selective though less active catalysts.

The specific action of transition metals and their compounds that has secured them their outstanding role as catalysts is explained by the fact that they have a large spectrum of states with varying degrees of oxidation (e.g., Mo^0, Mo^{2+}, Mo^{3+}, Mo^{4+}, Mo^{5+}, and Mo^{6+}) which differ little in energy and in which electron transitions are easy. Also of great importance is the fact that the atoms of transition metals have a large number of d orbitals of different symmetry which facilitate interaction with the reactants, and that the d shell is diffuse in comparison with the s, p shell. The latter property of transition metals is crucial for the formation of molecular complexes in which bonds involving electrons from the d shell are rearranged. For instance, H_2-D_2 exchange in the presence of an atom of transition metal M can proceed via formation of a four-member complex:

$$H_2 + D_2 \xrightarrow{M} \begin{matrix} H\cdots H \\ \vdots\ M\ \vdots \\ D\cdots D \end{matrix} \rightarrow 2HD \quad .$$

The specific ways in which d orbitals of different symmetry are occupied by electrons serve as a basis for well-defined correlations between catalytic activity and the number of d electrons and position of the transition metal on the periodic table [2.209]. In a great number of redox-type catalytic reactions, for example, a so-called two-peak pattern of change can be observed in the sequence of oxide catalysts. The catalytic activity of the oxides of metals in the first great period reaches its maximum with oxides having electron configurations d^3 (Cr_2O_3, MnO_2) and $d^7 - d^8$ (CoO, NiO), and its minimum at d^0 (CaO, Sc_2O_3), d^5(Fe_2O_3, MnO), and d^{10}(ZnO, Cu_2O). This is shown in [2.207] to hold for hydrogenation and dehydrogenation of hydrocarbons, $H_2 - D_2$ exchange, oxidation of hydrocarbons, oxidation of H_2 and CO and decomposition of H_2O_2 and $KC\ell O_3$. In polymerization of olefins a high level of activity is displayed by cations with electronic configuration $d^1 - d^2$ ($TiCl_3$, VCl_3, etc.), while oxides with approximately the same cation configuration (Mo^{5+}, V^{4+}) are active in metathesis of olefins. In adsorption the molecule interacts with the transition-metal ion, and a donor-acceptor complex is formed. The coordination bond between molecule (ligand) and metal is the most common type of bonding to facilitate the activation of a molecule for catalysis. According to DOWDEN [2.210], the role of d electrons in catalysis is so specific that one is tempted to reject the now accepted classification of catalysts into metals, semiconductors, and insulators, and distinguish instead between catalysts with ions or atoms in d states and catalysts with ions in s, p states.

Predictions on the catalytic activity of s, p semiconductors in redox reactions are at present based on much less precise correlations than those for compounds of transition metals. As is shown in [2.207,211], one commonly employed correlation is the connection between catalytic activity k and the width of the forbidden zone of the semiconductor, E_g:

$$\log k = A - BE_g \quad , \tag{2.75}$$

where A and B are constants. Figure 2.8 illustrates the experimentally obtained dependence of the logarithm of the rate of dehydrogenation of *iso*-C_3H_7OH in the adsorbed layer at 500°C upon the width of the forbidden zone of a semiconductor with a wurtzite structure.

The "solid-state specificity" is not so pronounced in acid-base catalysts as in metallic and semiconductor catalysts. In [2.212] it was shown that if a reaction in the liquid phase is boosted by the addition of an acid or a base, it will also be accelerated by a solid having acid or basic properties. One of the most reliable and well-founded correlations for acid catalysts is

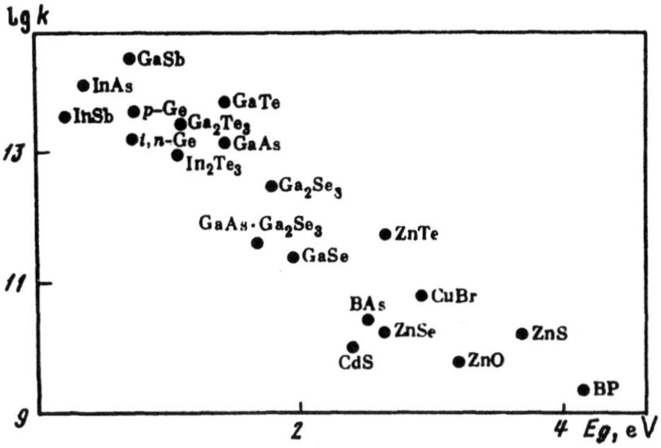

Fig.2.8. Dependence of the dehydrogenation rate constant of *iso*- $-C_3H_7OH$ upon the width of the forbidden band of the semiconductor [2.207]

the correlation with the acidity of the cation (Lewis acidity), which can be expressed through its electronegativity, ionization potential, or charge-to-radius ratio e/r. The greater the ionization potential, the greater is the Lewis acidity. Conversely, the basicity of an oxide is characterized by a low electronegativity value for the cation and a low value of e/r. The acidity of a solid surface can be due to the presence of mobile protons (H^+) on the surface, while basicity can be explained by the existence of hydroxyl groups. In the beginning of each period the M-OH bond is weaker than MO-H; the reverse is true for the elements closing the period. Correspondingly, oxides and hydroxides of metals opening the period accelerate the reaction via the basic mechanism, while those of metals at the end of the period boost the reaction via the acid mechanism.

Some industrial catalysts are made by depositing a small amount of transition-metal oxide on a dielectric non-transition-metal oxide with a well-developed surface. For instance, in hydrocracking and hydrodesulfurization, use is made of Co-Mo-Al_2O_3 catalysts obtained by depositing small amounts of cobalt and molybdenum oxides on an Al_2O_3 carrier. In metathesis of olefins, molybdenum oxide reduced to Mo^{4+} or Mo^{5+} and deposited on Al_2O_3 is used. Chromium oxide deposited on SiO_2 or aluminosilicate is employed in the polymerization of ethylene.

Such systems are of interest from a theoretical point of view as well, since they could help to determine whether local or collective interactions are most important in catalysis—in other words, whether the free charge carriers of the solid play the predominant role ("collective action"), or the interaction between adsorbed atoms or molecules and one active surface center ("local action"). Many researchers have tested these effects by investigating the catalytic and chemisorptive properties of diluted solid solutions of transition-metal oxides in the oxides of nontransition metals. These investigations showed that at high levels of dilution (< 1%), the behavior of the system depends on the reaction. In some reactions the active centers are isolated atoms of transition metal (local action), while in others the role of active centers is played by binary or even more complex centers with overlapping of the d orbitals of neighboring atoms, and possibly with electron transitions (collective action). For instance, the active centers in CO oxidation solid solutions of $Ni_xMg_{1-x}O$, where Ni^{2+} ions partly replace the Mg^{2+} ions in the MgO lattice, are both isolated Ni^{2+} ions and clusters of 50 to 100 Ni atoms, joined by oxygen bridges. The activity of clusters here, as in many other cases, is much greater than that of isolated ions [2.213].

An exact knowledge of the mechanism of catalysis is not absolutely necessary when choosing catalysts for reactions of practical interest. Whether they admit it or not, almost all researchers concerned with the theory and practice of catalytic action employ correlation methods. If out of 100 catalysts tested, 50 turn out to be active, and 40 of these are oxides of transition metals, then it is only natural to choose the 101st test sample from the same group of oxides. In some cases correlations encompass large groups of catalysts and reactions.

According to the BALANDIN multiplet theory [2.214], in many-point adsorption the atoms of the catalyst should follow the same geometric pattern as the atoms in a molecule of the reactant. This implies that there is a connection between catalytic activity and the geometric structure of the substance. Indeed, correlations between catalytic activity and the lattice parameter of the transition metals and certain oxides have been found for certain reactions of hydrogenation and dehydrogenation, and dehydration of some organic substances. An analysis of the structure of adsorption layers by the method of slow electron diffraction [2.191] revealed that the real structure of the adsorbed layer is not so simple as had been supposed in the multiplet theory. The adsorbate can form different structures on one and the same surface, depending on the occupancy.

For a number of reactions of catalytic oxidation, good correlations have been found between the catalytic activity and the bonding energy of the surface oxygen [2.175,215].

It is a common mistake, however, to use such correlations as a basis for speculations on the mechanism of catalysis. The mere existence of a correlation between certain quantities is not necessarily an indication of a causal connection between them.

The catalytic activity of a solid is determined by a large number of different factors. Each of these factors may correlate with some property of the solid. For this reason the real connections between catalytic activity and the properties of the catalyst are usually masked by other, often secondary, factors. Catalytic activity can almost never be related to a certain property of the substance and represented as a function of it, and therefore catalysts are usually chosen by statistical methods. In recent years the interest in correlations has dropped markedly, and greater attention is being focused on the experimental investigation of elementary acts of catalysis.

3. The Nature of Semiconductor and Dielectric Surfaces

Both macroscopically and electronically, the surface represents one of the greatest defects of a three-dimensional crystal structure. The termination of the periodic lattice structure causes changes in the coordination spheres of surface atoms and leads to rehybridization of bonds. This, in turn, results in changes in the effective charge of the surface atoms and in the spacing and order of the atomic pattern on the surface. The crystallographic structure of the surface should differ basically from that of the bulk, but since a distorted structure cannot abruptly turn into a periodic bulk structure, it is more correct to speak of a finite three-dimensional transition region (phase) [3.1]. In both the macroscopic and the microscopic sense, the crystal surface is always a three-dimensional problem. In Sect.2.1.3 it was pointed out that all the thermodynamic surface parameters, including the surface tension and surface energy of the solid, relate not to the geometric boundary, but to a surface layer of finite thickness. The modern physical chemistry of surface phenomena is characterized by a departure from a Gibbsian-Laplacian two-dimensional treatment of the surface and a move towards a three-dimensional approach [3.2-4].

Every real surface in nonuniform, a circumstance arising from a number of interconnected factors. Geometrical nonuniformity results from the different crystal faces that form the surface of the specimen, and their different adsorptive and chemical activity. Naturally, the destruction of the surface will also differ for different faces. Further geometrical nonuniformity results from unavoidable macroscopic defects: cracks, steps, pores, dislocation exits, etc.

In addition to macroscopic defects, a rather high concentration of various point defects is always present on the surface: vacancies (Schottky defects), interstitial atoms (Frenkel defects), replacement or implantation of atoms, etc. Due to their high mobility, surface defects can aggregate into "defect ensembles" or clusters of separate atoms. The concentration of certain types of defects is temperature dependent, and tends to zero at $T \to 0\,K$ (so-called thermal defects). Other types of defects remain in equilibrium with the crystal at 0 K ("biographical defects") [3.5]. In real crystals with a nonequilibrious cut, the concentration of defects increases exponentially with a decrease in the dimensions of the crystal. For high-dispersion crystals the concentration of defects is close in order of magnitude to that of the active centers of catalysis, and is sometimes even greater [3.4].

The real surface we are concerned with in adsorption and catalysis is always chemically nonuniform. Due to interaction between molecules of the environment and surface atoms which are unsaturated in terms of coordination or valence, the surface is always covered with various chemical compounds that are tightly bound to it. These complexes interfere strongly with adsorptive and catalytic surface processes.

Finally, the surface is nonuniform in the electronic sense. Disturbances in the periodicity of the surface potential, caused by the termination of the periodic lattice structure, by various macroscopic and microscopic defects, and by the presence of different surface chemical compounds, give rise to a wide spectrum of surface states which can act as centers that capture or recombine electrons or holes of the semiconductor. In the energetic spectrum of the crystal these states are represented by a set of local levels. The spatial and energetic nonuniformity of surface states results in an apparently continuous distribution of them in the forbidden zone of the crystal [3.6].

Naturally, all macroscopic and electronic properties of the surface ought to be considered in connection with its interaction with the medium; it is the surface that adsorbs molecules from the environment. But at the same time the adsorbed atoms and molecules often change the properties of the surface and its macroscopic and electronic parameters. So for the sake of simplicity we shall restrict ourselves in this chapter to the most general properties of solid-state surfaces that are in equilibrium or quasi-equilibrium with a vacuum.

The investigation of the nature and properties of surfaces involves great experimental difficulties. All the electrical-physical parameters should be measured on the surface of monocrystals, as it is only in this way that one can determine the quantitative relationships between surface and bulk electronic parameters. At the same time, in order to study the nature of the interface by modern adsorption, spectroscopic, and radiographic methods, specimens with well-developed surfaces are necessary. Consequently, measurements are usually carried out on high-dispersion powders to account for the real sensitivity of the existing investigation techniques. Of course, there is a great difference between the surface of a powder and that of a single crystal; the dissimilarity depends on the manner in which the powder has been obtained and processed, and varies for different aspects of the surface. It is possible, however, to apply the information obtained for powders to single crystals though such comparisons are primarily qualitative, and their accuracy is limited. To obtain quantitative results, the accuracy of adsorptive and spectroscopic methods must be improved so as to permit all measurements to be made on the surface of a monocrystal. Some promising approaches are discussed later.

There are two general methods of investigating a real nonuniform solid-state surface. In one of them, the ultrapure surface is obtained under high vacuum, with controlled formation of certain surface complexes which, to some extent, model a real surface. Analysis of the surface conditions during adsorption yields valuable information on the genesis of real surfaces and the nature of the adsorption and catalysis centers. In the second approach, the solid-state surface is studied directly by combined adsorptive, chemical, and spectroscopic methods.

3.1 Atomically Clean Solid-State Surfaces

3.1.1 Techniques of Obtaining an Atomically Clean Surface

In investigations of the properties of atomically clean surfaces three basic experimental obstacles must be overcome:

1) finding a reliable method of creating an atomically clean surface;
2) maintaining the purity of the surface in the course of the experiments;
3) choosing the experimental criteria for purity of the surface.

Detailed surveys of the data published on these problems can be found in [3.7-10].

New opportunities of studying pure surfaces that are free of adsorbed gases were opened up in the 1960s by progress in ultrahigh vacuum technology. This was later supplemented by reliable methods of growing good-quality monocrystals of metals and semiconductors, and by experimental techniques based on Auger spectroscopy and slow electron diffraction which made it possible to analyze both the chemical composition and crystallographic structure of the surface.

While no general agreement has been reached on what is meant by "ultrahigh vacuum," it is usually characterized by pressure below 10^{-9} torr [3.11]. Today, vaccums of 10^{-10} or even 10^{-14} torr are easily achieved in the laboratory. The number of collisions of gas particles A with molecular mass M at pressure p and temperature T per second and cm^2 of surface is given by

$$n = pN_A(2\pi MRT)^{-\frac{1}{2}} \quad . \tag{3.1}$$

At room temperature, $n = 2 \cdot 10^{21} \cdot pM^{-\frac{1}{2}}\ cm^{-2} \cdot s^{-1}$ (pressure measured in torrs). A square centimeter of any surface contains about 10^{15} atoms. If these atoms are involved in chemisorptive activity with a sticking coefficient close to 1, it is easy to estimate the time in which the surface will be covered with a monolayer of gas molecules (with M equal to 20 - 30). At 10^{-8} torr this time will be about 10^2 to 10^3 s, while at 10^{-10} torr it will be about 24 hours. Thus a vacuum of 10^{-9} torr is the boundary for dealing with really pure surfaces.

Since the properties of atomically pure surfaces depend greatly on the technique by which the surfaces are prepared, we shall briefly discuss the principal methods.

High-Temperature Treatment

In this method a real surface is cleared by evaporation and thermal decomposition of surface chemical compounds in a vacuum or a reductive (oxidative) atmosphere. For substances with a low melting point this method naturally has only limited applicability. High temperatures also induce changes in the distribution of impurities and defects in the specimen, and certain impurities may diffuse onto the surface and form nonvolatile surface compounds with residual gases. Contamination of the surface can also be caused by gases dissolved in the specimen which are freed by heating. High-temperature treatment often leads to reconstruction of the surface.

Deposited Films

Material purified in an ultrahigh vacuum is evaporated and deposited on the carrier surface or the walls of the vacuum chamber, which are kept at a lower temperature. Special care must be taken to ensure that the carrier surface is pure, so that gases desorbed from it will not contaminate the deposited film. In the deposition of binary and multicomponent compounds, it is difficult to conserve the stoichiometry of the deposited film. Epitaxial films are of especial interest here [3.12].

Ion Bombardment

This method was first introduced by FARNSWORTH and co-workers [3.13]. All surface impurities and several upper layers of the lattice are removed by a beam of inert gas ions having energies of several hundred electron volts. Then the specimen is subjected to annealing to heal the defects caused by bombardment and to remove the adsorbed atoms of inert gas. This heating leads to the same complications as in the first technique. Ion bombardment is widely employed with metals and semiconductors to obtain clean surfaces of monocrystals [3.14, 15].

Autoemission

In this method, surface atoms are removed from the tip of a metal or semiconductor in a strong (ca. 10^9 V · cm^{-1}) electric field. Such a field gives rise to intense autoelectronic emission from the tip. The widely employed field-emission electron microscope (FEEM) constructed by MÜLLER in 1937 [3.16] is based on this phenomenon. It consists of a very sharp metal tip (cathode) which, while remaining cold, emits electrons in a high vacuum due to a strong electric field maintained between the tip and the anode (Fig.3.1). The tip has a curvature radius of about 1000 - 3000 Å. The emitted electrons reach a luminescent screen positioned at a distance R of 3 - 10 cm. The image of the tip appears on the screen with a magnification of $R/r \sim 10^5 - 10^6$ and a resolution of about 20 Å [3.17-19]. Good images are obtained when the potential of the tip is $10^3 - 10^4$ V. A typical image of a tungsten tip is shown in Fig.3.2.

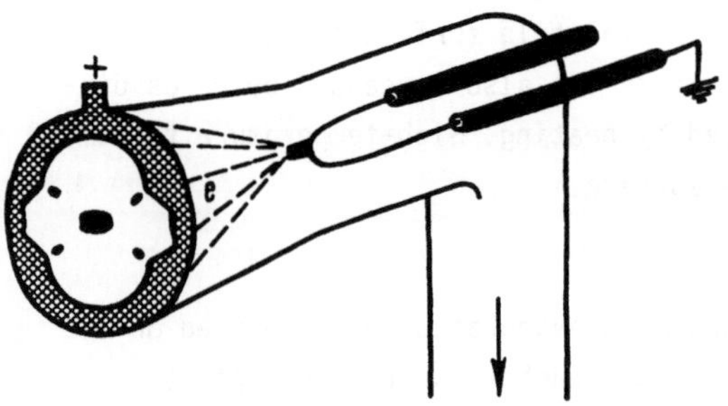

Fig.3.1 Field-emission electron microscope

The numbers in the photograph are the crystallographic indices of faces emerging onto surface W. The bright patches correspond to faces with a small thermionic work function φ_{th}, and the dark patches to faces with a high work function.

In autoelectronic emission the electrons tunnel through the potential barrier on the cathode surface. The nature of this barrier is evident from Fig.3.3. Curve 1 depicts the potential energy of the electron in the absence of an external field.

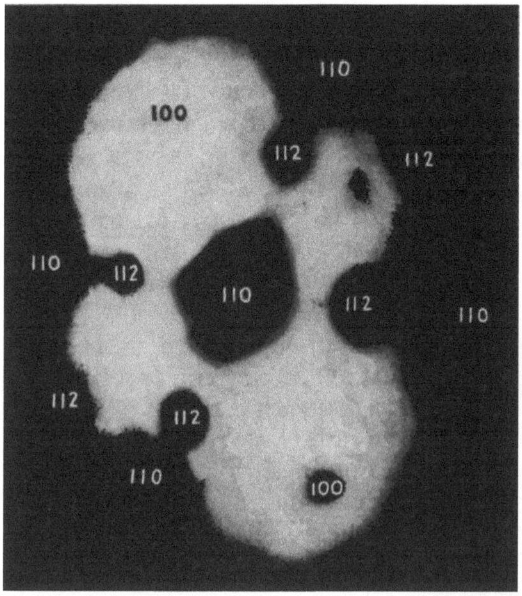

Fig.3.2. Image of a tungsten tip in a field-emission electron microscope

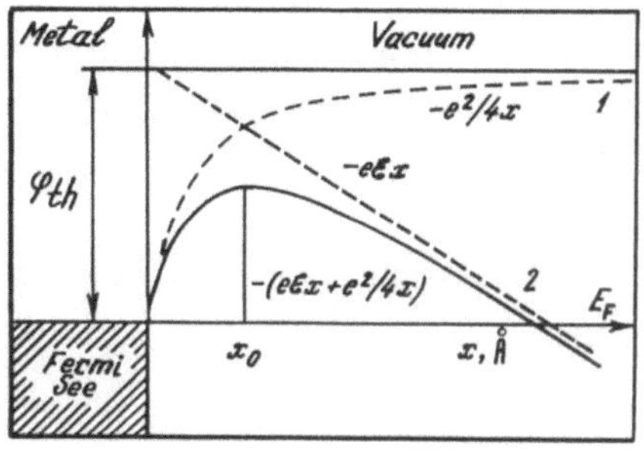

Fig.3.3 Shape of the potential barrier for electrons on a metal surface in the absence (*1*) and the presence (*2*) of field E normal to the surface

The autoemission current density j is given by the Fowler-Nordheim equation:

$$j(\mathrm{A\cdot cm^{-2}}) = 6.2\cdot 10^6 \,\frac{(\varphi_{th}/E_F)^{1/2}}{\varphi_{th}+E_F}\,\varepsilon^2 \exp\left[-6.8\cdot 10^7\,\frac{(\varphi_{th})^{3/2}}{\varepsilon}\right] \quad , \qquad (3.2)$$

where φ_{th} is the thermionic work function, E_F is the Fermi level, and ε is the field strength in $\mathrm{V\cdot cm^{-1}}$. Knowing the dependence of $\log(j/\varepsilon^2)$ upon $1/\varepsilon$, one can measure the work function. Measured data on the thermionic work function for different faces on tips made of various metals and semiconductors abound in the literature.

Autoelectronic emission was recently employed successfully as a source of electrons in an electron scanning microscope. CREVE [3.20] improved the resolution of the microscope to 2 or 3 Å in experiments with a penetrating beam which enabled him to observe the movement of isolated heavy atoms on an inert carrier.

Greater resolution (2 - 3 Å) is achieved in the field-ion microscope (FIM). It employs ionization of surface atoms, which are accelerated by an external field and create the autoionic image of the tip [3.21-23]. A great advantage of the ion microscope is that the individual atoms on the surface of the metal, can be monitored. Its drawback is that the image may be obtained during bombardment of the surface by He ions in strong electric fields ($\sim 10^9\ \mathrm{V\cdot cm^{-1}}$), and for very sharp tips having curvature radii that are quite small. This causes rapid healing of vacancies and dislocations, and removes extruding atoms, which results in an equilibrium image with a regular atomic pattern (Fig.3.4). Such an image, of course, scarcely resembles the surface as it was before the experiment. In such experiments, different faces of the monocrystal—faces with high and low indices—can be discerned. Use of the FIM together with a mass spectrometer made it possible to identify the ions desorbed from a given atomic patch, and thus a unique "atomic sonde" (AS) became available [3.21].

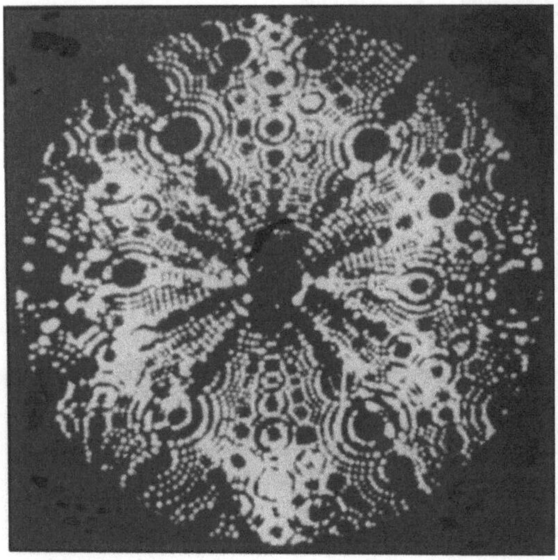

Fig.3.4 Image of a tungsten tip in a field-ion microscope at 12 K. The dots correspond to the positions of individual atoms [3.17a]

Autoemission techniques have provided valuable data on the mobility of atoms on the surfaces of metals, semiconductors, and dielectrics. A detailed survey of various techniques and the results obtained with them can be found in [3.17-23]. However, autoemission methods have serious limitations when it comes to electrophysical and adsorptive experiments, mainly because of the minuteness of the patch of clean surface.

Cleavage of a Single Crystal in Vacuum

This method is free of the drawbacks arising from the necessity of subsequent heating. However, cleavage releases a certain quantity of dissolved gas, which can contaminate the surface if the area is small. It is therefore necessary to subject the specimen to a prolonged high-temperature treatment prior to cleavage; this causes a redistribution of defects and impurities. With regard to electrophysical measurements this method has two important disadvantages:

cleavage gives only one clean surface, and that surface has rather a large number of steps with heights of about 100 Å [3.24]. Furthermore, LEMKE and HANEMAN warn that microcracks may form during cleavage of a silicon crystal [3.25].

Methods of cleavage have also been tried in liquid inert media [3.26,27]. To keep the surface clean, however, special precautions should be taken to remove all traces of oxygen and other impurities from the liquid. The cleavage technique has been widely employed to analyze the electrophysical properties of atomically clean surfaces of semiconductor monocrystals.

Crushing

In order to investigate adsorption on atomically clean surfaces one must deal with disperse powders obtained by crushing the initial specimen in a high vacuum. To improve the accuracy of adsorption measurements and reduce the contamination of the surface by chemisorbed residual gases, it is desirable that the dispersity of the particles be increased. However, this leads to two important complications. The first of these is connected with the contamination of the powder by foreign particles worn off the crusher rollers. The activity of these particles should therefore be investigated in advance. The second complication arises from aggregation of the powder particles during crushing, which renders quantitative measurements of the specific surface of the particles impossible. Unfortunately, the effect of aggregation has been overlooked in the overwhelming majority of investigations on the surface of fine powders. Since this effect can cause the results of certain experiments to be misinterpreted, we shall discuss it here in more detail.

The effect of aggregation was first discovered by KHODAKOV and REBINDER in 1959 [3.28,29]. They discovered that during air-dry crushing the specific surface of quartz powder, which is determined by low-temperature nitrogen adsorption, may cease to increase, and even start to decrease. This effect was attributed to the aggregation of quartz particles and the formation of an ultraporous structure whose surface is partly inaccessible to the nitrogen molecules. The aggregates formed in dry-air or vacuum crushing cannot be undone by subsequent treatment with water or heat, though their mechanical strength is rather low. Several minutes of grinding with an excess water towards the end of the crushing procedure, however, will destroy the aggregates and increase the specific surface. Conversely, if a dry- high-dispersion disaggregated powder is subjected to dry crushing, its specific surface will decrease markedly. Brief crushing with an excess water restores the initial value of the specific surface. Crushing with an excess of water usually produces disaggregated powders whose surface, of course, is far from being atomically clean. Systematic investigations of the effect of aggregation on the adsorptive properties of quartz can be found in [3.30,31].

Figure 3.5 shows the reversible isotherms of the adsorption of water vapor at 300 K and nitrogen at 77 K on disaggregated quartz powder Qu-4 produced by crushing quartz with excess water, and on aggregated powder Qu-4A obtained by 16 minutes of additional air-dry crushing of the initial specimen Qu-4. From the figure it is clear that the adsorption isotherm for nitrogen on Qu-4A is lower than that on Qu-4. The calculated value of specific surface s is reduced from 53 $m^2 \cdot g^{-1}$ for Qu-4 to 35.4 $m^2 \cdot g^{-1}$ for Qu-4A due to the aggregation. At the same time, the isotherm of the adsorption of water is higher for the Qu-4A specimen, indicating an increase in the total surface accessible to small (compared to nitrogen) molecules of water, due to additional crushing of the specimen.

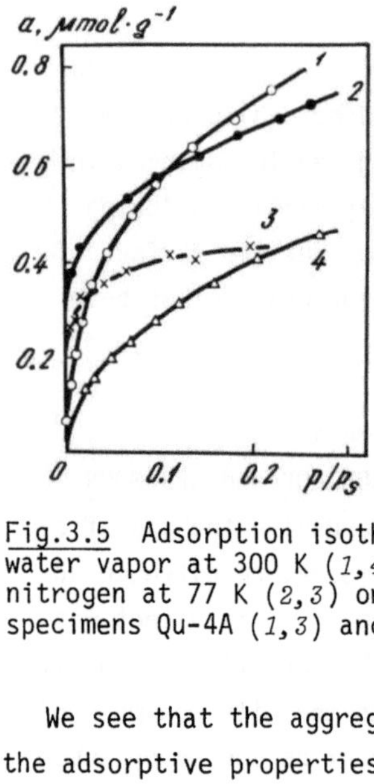

Fig.3.5 Adsorption isotherms of water vapor at 300 K (*1,4*) and nitrogen at 77 K (*2,3*) on quartz specimens Qu-4A (*1,3*) and Qu-4(*2,4*)

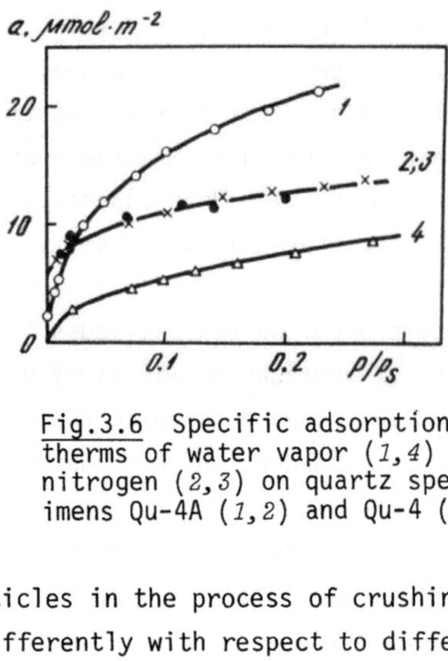

Fig.3.6 Specific adsorption isotherms of water vapor (*1,4*) and nitrogen (*2,3*) on quartz specimens Qu-4A (*1,2*) and Qu-4 (*3,4*)

We see that the aggregation of particles in the process of crushing affects the adsorptive properties of quartz differently with respect to different molecules (in our case H_2O and N_2). Naturally, the specific isotherms of water adsorption a [$\mu mol \cdot m^{-2}$], calculated for the unit of surface area found from the nitrogen adsorption isotherms (s_{N_2}), will be quite different (Fig.3.6). This difference, however, is due not so much to the different adsorptive activity of the surfaces, but to the fact that part of the surface of the aggregated powder, though inaccessible to nitrogen molecules, is nonetheless accessible to water molecules, i.e., $s_{N_2} < s_{H_2O}$. The reversibility of the effects of aggregation proves that the resulting ultraporous structure is not intrinsic to the initial quartz particles, but is of secondary origin [3.32].

The effect of aggregation on the adsorptive activity of a solid is apparently greater, the smaller the particles are. For instance, during 20-minute crushing of dry silica gel, its specific surface s_{N_2} is irreversibly reduced from 280 to 22 $m^2 \cdot g^{-1}$ [3.33]. Analysis of the isotherms of water adsorption on such a sample reveals that the porous structure of the silica gel is completely destroyed.

As yet there is no conclusive explanation for the phenomenon of aggregation or for the nature of the ties between the particles in an agglomerate. The very existence of these effects is a serious obstacle to quantitative measurements of adsorption on atomically clean surfaces. Neglect of this effect has led to great discrepancies in the data published on the adsorptive properties of crushed powders of quartz, silicon, germanium, and other solids (for more details see [3.4,32]). The exceptions are solid bodies with a pronounced anisotropic structure. Analysis of the adsorptive properties of graphite powders does not reveal any appreciable aggregation of particles up to values of $s \sim 400 - 500$ $m^2 \cdot g^{-1}$ [3.34,35]. A slight aggregation due to sticking of the basic faces of particles has little effect on adsorption upon reactable prismatic planes, and aggregation has not been observed in the crushing of mica, boron nitride, or talc, all of which also have a laminated structure [3.29].

Aggregation of particles during crushing is apparently connected with the conditions under which the process takes place, in particular the kinetics of crushing. In experiments conducted by STRELETSKY and BUTYAGIN [3.36] on the crushing of fused quartz with an initial specific surface of $5 \cdot 10^{-2}$ $m^2 \cdot g^{-1}$, the specific surface increased continuously with no signs of aggregation. The rate of increase of the specific surface did not depend on temperature or on the composition of the gaseous media (vacuum, hydrogen). Even subsequent crushing in a surface-active medium (water) did not lead to any further increase in the surface.

BUTYAGIN performed a detailed study of the crushing of solid bodies [3.37]. The formation of the surface was shown to be accompanied by dissipation of excess energy in the form of heat, emission of electrons, ions, and neutral particles, luminescence, etc.; often the released energy was channeled to the initiation of chemical reactions.

We now turn to the question of how atomic purity of the surface is maintained in the course of an experiment. The "cleanness" of the experiment can be described by the contamination coefficient $\varkappa$, which is defined as the ratio of the number of residual gas molecules in the vacuum chamber to the total surface area of the sample being studied [3.38]. In early investigations of the electrophysical properties of atomically clean surfaces of semiconductor monocrystals in an ultrahigh vacuum, the value obtained for $\varkappa$ was $\sim 10^8$ - 10^9 molecules · cm^{-2} [3.38]. Later, $\varkappa$ was reduced by further improving the vacuum or increasing the surface of the samples. The use of fine powders made it possible to reduce $\varkappa$ to about 10^5 molecules · cm^{-2}. It should be noted that in fact $\varkappa$ is much lower, since the partial pressure of the most active gases is lower than the total pressure of 10^{-9} - 10^{-10} torr by more than an order of magnitude, and their sticking coefficients are also quite small (Chap.4).

3.1.2 Methods of Studying the Structure and Composition of an Atomically Clean Surface

Various diffraction techniques, as well as methods based on the interaction of corpuscular and electromagnetic radiation with the surface, are widely used to study the purity and structure of atomically clean surfaces of single crystals. A very common technique is low-energy electron diffraction (LEED), a method based on the classical works of Davisson and Germer and modified by FARNSWORTH and others [3.39-42] to meet the needs of surface analysis. It analyzes the distribution of diffractional maxima of slow electrons elastically scattered by the surface over angles and intensities (ELEED). According to the "kinematic" theory of diffraction, at normal incidence (along the z axis) of a monochromatic beam of electrons of energy E on the surface, the diffraction will satisfy the following conditions:

$$m_x\lambda = d_x \cos\psi_x \quad , \quad m_y\lambda = d_y \cos\psi_y \quad , \tag{3.3}$$

where λ is the electron wavelength (according to de Broglie, $\lambda = h/\sqrt{2mE}$); m_x, m_y are the orders of diffraction; d_x, d_y are the lattice constants; and ψ_x, ψ_y are the angles of diffraction in the directions of the x axis and y axis respectively.

The ELEED method usually employs beams with E at about 10 - 500 eV; the electrons in this case penetrate the crystal to the depth of several lattice periods. Foreign atoms on the surface can be detected if they form a periodic pattern, or if their presence lowers the intensity of the main diffraction maxima. In most cases the sensitivity of the method does not exceed 5% - 10% of the impurities of a monolayer. To improve its sensitivity, ELEED is often combined with measurements of the thermionic and photoelectronic work function. ELEED enables one to detect macroscopic surface defects (dislocations, steps, etc.) whose dimensions are comparable to the beam coherence radius ($\sim 10^2$ - 10^3Å)

[3.43]. If the size of a linear defect is assumed to be about 10^3 Å, then the concentration of surface atoms belonging to such defects is of the order of $2 \cdot 10^{+12}$ cm^{-2} [3.44]. These atoms are often considered to be the most active chemisorption centers and electron traps. A detailed analysis of disturbances in the diffractional pattern caused by point defects, steps, facets, and domains was recently carried out by HENZLER ([3.45] and [Ref.3.46, Chap.3]).

High-energy electron diffraction (HEED) is often used to detect macroscopic defects [3.40,47]. In this method a monochromatic electron beam with energies of 10 - 100 keV is directed at the surface at a small angle (of the order of several degrees), and the diffractional pattern of the scattered beam is analyzed. The reliability of this technique is improved by supplementary observations of the decorated surface with a scanning electron microscope [3.48].

The simplest version of the kinematic theory of diffraction, on which ELEED is based, considers only single elastic collisions between incident electrons and the atoms of the lattice. In reality, however, due to a relatively large scattering cross section ($\sim 10^{-15} - 10^{-16}$ cm^2), multifold reflections become quite probable, contributing to the intensity of the reflected beam [3.49]. This effect can be accounted for, to some extent, by carrying out measurements at various angles of primary beam incidence. Much greater difficulties are encountered when one tries to account for the contribution of inelastic collisions [3.49,50]. The energy released in inelastic collisions can be expended on the excitation of plasmons, on transitions within one band or between bands (sometimes accompanied by emission of Auger electrons), and on the excitation of phonons, provided the energy of primary electrons is great enough [3.50-53]. HIROBAYASHI points out that the diffractional pattern can be influenced by the resonance in the surface states [3.54].

While before the 1970s the above effects were merely considered a nuisance in the interpretation of data obtained by ELEED, in the past decade they have become a subject in their own right in the dynamic theory of diffraction which studies the mechanism of the elementary act of interaction between the incident electron and the surface atoms either in the zone theory approximation, or by the methods of many-body theory [3.46,50-52].

One promising technique of surface spectroscopy is based on measurements of electron energy loss during inelastic interactions. Energy loss spectroscopy (ELS) has already supplied valuable data on surface electron states, including those of a chemisorptive nature, as well as on plasmons, surface phonons (surfons), and the oscillatory modes of adsorbed molecules [3.53]. Because of its very shallow penetration ($\sim$1 nm at 100 eV), the ELS method is extremely surface oriented ([3.52] and [Ref.3.46,Chap.3]).

In the past decade emission methods of surface analysis have gained considerable ground. The emission is stimulated by heating of the specimen, creation of a transverse electric field ("field emission"), and irradiation by electrons, ions, photons, and molecular beams. The information is obtained from analysis of secondary (i.e., emitted by the surface) photons, electrons, atoms, molecules, and ions. It would be impossible to describe even briefly here the entire range of modern emission techniques employed to investigate the structure and composition of surfaces; they have long numbered over 50 ([Ref.3.46, Chap.6] and [3.55,56]). To remain within the framework of this book, we shall discuss only photoelectron, absorption, ion and Auger spectroscopy. The data obtained by these methods will be of help in our analysis of adsorptive and electronic processes on semiconductor and dielectric surfaces.

The methods of photoelectronic emission are based on photoexcitation of the electrons of atoms in the surface region of the crystal [3.57]. Depending on their energy, the incident quanta will knock out electrons of the valence band (Ultraviolet Photoelectron Spectroscopy, or UPS [3.58] or electrons from inner shells (X-ray Photoelectronic Spectroscopy, or XPS, also often called Electron Spectroscopy for Chemical Analysis, ESCA [3.59-62]; see Fig.3.7a). To illustrate, we shall briefly discuss the electronic phenomena that occur in a solid under the influence of X-rays. Commercial electron spectrometers (ESCA) employ as the source of spectrum excitation either the K_α line of aluminum (E = 1487 eV) or the K_α line of magnesium (E = 1253 • 6 eV). The resolution on the energy scale is then about 0.5 eV.

The energy of the electron level (also called the electron-binding energy) is given by the equation:

$$E_b = h\nu_0 - E_{kin} - \varphi_{sp} \quad , \tag{3.4}$$

where $h\nu_0$ is the energy of the incident photon; E_{kin} is the kinetic energy of

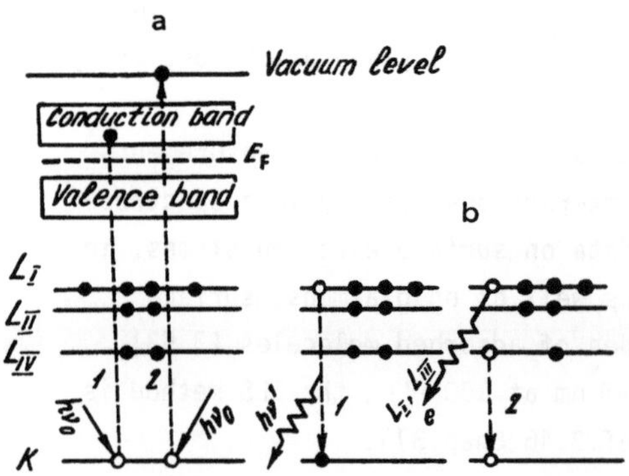

Fig.3.7 a,b. Electron transitions in excitation of a photoelectron (a) and the Auger effect (b)

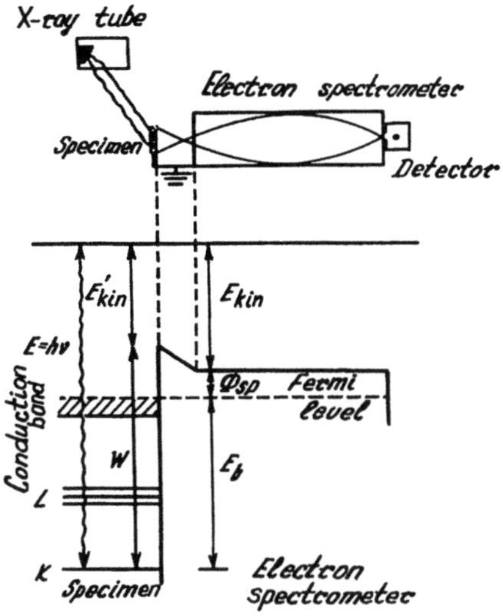

Fig.3.8 Electron transitions in a solid in X-ray photoelectron spectroscopy [3.62]

the electron after photoionization, measured in the course of the experiment; and φ_{sp} is the work function of the spectrometer. Figure 3.8 is a schematic diagram of the energy levels in a spectrometer for the case of metal (the Fermi level of the sample is balanced with the potential of the spectrometer). Samples of dielectrics and semiconductors with a wide exclusion band get charged in the course of experiment, causing additional complications in determining E_b. In such cases one usually employs a reference line with known E_b, and determines the E_b of other electrons with respect to this line. Most often the 1s carbon line belonging to the adsorbed organic substances of vacuum grease is employed.

The depth from which the photoelectrons are knocked out depends on the element's atomic number Z and on the kinetic energy of the electrons. Generally speaking, high-energy electrons from substances with high Z numbers come from the bulk, while low-energy electrons from substances with low Z numbers come from a thin surface layer. According to [3.63], the average depth from which electrons with a kinetic energy of 72 eV come out of silver is just 4 Å. For electrons with an energy of 362 eV, the depth is a little greater – about 8 Å, while electrons with an energy of 800 eV coming out of iodostearic acid are knocked out from a depth of more than 100 Å. If one takes measures to lower the angle of departure of photoelectrons with respect to the specimen surface, then the effective analyzed area will increase, and the depth of sondage will decrease [3.61]. The low-angle ESCA technique can greatly improve the intensity of peaks due to surface atoms [3.64].

Any changes in the valence shell (change of valence, coordination, or formation of chemical bonds) lead to "chemical" shifts in the X-ray electron spectra. For the crystal lattice the chemical shift ΔE_{chem} can be calculated on the basis of an electrostatic model [3.59-61]:

$$\Delta E_{chem} = c\left(\frac{1}{r} - \frac{\alpha}{R}\right)n \quad , \tag{3.5}$$

where r is the radius of the valence orbital; R is the distance between the nucleus of the atom being studied and the orbital of the atom which has accepted the electron (to a first approximation the distance between the nuclei); α is the Madelung constant; n is the number of migrant electrons; and c is a constant close to 1.

If one assumes the average radius of the valence orbital to be about 1 Å, then the chemical shift resulting from the removal of one electron will be about 14 eV [3.59]. The shifts observed are usually much smaller. The ESCA method is widely employed to investigate the zonal structure of solids, the effective charges of surface atoms, and the electronic structure of surface complexes.

Photoelectron spectroscopy provides valuable information on the occupied electronic states [3.57]. In order to obtain a more complete picture of the energy spectrum of the surface, including chemisorptive states, one usually combines the UPS method with ELS or Auger spectroscopy, and sometimes all three are used together. Improvements made in the monochromatization of excitation sources, and the utilization of polarized radiation have made it possible to study the angular distribution of electronic spectra in detail ("angle-resolved UPS" or UPS AR; ELS AR, XPS AR). The data obtained in this way have greatly broadened our conceptions of the structure, and in particular the symmetry, of surface states (see [3.65] and the literature cited there), and provided valuable information on the structure and orientation of adsorbed molecules ([Ref.3.58, Chaps.6,7,12] and [3.66]).

The use of synchrotron radiation as the source of the exciting beam has proved very fruitful in the study of the structure of solid-state surfaces. The high monochromacity of this type of radiation has made it possible to obtain valuable information not only from photoelectronic spectra [3.66,67], but also from absorption spectra. High-resolution absorption spectra of soft X-rays (SSXA) are determined by the excited states near the boundaries of bands. As was first shown by LUKIRSKY and co-workers [3.68], there is a striking parallel between the knock-out of photoelectrons and the absorption spectrum. The SSXA method is applied to the study of excitons and local disturbances on the surface caused by adsorbed atoms [3.69]. Additional information is gained from analysis of interatomic Auger transitions [3.70]. Lately, much attention has been directed towards the study of extended X-ray absorption fine structure (EXAFS, or SEXAFS when applied to the study of the surface [3.71]). SEXAFS rivals ESCA in its ability to provide information on the nature of catalysis and absorption centers [3.72]. In particular, the SEXAFS method makes it possible to study the local geometry of active surface centers.

Investigations by the SSXA and SEXAFS methods require sophisticated synchrotron accelerators, which is a rather grave obstacle. For this reason Auger spectroscopy (AES) has found greater recognition in surface studies [3.73]. It consists basically of the following. If external ionization creates a vacancy in one of the inner orbitals of an atom (e.g., K in Fig.3.7a), then once the atom has returned to the ground state this vacancy can be occupied by an electron from a higher level (e.g., L_I in Fig.3.7b; the Roman-numeral subscripts denote the multiplet values of the total angular momentum j). The energy released in such a transition can either be emitted as a quantum of characteristic X-ray radiation $h\nu'$ (X-ray fluorescence; see Part 1 of Fig.3.7b); or transferred to another inner electron, e.g., L_{III}(Part 2 of Fig.3.7b), thus causing the escape of the latter (Auger emission). The intensity of emitted Auger electrons for light atoms is greater than the intensity of fluorescence. For heavier atoms this relationship is reversed; e.g., for the K shell of atoms with $Z > 33$ (As) the fluorescence dominates over Auger transitions.

Modern X-ray spectroscopy of fluorescent radiation is capable of detecting as little as 10^{-14} g of a substance. However, the great penetrating power of the exciting X-ray radiation allows judgments to be made about the structure of a rather thick surface layer ($\sim 10^2 - 10^3$ A) [3.74]. Vast amounts of data on the surface itself are also obtained from Auger spectroscopy. By employing glancing electron beams with an energy of $10^2 - 10^3$ eV for the excitation of Auger electrons, one can achieve a sensitivity to foreign atoms of several units over 10^{12} atoms $\cdot$ cm^{-2} in a very thin surface layer, i.e., about 0.1 % of a monolayer [3.73]. If ELEED and Auger spectroscopy are combined in one experimental unit, the structure and composition of the surface can be studied at the same time. This combined technique is often used to investigate atomically clean surfaces of metals and semiconductors. In addition to conventional Auger spectroscopy, the AESAR (Auger electron spectrum angle resolution) method is gaining growing recognition [3.75].

Another class of techniques in surface physics is based on analysis of surfaces with the aid of ionic [3.56,76] and molecular [3.77] beams. Of the numerous ion spectroscopy methods, the one most often employed is secondary ion mass spectroscopy (SIMS) [3.78], which studies the ions knocked out from the surface layer by the initial ion beam [Ref.3.79,Chap.6]. This method features ion etching and impurity analysis, and is widely used in the study of thin dielectric layers in MIS structures and metallic coatings in microelectronic technology [Ref.3.79,Chap.7]. Molecular beams have only recently begun to be used for surface study, but have already provided valuable data on the structure of ordered adsorption layers, the degree of surface disorder, and the elementary mechanism of certain adsorptive and catalytic processes that take place on the surface of a monocrystal [3.57,77].

Incremental refractometry [3.80], ellipsometry [3.81], and EPR [3.82] are used widely to investigate atomically clean surfaces of semiconductors and dielectrics.

3.1.3 The Structure of Atomically Clean Surfaces and the Dynamics of Surface Atoms

In the 1920s E. Madelung, G. Bimueller, and later J. Lennard-Jones, B.M. Dent, I. Stransky, and other research workers (see the surveys in [3.1,83]) had shown that the distribution of ions on the surface of an ionic crystal differs from that in the bulk. Based on Fayence's concept of screening a cation, WEYL [3.1] surmised that a double electric layer (Fig.3.9) ought to exist on the surface of a NaCl crystal; it would ensure better screening of Na^+ ions and result in a decrease in the surface energy. The anions of this double electric layer would be polarized. As was already mentioned, disorder on the surface cannot abruptly turn into the orderly distribution in the bulk, and

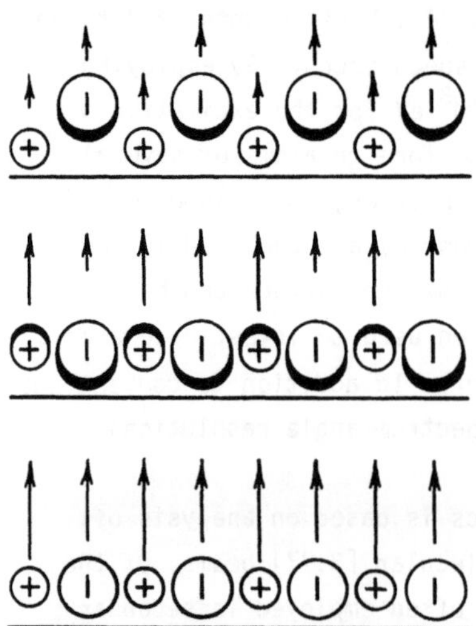

Fig.3.9 Scheme of a reconstructed surface of NaCl

therefore one should speak of reconstruction not of the surface, but of a surface layer of finite thickness. Theoretical speculations by BENSON and YUN [3.83] have shown that the first five subsurface layers are reconstructed in the <100> plane of the NaCl lattice. Similar results have also been obtained for other alkaline halides [3.84].

ELEED continues to be the most informative method of investigating the surface structure of a monocrystal. The first investigations of slow electron diffraction on atomically clean surfaces of semiconductor elements (Si and Ge) [3.85] showed that together with reflexes typical of the "diamond" structure of the bulk, the diffraction pattern displays reflexes of a fractional order which point to the existence of a surface superstructure. Surface superstructures have since been discovered in a number of semiconductor and metallic crystals: Si, Ge, C (diamond), Te, GaAs, InSb, GaSb, GaP, CdS, ZnO, ZnTe, Pt, Au, Bi, Sb, etc. (see [3.9,40,41,86,87] and the references cited in [3.42, 88]). Some of these data are presented in Table 3.1, which was taken from [3.41]. The numbers in angular brackets denote the indices of the face, and those in parentheses, the number of times which the period of the surface superstructure exceeds that of the bulk lattice. Superstructures have not been found on certain surfaces of metals and dielectrics, e.g., the planes <110>, <100>, and <211> of tungsten and nickel, and certain planes of aluminum, ionic crystals, and compensated binary compounds [3.41].

Table 3.1. Some surface structures of semiconductors and metals [3.41]

Crystal	6th-order axis	4th-order axis	2nd-order axis
Si	<111> - (2 × 1); (7 × 7)	<100> - (2 × 2); (4 × 4)	<110> - (5 × 2)
Ge	<111> - (2 × 1); (2 × 8)	<100> - (2 × 2); (4 × 4)	<110> - (2 × 1)
Diamond	<111> - (2 × 2)	<100> - (2 × 1)	
Te	<0001> - (2 × 1)		
GaAs	<111> - (2 × 2); (3 × 3)		
GaSb	<111> - (2 × 2); (3 × 3)		
InSb	<111> - (2 × 2)	<100> - (2 × 1)	
CdS	<0001> - (2 × 2)		
Pt		<100> - (5 × 1)	
Au		<100> - (5 × 1)	<110> - (2 × 1)
Ir		<100> - (5 × 1)	
Pd		<100> - (5 × 1)	
Bi		<100> - (1 × 1)	
Sb			<1120> - (2 × 10) <1120> - (6 × 3)

The data obtained by ELEED show that the observed superstructures are stable in a restricted temperature range and depend significantly on the method by which the surface is prepared. Consider, for example, the surface of the <111> plane of Ge. Immediately after cleavage the <111> face displays a (2×1) superstructure which at $T=420$ K turns into a (2×8) or (8×8) structure [3.89]. The (2×8) superstructure is also formed during ionic bombardment of the <111> face with subsequent annealing of defects. PALMBERG and PERIA [3.90] assume the observed diffraction pattern (8×8) to be a superposition of diffraction reflexes from domains with a (2×8) structure, oriented in three different directions.

In addition to the irreversible phase transitions, reversible phase transitions of the order-order type have been discovered, for instance, on certain faces of Ge and Si [3.91,92]. The diffraction pattern during heating reflected the formation of transition phases.

Phase transitions in surface superstructures are found to be very sensitive to various surface imperfections and adsorbed atoms. In [3.91] it is pointed out that the phase transition on a cleavage <111> surface of Ge occurs at lower temperatures than on annealed surfaces cleansed by ionic bombardment. PALMBERG [3.93] found out that the presence of Na^+ ions on a cleavage surface of Ge leads to destruction of the superstructure even at 77 K. In some other cases, in contrast, the presence of foreign atoms stabilizes the superstructure. The precise role of impurities in the formation of superstructures is not yet clear [3.94,95]. On the one hand, investigations by ELEED and AES reveal a total absence (within the accuracy of AES, i.e., about 10^{12} atoms $\cdot$ cm^{-2}) of impurities on the surface of a Si crystal heat-treated at 1500 K [3.96]. On the other hand, even a minute amount of impurities may, due to collective effects, stimulate a phase transition in superstructures. With binary surface compounds the formation of a superstructure strongly depends on stoichiometric deviations. For instance, the <0001> face of alumina initially has no superstructure. After reduction in a vacuum, a superstructure $(\sqrt{31}\times\sqrt{31})$ is formed and disappears after reoxidation. In [3.97] it has been emphasized that nonstoichiometric aluminium oxides Al_2O or AlO, which are thermodynamically unstable in the bulk, become stable on the surface.

Opinions on the reconstruction of surfaces of ionic and partially ionic crystals differ. Superstructures were not found on the <100> plane of NaCl [3.98] or on the LiF surface [3.99]. However, the interatomic distances were observed to change significantly in the direction normal to the surface, which supports theoretical estimates by BENSON [3.83]. As was mentioned in Sect.3.1.2, the ELEED method is not very sensitive to impurities (accuracy about 10%) or stoichiometric deviations. The fact that reconstruction is strongly influenced by these factors might be one of the reasons for the existing controversy on the surface structure of binary crystals. By combining ELEED with Auger and photoelectron spectroscopy, FIERMANS and co-workers [3.100] showed that the superstructure on the <0001> face of ZnO appears on those patches of the surface which contain very small amounts of impurities not detected by ELEED.

In recent years, MARK, DUKE, and co-workers [3.41,101] conducted detailed studies of the polar and nonpolar faces of tetrahedrically coordinated semiconductors (ZnO, CdS, GaAs). They reached the important conclusion that nonpolar planes are never reconstructed, whereas polar planes are almost always reconstructed, and display steps and a faceted structure.

A number of models have been proposed to explain the additional reflexes coming from surface superstructures. Consider, for example, cleavage along the <111> face of crystals with a diamond structure, e.g., Si and Ge. The atoms of Si and Ge in the bulk are in sp^3 hybridization, and the broken bondings appear on the cleavage surface <111>. LANDER and MORRISON [3.102] considered the possibility that the valence state is changed by a number of surface atoms, resulting in the displacement of some atoms in the direction normal to the surface (in addition to possible displacements in the surface plane; see Fig.3.10). The atoms next to the vacancies which form are assumed to establish conjugated $\pi\sigma$-bonds. SEIWATZ [3.103] also considers it possible that conjugated double bonds form on the surface of Si and Ge, and concludes that superstructures with fractional unit vectors $\sqrt{7}$, $2\sqrt{3}$, etc. may form which, in principle, can answer for any diffractional pattern ever observed by ELEED methods. However, there are no independent confirmations of the existence of double bonds on the surface of Si and Ge.

HANEMAN [3.104] considered another probable model (Fig.3.11) according to which two groups of atoms, ordered in rows in the <01> direction, exist on

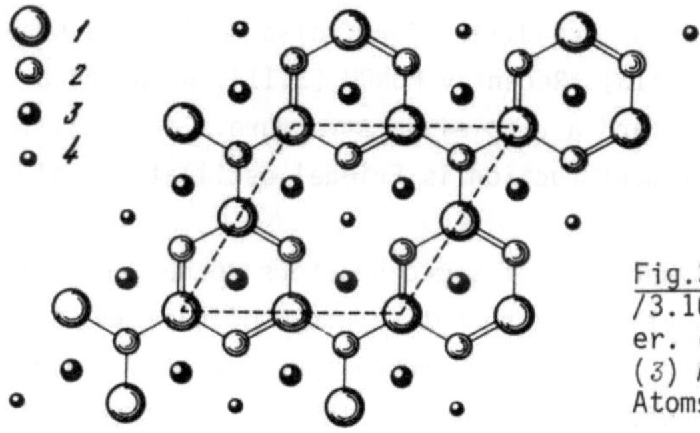

Fig.3.10 Model of <111> Ge surface /3.102/. (*1*) Atoms of the upper layer. (*2*) Atoms of the second layer. (*3*) Atoms of the fourth layer. (*4*) Atoms of the sixth layer

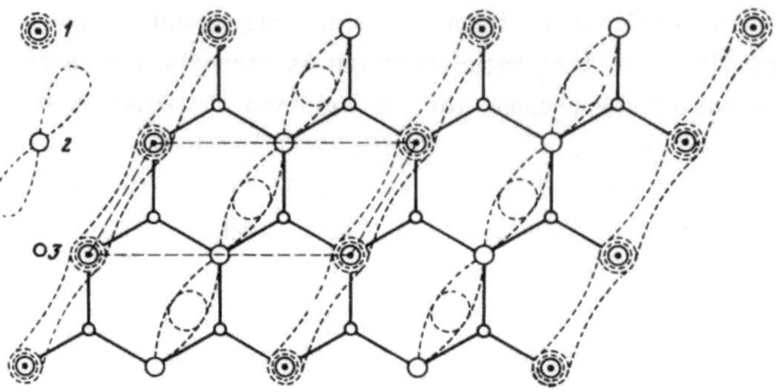

Fig.3.11 Model of <111> Si surface /3.104/. (*1*) Elevated atoms of the upper layer. (*2*) Lowered atoms of the upper layer. (*3*) Atoms of the lower layer

the <111> Si surface. One group of atoms is shifted 0.17 Å upwards; the other 0.13 Å downwards. The elevated atoms have dangling s-bonds which may partly overlap. The lower atoms have dangling p-bonds (of the π-bond type) which strongly overlap.

The formation of long-period structure [of type <111> Si(7 × 7), <111> Ge (2 × 8), etc.] has been attributed to the formation of vacancies and surface defects and to rehybridization of chemical bonds of the surface atoms [3.102, 105-107]. Recently CHADI and co-workers [3.92] presented convincing proof that Si(7 × 7) and Si(2 × 1) reconstructions are similar in character. They proposed a new model of (7 × 7) surface based on ringlike arrangements of raised and lowered atoms. HARRISON [3.107] examined the possibility that two-electron states might originate on atomically clean surfaces of Si. MARK and co-workers proposed a ripplelike model of <111> Si(7 × 7) [3.108,109], <111> Ge(2 × 8) [3.110], and <110> GaAs [3.111] surfaces. In this model the upper two layers of atoms are deformed and have a ripplelike structure. The deformation is assumed to be due either to a charge-density wave (Peierls) instability, or to tensions resulting from rehybridization of surface bonds. Some authors have considered the possibility that surface dimers form, disordering the subsurface region of the crystal [3.112,113]. Recently MÖNCH [3.114] proposed a hybrid model including both vacancies and a ripplelike structure. For metallic surfaces another possible cause of reconstruction is Friedel oscillations of the surface potential [3.115].

These models of atomically clean surfaces of semiconductors do not, however, provide a basis for a general explanation of the entire range of experimental data provided by the ELEED method and data on the energy spectrum of surface states obtained by UPS and UPSAR. A vivid comparison of theory and experiment on the Si <100> face is given in [3.80]. Unfortunately, in many papers on ELEED and UPS, surface models are discussed without reference to EPR data, which directly demonstrate the hybridization of dangling bonds on the cleavage surface. (These questions are discussed further in Sect.3.1.4.)

Many theoretical concepts of adsorption and surface states on a reconstructed atomically clean surface are based on a uniform two-dimensional surface superlattice; it is in this approximation that the energy spectrum of surface states is considered. There is strong evidence, however, that such surfaces have a cluster structure [3.4]. The diffraction pattern observed in ELEED is a superposition of diffraction reflexes from such a structure. Recent ellipsometric investigations [3.81] of a Ge surface cleansed by ionic bombardment indicate the existence of an amorphous layer about 4 Å thick. According to [3.81], reconstruction proceed not uniformly but in patches. Such a surface can be studied to some extent by methods used in the electronic theory of disordered systems. The question of the adequacy of the diffraction pattern to represent the actual surface structure has not yet been solved. A detailed discussion of these problems was recently given by JONA and SHIH [3.116].

So far we have been speaking of the quasi-equilibrium properties of a crystal. Important requirements in investigations of the adsorptive, catalytic, and electrophysical characteristics of surfaces are data on the dynamics of the surface atoms: phonon spectra and the diffusion processes of surface atoms and defects. The coordination of atoms on the surface differs from that in the bulk, causing changes in the dynamic constants of bondings with neighboring atoms, and consequently alterations in the amplitude and isotropy of their oscillations [3.117].

Lord Rayleigh, while investigating the elastic properties of a continuum, suggested that the introduction of an interface leads to the appearance of localized oscillation modes; these were later named Rayleigh surface waves. LIFSHITZ and ROSENZWEIG [3.118] showed on the basis of solid-state microscopic dynamics that if the surface is taken into account, the localized surface modes are split from the acoustical and optical branches allowed for the bulk. The low-frequency mode in the limit turns into Rayleigh waves; the high-frequency mode turns into the optical modes of the surface phonon spectrum. Various methods were later used to calculate the surface modes of oscillations for primitive, face-centered, diamond-type, and more complex lattices [3.117, 119-121].

Theory shows that for the mean-square amplitude of atomic oscillations on the surface, $\langle x_S^2 \rangle$ is greater in all cases than that in the bulk, $\langle x_V^2 \rangle$; the atomic oscillations on the surface can be anisotropic even if they were isotropic in the bulk. Qualitative relationships have been established between the character of chemical bonds and the magnitude of $\langle x_S^2 \rangle$. The influence of the surface on the spectral distribution is most vividly demonstrated by the calculations carried out in [3.122, 123]. From Fig.3.12 it is easy to see that the transition from an infinite crystal to thin films in the <110> plane causes a shift of the spectral density maximum towards the low-frequency region.

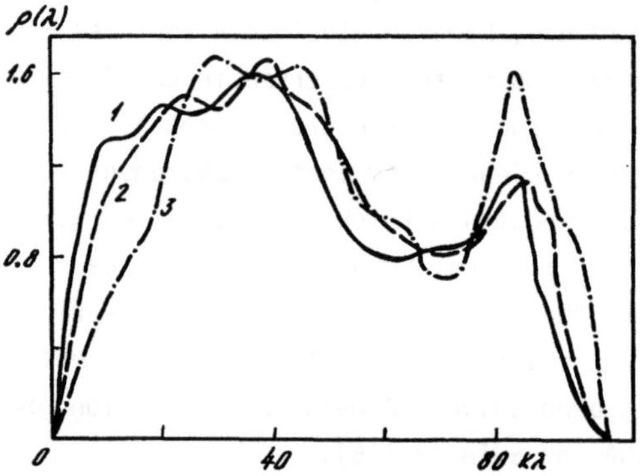

Fig.3.12 Spectral density distribution of thin films /3.122/. (*1*) Film of seven atomic layers. (*2*) Film of four atomic layers. (*3*) Infinite crystal

The majority of calculations have been carried out for highly idealized model surfaces, for which it is assumed that the symmetry of surface atoms and their dynamic bond constants remain unchanged. WALLIS [3.124] pointed out that the dynamic bond constants on the surface can be 25% - 30% smaller than those in the bulk. Thus, at present one can speak only of qualitative agreement between theory and experiment.

Valuable information on the mean-square amplitudes of oscillations of the surface atoms is obtained by measuring temperature-dependent intensities observed in ELEED. The intensity of diffracted beam I (multiple reflections being disregarded) is determined by the Debye-Waller factor [3.41,124]:

$$I_{h,k,\ell} = A \exp(-B\langle x_S^2 \rangle) \quad , \tag{3.6}$$

where A and B are derived from experiment. In the harmonic oscillator approximation, at temperatures above the Debye temperature θ_D,

$$\langle x_S^2 \rangle = \frac{3h^2}{mk\theta_D^2} T \quad , \tag{3.7}$$

where m is the atomic mass. From measurements of I(T) one can calculate, according to (3.6,7), the values of $\langle x_S^2 \rangle$ and θ_D for surface atoms, which determine the diffraction pattern. Such measurements for a number of metals have shown that the normal components of the amplitude of atomic oscillations $<x^2_{S\perp}>$ are twice as high as the amplitudes in the bulk $<x^2_V>$. Consequently, the ratio of the Debye temperatures for the surface and the bulk, θ^S_D / θ^V_D, is about 1/2 [3.41,125,126]. For the silicon <111> face, $<x^2_{S\perp}> \sim 6<x^2_V>$, while for tangential components, $<x^2_{S\|}> \sim <x^2_V>$. For the same face of germanium, $<x^2_{S\perp}> \sim <x^2_{S\|}> \sim <x^2_V>$ [3.127]. Certain diversions from the theoretical calculations are attributed to the fact that superstructures on the Ge<111> and Si<111> faces are disregarded.

The ELEED method makes it possible to estimate not only the alterations in the oscillation amplitudes of surface atoms, but also the coefficients of thermal expansion $\alpha_{S\perp}$ and $\alpha_{S\|}$, which are proportional to the coefficient of anharmonicity of the dynamic bond constants. For GaAs, Ge, Si, and certain metals it was found that $\alpha_S > \alpha_V$ [3.128], i.e., that the anharmonicity of oscillations is greater for the surface atoms. However, the above-mentioned difficulties in determining the absolute values of the intensities of diffraction reflexes, and the fact that inelastic collisions are not always accounted for, demand that one be quite cautious in relying on the Debye-Waller factor when calculating the actual dynamic characteristics of surfaces. Very interesting in this connection is the work of IBACH [3.129], who employed electron beams with energies of 1 - 100 eV and a resolution of $2 \cdot 10^{-2}$ eV to find that the electron energy loss upon reflection from quite different surfaces is about the same: $\Delta E = 69 \cdot 10^{-3}$ eV for ZnO; $\Delta E = 55 \cdot 10^{-3}$ eV for an atomically clean Si<111> face; and $\Delta E = 48 \cdot 10^{-3}$, $90 \cdot 10^{-3}$, and $125 \cdot 10^{-3}$ eV for an oxidated Si<111> face. IBACH attributed this to the interaction between the incident electrons and the surface phonons, which have a wavelength of about 100 Å. In view of the handicaps inherent in the ELEED method, data on the dynamic parameters of the lattice obtained by ELEED should be supplemented with information gained by other means, e.g., Mössbauer spectroscopy and atomic beam scattering.

The increased amplitudes of the oscillations of surface atoms must have an effect upon the magnitudes of surface diffusion and self-diffusion, as well as upon the surface melting. The coefficient of surface diffusion is known to be $D_S = D_0 \exp(-E_a / kT)$, where $E_a = E_D + E_0$ [3.41,130]. In the case of an atomically clean surface, the quantity E_0 characterizes the activation energy of the birth of a mobile particle (atom or vacancy); E_D is the activation energy of diffusion of this particle along the surface. Because of the nonuniformity of the surface, which is covered with a high concentration of macroscopic and point defects, E_0 and E_D are only average values. The value of D_S is obtained by various techniques: with isotope tracers, ion projectors (field emission ion microscope—FEIM), etc. [3.130]. The results obtained for metals [3.41] show the activation energy of surface diffusion, E_a, to be about an order of magnitude lower than that for the bulk. NESTERENKO [3.128] points out that the greater the value of $\langle x_S^2 \rangle$ from ELEED measurements, the lower E_a is for different faces of a crystal (e.g., Ni).

In 1910 Lindemann, [3.131], suggested that a solid body starts to melt when the mean-square amplitude of the oscillations of its atoms builds up to a certain critical value $\sqrt{\langle x_V^2 \rangle} \sim \alpha d$, where d is the mean distance between adjacent atoms. In the Debye approximation according to (3.7), the Lindemann melting temperature T_L is [3.131]:

$$T_L \sim (\alpha d)^2 \frac{mk\theta_D^2}{3\hbar^2} \quad . \tag{3.8}$$

Depending on the type of lattice, α takes on values from 0.1 to 0.2 [3.132]. A sufficiently exact relationship between T_L and the lattice parameters was obtained by GILVARRY [3.133]. Since θ_D^S, as was already pointed out, is much lower than θ_D^V, the melting temperature for the surface will be lower than that for the bulk. Because melting is accompanied by expiration of the long-range order, the sensitivity of the ELEED method in investigations of this kind is unfortunately rather low.

An extreme manifestation of the peculiarities of the structure and dynamic properties of the interphase is observed in disperse solid-state particles, where the ratio of surface atoms to bulk atoms is greatly increased. It has long been known that polishing or honing a metal or a dielectric creates an amorphous surface layer (Bailby layer). Similar layers are also created when a crystal is crushed. Since disperse materials are most commonly employed in investigations of catalytic and adsorptive processes, it is worth going into greater detail here.

The best-studied case of surface amorphization is in crushing of α-quartz. It has been proven beyond doubt by diverse techniques (X-ray analysis [3.29, 134], IR spectroscopy [3.135,136], measurements of the solubility of quartz powders [3.137], and differential thermal analysis of the fraction of the crystal quartz phase undergoing the $\alpha \rightleftharpoons \beta$ transition at 874 K [3.138]) that on the surface of crushed particles there is always an amorphous layer of SiO_2 with a density of about $2.25 - 2.59\ g \cdot cm^{-3}$, i.e., much lower than the density of the original crystal ($2.66\ g \cdot cm^{-3}$). Taking the effect of aggregation into account (Sect.3.1.1), KHODAKOV [3.29] estimated the thickness of the amorphous layer and investigated the influence of crushing and of the surface-active media on the degree of amorphization.

Other alterations in the crystal structure have also been observed to be caused by crushing. For instance, the formation of an aragonite phase during the crushing of calcite has been proved by X-ray analysis [3.29] and in high-dispersion graphite, traces of its rhomboid modification [3.34] have been found. NESTERENKO and co-workers, while analyzing the scattering of slow neutrons (25 MeV) on silicon, observed noticeable changes in the spectral distribution function upon transition from coarse to fine powders of Si [3.139]. Based on the data presented in Fig.3.12, NESTERENKO and GORBACHEV [3.139] attributed these changes to an increased contribution by the phonon spectrum of the surface.

The existence of an amorphous layer on the surface of particles must affect the melting temperature as well. Usually a decrease in the melting temperature of small particles, ΔT, as compared with a bulk sample (T_∞) is related to the dependence of the interphase surface energy upon the radius of curvature. For instance, GLADKIKH and KHODKEVICH [3.140] employed the following equality for metallic paricles of radius r:

$$\frac{\Delta T}{T_\infty} = \frac{3}{\lambda\rho}\left[\sigma_S + \sigma_L\left(1 + \frac{A}{3}\right)\right]\frac{1}{r} \quad , \tag{3.9}$$

where λ is the latent melting heat; ρ is the solid phase density; $A = (V_L - V_S) / V_S$; and σ_S and σ_L, V_S and V_L are, respectively, the specific surface energies and volumes of the liquid (L) and solid (S) phases. However, the parameters λ, ρ, σ for disperse particles, which enter into relations of this kind, are usually not known. They are assumed to be equal to those for the bulk, which naturally leads to considerable errors. With small particles, the principal factor causing a decrease in the melting temperature comes from the amorphous layer, where the mobility of lattice elements is much greater than that in the bulk. Since $\theta_D^S \approx 1/2\theta_D^V$, the Lindemann temperature (3.8) for the surface is much lower than that for the bulk, as was mentioned above. For small particles the effect of "surface melting" will be the predominant one.

KHODAKOV and REBINDER [3.29,138] attributed the formation of an amorphous layer on the surface of particles of brittle and rather high-melting substances to plastic deformations which result from crushing. Amorphous layers, however, are also present on particles obtained by other techniques, such as condensation of vapors and sedimentation. It appears that the disordering of the surface phase is due to the rather high mobility of the lattice elements in the subsurface region of a small particle. The starting point for the mobility in a solid-state lattice is often described by the Tamman temperature (T_T). For monocrystals it is usually assumed that $T_T \sim (0.3 - 05)T_{melt}$. For the surface phase of small particles T_T should be even lower. A remarkable phenomenon was observed in dispersed ice [3.141]: a liquid phase (water) was detected by NMR at a temperature of about 250 K, which is 23 degrees lower than the melting point of ice.

In a low-melting molecular crystal (ice) the amorphous phase is already in a quasi-liquid state. The coefficient of self-diffusion for the surface water phase (10^{-6} - 10^{-7} $cm^2 \cdot s^{-1}$) proved to be just about two orders of magnitude lower than that for common water. Recently the formation of a quasi-liquid film on the surface of an ice monocrystal was demonstrated once again by measurements of the thermal expansion coefficient [3.142]. A theoretical discussion of this phenomenon had been carried out earlier by FLETCHER [3.143].

It is interesting to recall in this connection the work of HENZLER and CLABES [3.144], who observed the disappearance of steps on atomically clean vicinal surfaces of monatomic semiconductors during heat treatment at about 1/2 T_{melt}. The transition from one type of superstructure to another depends on the density of steps on the surface.

Data obtained by ELEED and other methods show that reconstructed surfaces of certain monocrystals, as well as the amorphous surface phase of disperse particles, can exist in quasi-stable equilibrium with the ordered structure of the bulk. This is probably connected with a significant change in the entropy of the surface phase, η^ω, as compared with the entropy of the bulk crystal, see (2.32b and 47): $\eta^\omega = \eta_{tr} + \eta_{rot} + \eta_{vib} + \eta_{con}$. For high-melting materials the increase in η^ω is due primarily to an increase in the vibrational degrees of freedom of the surface atoms (η_{vib}) and an increase in the configurational component of entropy (η_{con}). For materials with a lower melting point (e.g., molecular crystals) an important part is played by η_{tr} and η_{rot}. Turning from single crystals to powders, we can observe the whole range of surface disordering: from domains with a reconstructed yet ordered structure to a totally amorphous surface phase.

ELEED only provides data on the ordered elements of atomically clean surfaces, and gives merely indirect hints regarding the closure and rehybridization of surface chemical bonds. More straightforward information is obtained with the aid of spectroscopic and radiospectroscopic methods.

3.1.4 Hybridization of Surface Atoms

Let us consider first the ideal cleavage surface of a crystal with a diamond-type structure (Fig.3.13). The <111> plane dissects one of the strongest

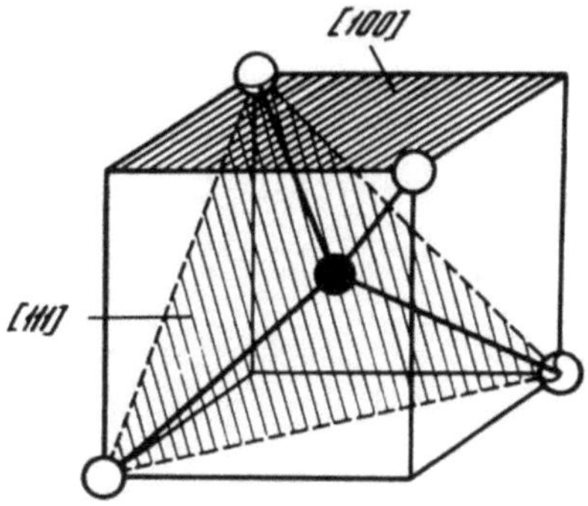

Fig.3.13 Diamond crystal cell

bondings which result from interaction of the sp^3 orbitals of two nearest-neighbor carbon atoms. The breakage of this bond gives rise to a free valence directed outwards normally to the <111> plane. The hybridization of the surface atom is retained, so the surface of the <111> face should behave like a polyradical. The <100> plane dissects two bonds at once. These bonds interact strongly with the sp^3 bond of the central atom, which leads to rehybridization of bonds.

Direct information on broken surface bonds is obtained by EPR. The most extensive data are on EPR signals from atomically clean surfaces of silicon and germanium, which display a diamond-type structure. In pioneering works carried out as early as the 1950s (see the review in [3.4]), silicon monocrystals which had undergone mechanical treatment or crushing were observed to give an EPR response which was attributed either to dangling Si bonds, defects in the subsurface region, or the oxygen complexes on the surface. Since the crushing in these experiments was performed in air or in a low vacuum, one can hardly draw meaningful conclusions regarding the nature of the observed paramagnetic centers.

The first detailed investigations of EPR signals from atomically clean surfaces of silicon in a vacuum of 10^{-9} torr were carried out by HANEMAN and his colleagues who analyzed both powders [3.145] and single crystals [3.104, 146] of silicon. In both cases a single isotropic line was observed (Fig.3.14) whose parameters (the width of the line ΔH and g factor) are presented in Table 3.2. The intensity of response from <111> faces,[1] N_S, was about 10^{14} spin $\cdot$ cm^{-2}. The parameters of the signal were not influenced by various impurities (B and P [3.157]) or by oxygen dissolved in the silicon [3.148,

---

1 The surface of Si and Ge powders obtained by crushing consists largely of <111> faces [3.4].

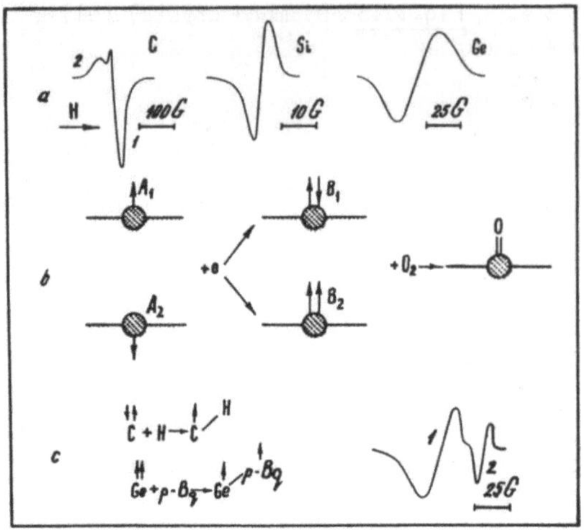

Fig.3.14 (a) EPR signals from atomically clean surfaces of graphite, germanium, and silicon. (b,c) Schemes of elementary acts of adsorption

Table 3.2. Parameters of EPR signals from atomically clean surfaces of semiconductors and dielectrics

Material	Prep. atomically clean surface	N_S[cm^{-2}]	ΔH[G]	g factor	Temp. anneal.	Ref.
Graphite	Crushing at 300 K	$<10^7$			>800	[3.147]
	at 77 K	10^{13}	1[a] -12	2.0270		[3.148]
			2[a] -55	2.0036		[3.147]
Ge	Crushing at 300 K	$<10^9$			>550	[3.4,149]
	at 77 K	10^{13}	35	2.0240		[3.4,149]
	at 300 K	$5\cdot10^{12}$	30	2.02		[3.150]
Si	Crushing at 300 K	10^{14}	6	2.0056	>700	[3.148]
	at 700 K	10^{12}	6	2.0056		[3.148]
	Cleavage at 300 K	$8\cdot10^{13}$	6	2.0055		[3.104]
GaAs	Coarse Crushing at 300 K	$5\cdot10^{13}$	9 - 10	2.0038		[3.151]
GaSb	Coarse Crushing at 300 K	few	9	2.0038		[3.151]
InAs	Coarse Crushing at 300 K	0				[3.151]
CdS,CdSe ZnSe, PbS	Coarse Crushing at 300 K	$\sim10^9$		1.90-2.10		[3.152,153]
SiO_2	Crushing at 300 K	$4\cdot10^{12}$	1	$g_{\parallel}=2.0017$, $g_{\perp}=2.0003$	>800	[3.154,155 156]

[a] see Fig.3.14a

158]. All this suggested that the EPR signals should be attributed to the dangling bonds on the Si<111> face [3.145,146,148,157,158].

The first experiments on the crushing of germanium in an ultrahigh vacuum (10^{-9} torr) at room temperature failed to reveal any spin centers (within an accuracy of 10^9 spin $\cdot$ cm^{-2}) [3.4,38,158]. This result was confirmed in [3.159]. However, when germanium was crushed in a high vacuum at the temperature of liquid nitrogen (77 K), and EPR measurements were taken at the same temperature, DEMIDOVICH [3.149] (cf. also [Ref.3.4,p.104]) observed a wide band (Fig.3.14a) whose parameters are presented in Table 3.2. Later a similar signal, though of lower strength (cf. Table 3.2), was detected with the aid of a storage amplifier during rough grinding of a Ge single crystal under high vacuum at 300 K [3.150], and during cleavage of Ge-Si alloys with an excess of Ge [3.160]. All these wide bands were interpreted as coming from free radicals on the Ge<111> face.

This series of investigations led to an important conclusion. It turned out that the signals obtained during cleavage of Si monocrystals [3.104] have practically the same parameters as those from disperse powders (with s up to 10^4 $cm^2 \cdot g^{-1}$) [3.148]. This makes it justifiable to compare data obtained for the surface of single crystals with those for powders.

A vast amount of data has been accumulated from EPR measurements on atomically clean graphite surfaces [3.147,148,161]. Due to the absence of the aggregation effect (Sect.3.1.1), graphite turned out to be a very advantageous object to investigate. Cleavage along the AA' line (Fig.3.15) in which two atoms with broken π-bonds appear close to each other on the surface, should not give rise to any polyradical properties [3.4]. However, a surface with alternating atoms with broken and intact bonds will display polyradical properties (line BB' in Fig.3.15). EPR measurements in the ranges 8 mm, 3 $\cdot$ 2 cm,

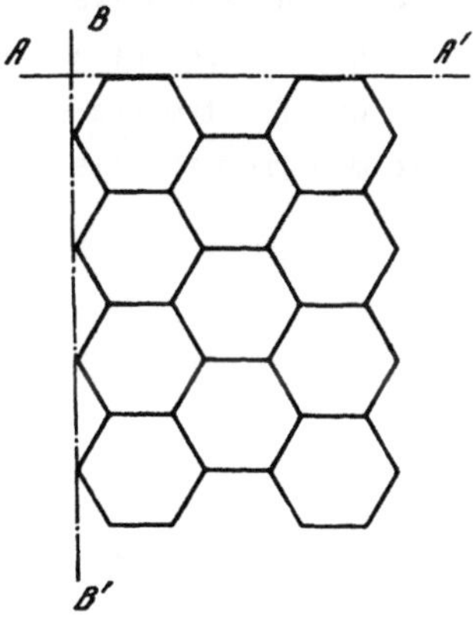

Fig.3.15 Cross section of a two-dimensional plane of graphite

and 1 m failed to show any response (within 10^7 spin • cm^{-2}) from graphite samples prepared by crushing a single crystal in a high vacuum at 300 K [3.148]. The EPR signals observed earlier [3.162] from coarse graphite particles apparently belong to bulk defects. However, samples prepared at 77 K display a wide asymmetric band composed of narrow (1) and wide (2) components (see Fig. 3.14a and Table 3.2). The wide component is attributed to broken σ-bonds of carbon; the narrow one to partially delocalized π-electrons [3.147,148]. Since only the prismatic faces of graphite are reactable (according to adsorption measurements, their area is about 1/18 of the total surface of the particles [3.4]), the concentration N_S for graphite in Table 3.2 refers only to these faces. For Si and Ge particles it refers to the total surface.

Comparison of the linewidths for graphite (σ-bonds), Ge, and Si (Table 3.2) shows good agreement with the corresponding estimates of spin-orbital bond constants made in [3.159]. This is a further indication that the signals presented in Table 3.2 belong to parent crystal atoms, and not to impurities. It is interesting to note here that in investigations on dielectric and semiconductor powders fired at high temperatures, a characteristic narrow signal ($\Delta H \sim 1.0 - 1.5$ G and $g = 2.0028$) was observed and conclusively proved to belong to carbon contaminants due either to decomposition of organic substances on the surface, or to diffusion of carbon atoms from the bulk [3.152,163]. These data were not, of course, included in Table 3.2.

HANEMAN and co-workers carried out EPR measurements from atomically clean surfaces of $A^{III}B^{V}$ [3.151], $A^{II}B^{VI}$, and $A^{IV}B^{VI}$ [3.153] compounds. From Table 3.2 it is easy to see that the concentration of spin centers for all these compounds is much lower than that for silicon. The value of N_S is somewhat greater for GaAs. The low density of paramagnetic centers was attributed by HANEMAN and coworkers [3.151,153] to the high ionicity of the bond in these compounds.

The cleavage surface of α-quartz has received considerable investigative attention [3.154-156,164-168]. According to WEYL [3.166], the cleavage of a quartz crystal in a vacuum (Fig.3.16) leads to the appearance of a free-radical surface, which in the plane of cleavage <1011> is formed from the elements

$$\begin{array}{c} O^{\cdot} \\ | \\ -O-Si-O- \\ | \\ O \\ | \end{array} \qquad\qquad \begin{array}{c} -Si^{\cdot}- \\ | \end{array} \qquad (3.10,11)$$

and in the plane of the hexagonal prism from the following elements [3.165]:

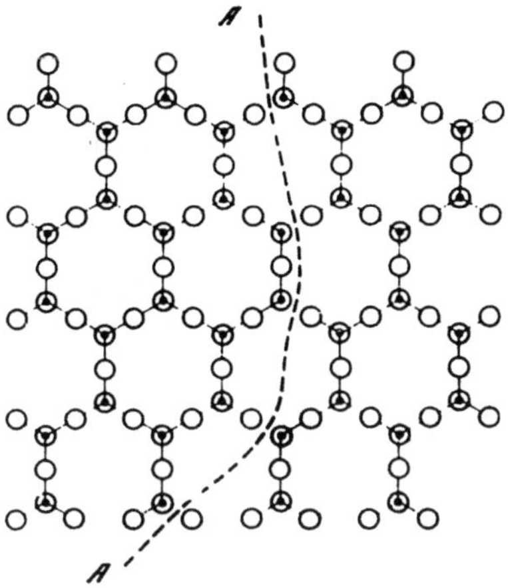

Fig.3.16 Cleavage of α-quartz crystal

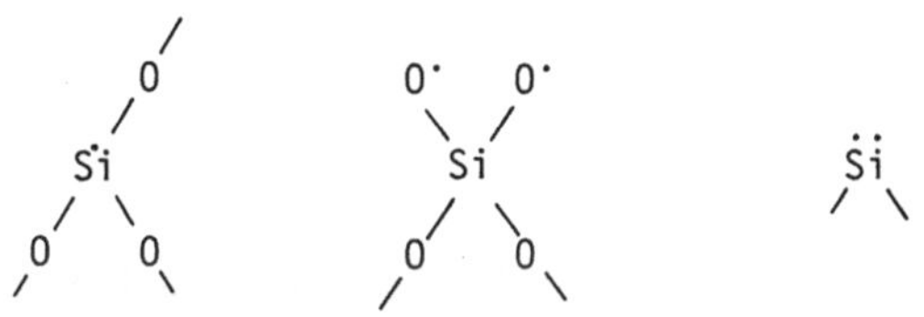

(3.12,13,14)

The mechanism by which active centers are formed during the crushing of quartz was investigated in detail by AKHMED ZADE and co-workers [3.154], HOCHSTRASSER [3.155] and RADZIG [3.156,167,168]. It was shown that crushing quartz (or quartz glass) in an atmosphere of inert gas or in an ultrahigh vacuum gives rise to a narrow asymmetrical EPR response (Table 3.2). This signal has an ultrafine structure resulting from the interaction of a non-paired electron with the nuclei of a ^{29}Si isotope ($I = 1/2$, natural content 4.7%). This signal corresponds to radicals of type (3.11). Radicals of type (3.10) cannot be detected by EPR because of the degeneracy of the ground orbital state; however, their number can be determined through chemisorption of active gases [3.156,167]:

$$\equiv Si - O^{\cdot} + CO \rightarrow \equiv Si - CO_2^{\cdot} \qquad (3.15)$$

$$\equiv Si - O^{\cdot} + O_2 \rightarrow \equiv Si - O_3^{\cdot} \qquad (3.16)$$

Radicals of types (3.15) and (3.16) are detected by EPR. The number of paramagnetic centers of type (3.10) is equal to the number of centers of type (3.11), and is about $(2-5) \cdot 10^{12}$ cm^{-2}, or 1% - 2% of the number of surface atoms of Si. In the course of cleavage and subsequent recombination of radicals, not only do centers of types (3.10) and (3.11) appear, but also stressed silicon bonds

```
    O
\  / \  /
 Si   Si                (3.17)
/  \ /  \
    O
```

which have augmented chemisorptive activity. In order to form new strong bonds, the adjacent silicon tetrahedrons must turn, and this process is hindered by their limited mobility. There are many more centers of type (3.17) than of types (3.10) and (3.11). The strength of the Si - O bond in centers of type (3.17) is not as great as that in the ordinary quartz lattice [3.168], and chemisorption irreversibly destroys centers of type (3.17). ZHDANOV and A. KISELEV [3.165] assume that there is rehybridization of bonds in centers of types (3.12) and (3.13) with the formation of a conjugated bond:

```
- O
   \
    Si = O              (3.18)
   /
- O
```

This should cause the basic crystallographic cell of silicates to be destroyed, which is rather doubtful [3.169-171].

An analysis of Table 3.2 reveals a number of characteristic properties of atomically clean semiconductor and dielectric surfaces. Some of these are now briefly discussed.

1) The concentration of spin centers is always lower than the concentration of surface atoms on the cleavage plane predominantly formed in crushing (0.1% - 10%). This can hardly be attributed to the destruction of some of the radicals due to adsorption. As mentioned in Sect.3.1.1, the contamination coefficients for disperse powders at 10^{-9} torr can reach 10^{5} molecules $\cdot$ cm^{-2}. One could presume that the majority of broken bonds is closed, for example, according to scheme (3.17) in the case of SiO_2. However, investigations of graphite powders have revealed that the chemisorptive activity of samples obtained at low temperatures (77 K) is about the same as that of samples obtained at 300 K, while their values for N_S differ by many orders of magnitude. Investigations of isochronous annealing of paramagnetic centers in a vacuum [3.172,173] show them to disappear at temperatures higher than the temperatures at which surface superstructures are reconstructed (as observed by ELEED; see Sect.3.1.3).

2) The spin concentration N_S for C and Ge depends on the temperature of crushing. It is interesting to note that in crushing at 300 K the value of

N_S for the semiconductors with the narrowest band (C, Ge, InAs, etc.) is extremely low compared to that for semiconductors with a wider forbidden band (Si and GaAs). If silicon is crushed at a higher temperature (700 K), the value of N_S will be smaller.

3) Despite the presence of free radicals on the surfaces of certain semiconductors, the sticking coefficients k_S are quite low, even for so active a gas as oxygen. For instance, at 300 K, k_S is about 10^{-2} for Si, 10^{-3} - 10^{-4} for Ge, and 10^{-5} for GaSb [3.4] (see also Sect.4.1.2). The lowest values of k_S ($\sim 10^{-12}$) are observed with compounds of type $A^{IV}B^{VI}$ [3.153]. The adsorption at 300 K of conventionally demoisturized oxygen does not bring about any changes in the EPR response from broken bonds of Ge and Si [3.4,173,174]. The absence of direct chemical interaction between O_2 and the spin centers of Ge and Si is an indication of the specificity of hybridization of the surface orbitals of unpaired electrons.

In [3.102-104], the chemical bonds on Si<111> and Ge<111> faces were qualitatively analyzed with the change in interplanar distance in the upper two layers of atoms being taken into account. SEIWATZ [3.103] investigated the breakage of not one, but three bonds at once; he assumed that double $\pi\sigma$-bonds form, which is not typical of Si and Ge (Fig.3.10). In the scheme proposed by HANEMAN (Fig.3.11) the energy of interatomic interaction is assumed to depend only on the distance, and the change in bond strength brought about by rehybridization is not taken into account [3.104].

An analysis of the behavior of the overlapping integral of atomic orbitals of Si in the upper two layers in the <111> plane revealed that the equilibrium distance d between the first and second layers is small compared to that in the bulk [3.173]. Due to the decrease in d, the parameter which characterizes the overlapping of orbitals becomes negative, which corresponds to the turnover of the orbital of the unpaired electron of the upper layer of Si atoms inwards. This turnover is accompanied by a loosening of the electron cloud and an increase in the overlapping integral of the orbitals of adjacent atoms in the first layer. The change in hybridization can manifest itself in a different shift of the atoms in the first layer, both normal to the surface and tangentially, which does not contradict ELEED data. Such a surface more or less resembles the ripplelike model of <111> Si(7×7) proposed in [3.108].

The turning over of dangling bonds and the corresponding rehybridization of surface atoms may follow the activation mechanism. For this reason the concentration of upturned bonds will depend on the temperature at which the crushing has been performed. ELEED data give evidence that the surfaces of silicon and germanium specimens are reconstructed in the course of firing after cleavage (Sect.3.1.3). This could explain certain discrepancies between EPR responses from specimens prepared by coarse crushing in a mortar [3.150] and those prepared on a vibrating mill [3.158,174]. The mechanical deformations and local heat buildups which are necessarily connected with vibrational crushing [3.29,31] cause some of the dangling bonds to upturn, and they become inactive with respect to adsorption.[2] In coarse crushing of a monocrystal, the mechanical deformations and local heat buildups are much smaller, so that the initial orientation can be retained by a greater number of orbitals of unpaired electrons, thus facilitating the interaction with O_2. HIGINBOTHOM and HANEMAN [3.150] noticed a small alteration in the EPR response due to adsorption of O_2 on an atomically clean surface of Ge.

In graphite the effect of the upturning of bonds appears to be quite small. Chemisorption of oxygen always causes a drastic change in N_S and leads to the appearance of EPR signals from the peroxide radical $>$C-O-O$^{\cdot}$ [3.161]. The same is true of atomically clean surfaces of SiO_2, where HOCHSTRASSER and ANTONINI [3.155], as well as RADZIG [3.156,167] observed radicals (3.15) and -Si-CO$^{\cdot}$ during adsorption of CO; (3.16) and $\geqslant$Si-O-O$^{\cdot}$ [3.167] during adsorption of O_2; (3.15) during adsorption of CO_2 [3.155,167]; $\geqslant SiON_2^{\cdot}$ during adsorption of N_2O; and Si - OH groups and H atoms during adsorption of H_2 [3.167].

Another attempt to explain the low sensitivity of EPR from dangling bonds during adsorption of hydrogen on semiconductors was made by LEMKE and HANEMAN [3.25]. They assumed that the majority of spin centers are situated on the surface of microcracks, which appear on the crystal surface in the course of cleavage or crushing, and on abraded specimens. The wave functions of unpaired electrons on the opposite walls of these microcracks overlap. Due to perturbation of the potential in such microcracks, Anderson localization takes place. A clean cleavage surface gives no EPR response, which according to [3.25] is due to the pairing of electrons on alternate atomic sites. Without going into the details of the adsorption mechanism, LEMKE and HANEMAN [3.25] assumed that the adsorbed molecules (e.g., O_2 or H_2) interact with the localized states of the electrons and influence the width of the EPR line. The same model is extrapolated for vacuum-cleaved Ge crystals, Ge - Si alloys, and even amorphous substances. Notice that deposited amorphous films of Ge and Si give an EPR response similar to that from crystal specimens (Table 3.2, and [3.175]).

In spite of its many advantages, the Haneman model fails to explain all of the experimental data on EPR signals from atomically clean cleavage surfaces

---

[2] It is interesting to note the absence of reversible broadening of the EPR lines from these centers in physical adsorption of oxygen.

of semiconductors. For example, small variations in the EPR line widths for single crystals (Table 3.2), coarse powders [3.176], and high-dispersion, vibrationally crushed specimens (Table 3.2) are still a puzzle. One may ask whether the crushing of a specimen in a vibrational mill does not proceed along the microcracks first. The formation of microcracks on the prismatic faces of graphite particles is difficult to explain (see Table 3.2). The Haneman model does not give an exhaustive explanation of the IWASAKI and NAKAMURA experiments [3.176]. In those experiments the adsorption of H_2, O_2, and Cl_2 molecules, which are quite different in size and electronegativity, were observed to increase the strength of the signal from Si (15% with O_2, 80% with H_2, 35% with Cl_2). The adsorption of atomic hydrogen, in contrast, worked to decrease the signal amplitude. The adsorption of O_2 upon atomically clean surfaces of Si is known to be irreversible, as it is connected with the formation of strong chemisorptive bonds.

The dangling bonds observed in EPR experiments with crushed graphite, silicon, and germanium are interpreted in [3.4,147,148] as acceptor surface states of the Shockley type (see Fig.3.14b, centers A_1 and A_2). In the case of graphite dispersed at 77K, the orbitals of the unpaired electron are directed outwards (A_1), while those of germanium and silicon are directed inwards (A_2). The capture of the free charge carrier by these traps leads to the formation of two-electron states of type B (Fig.3.14b) — occupied Shockley states. Neither singlet (B_1) nor triplet (B_2) electron states in these centers are observed in the EPR spectra [3.4]. A similar model of traps for the surface of silicon has been proposed by CHUNG [3.157].

In a later work [3.25] HANEMAN also employed the concept of one- and two-electron traps (A and B in Fig.3.14) on the surface of silicon. We have already mentioned that the character of hybridization of surface atoms is closely connected with the variation in the distance d between the upper two layers of the lattice [3.173]. HARRISON [3.107] has shown that depending on the value of d, not only A and B states can appear on the surface, but empty states as well ($\geqslant Si^+$).

One explanation of the high chemisorptive activity of graphite crushed at 300 K(C_{300}), whose surface lacks a noticeable concentration of radicals, is based on the assumption that the occupied B states play the role of adsorption centers [3.147,161]. The transition from state A_1 to state B_1 seems to be connected with the surmounting of a rather high potential barrier by the charge carriers. This perhaps explains the dependence between the concentration of spin centers and the temperature of crushing (Table 3.2). The same model of surface centers has been applied to Ge and Si [3.148,174]. In order to explain the high values of oxidation heat for atomically clean surfaces of Ge and Si, and the absence of an EPR response from adsorbed oxygen, DEMIDOVICH, KISELEV and co-workers [3.148,174] assumed that the adsorption of O_2 proceeds on surface centers of type B first (Fig.3.14b), and that the upturned orbitals of the unpaired electron (A_2) then become covered with an oxide film which can only be removed by etching.

Direct proof of the presence of centers of type B on graphite surfaces was obtained in measurements of the static paramagnetism of graphite [3.177] and experiments on the adsorption of atomic hydrogen and deuterium [3.148]. COULSON [3.178] considered three possible states of the peripheral atoms of carbon:

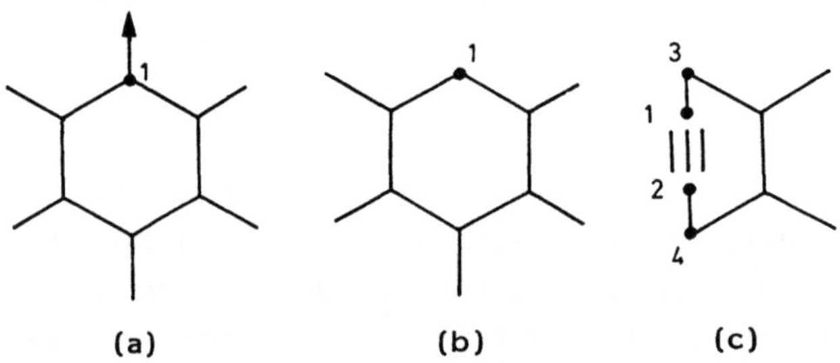

(3.19)

State (3.19a) corresponds to an unsaturated valence, and the carbon atom (1) retains its $sp^2\pi$ hybridization. In state (3.19b), the corner atom (1) goes from the $sp^2\pi$ state to the s^2p^2 state (it is assumed that the state of a free atom of bivalent carbon is a triplet one). In state (3.19c), provided 1 and 2 are boundary atoms, the elementary cell can undergo deformation, and these atoms may come close to the line of the bond 3 - 4. This results in hybridization close to the sp state, which in the limit (when atoms 1 to 4 all line up) becomes acetylene bond. Magnetochemical measurements have shown [3.177] the majority of two-electron centers B on the graphite surface to be in the singlet state, i.e., to be nonparamagnetic (Fig.3.14b, center B_1).

In experiments on the adsorption of hydrogen atoms on the nonradical surface of graphite C_{300} (Table 3.2), the authors of [3.148] detected an EPR signal with $g = 2.0036$ and $H = 11 \pm 1$ G, which in terms of parameters is similar to the signal from a π electron partially localized near the newly formed bond $>$C - H (Fig.3.14c). Since the linewidth (with blurred SFS) is determined by the magnetic moment of the nucleus μ_H, the adsorption of deuterium atoms ($\mu_H / \mu_D = 3.25$) was investigated. In agreement with theory, the signal coming from electrons localized near the C - D bonds was indeed three times as narrow ($g = 2.0036$, $\Delta H = 4.0 \pm 0.5$ G). A similar observation of "dispairing" of B states was made during adsorption of acceptor molecules of p-benzoquinone (p-Bq) on a nonradical surface of Ge [3.174]. The adsorption brought about a complex EPR signal (Fig.3.14c): the narrow component is connected with the formation of the anion radical $(p\text{-Bq})^-$; the wide component ($g = 2.02$, $\Delta H \sim 22$ G) is close in parameters to the signal from Ge. A certain difference in ΔH is due to the electric field arising from a redistribution of the charge within the complex. It is possible that the increase in the strength of the EPR signal observed in [3.176] during adsorption of O_2, H_2, and Cl_2 molecules is also connected with partial dispairing of two-electron states B.

The position of local levels of broken bonds in the energy spectrum of a crystal is not entirely clear. If one adopts the model of activated occupation of traps (i.e., in which the transition from centers of type A to centers of type B is facilitated by considerable mechanical deformations and local heat build ups in the course of crushing), then it can be assumed that the energy levels of these traps fall within, for example, the allowed zones. The barrier that separates these states from the allowed zone could be connected with a different symmetry of the wave functions. It is interesting to recall in this connection the paper by CHUNG [3.157] in which it was shown that the concentration of spin centers on the surface of silicon does not depend on doping of the bulk crystal. The transition from degenerated n-type to p-type silicon, when the Fermi level moves across the whole width of the forbidden zone, does not change the value of N_S in any considerable way. It must be noted, however, that these data say nothing about the alteration of the position of the Fermi level on the surface, which in fact determines the population of traps in the forbidden zone of the crystal.

The data on the temperature dependence of N_S for graphite and silicon [3.148] do not contradict the model of levels in the allowed zones. The dependence $N_S(T)$ obeys Curie's law; i.e., the value of N_S in both cases is determined by the population of Zeeman levels, and not by the position of traps in relation to the Fermi level in the forbidden zone. On the other hand, in the case of germanium, Curie's law is not valid [3.160]. It is possible that with germanium some of these levels fall within the forbidden zone, so that the signal coming from them strongly depends on the temperature. The rest of the levels are separated from the allowed zones by the activation barrier, and electrons in these states are responsible for the chemisorption of p-Bq with the formation of radicals $(p\text{-Bq})^-$.

It is interesting to note that according to NISHIDA [3.179], the hybridization of dangling bonds on a relaxed surface of Ge differs greatly from that on an ideal and relaxed surface of Si. According to estimates made in [3.180], the positions of levels of empty, one-electron and two-electron traps (centers

A and B in Fig.3.14) in the energy spectrum of the surface depend on the constants of electron-phonon interactions, and on interactions between electrons.

At present it is hardly possible to probe further into the hybrid states of surface atoms. In the course of mechanical crushing, one can observe not only considerable local deformations and heat build ups, which can result in substantial reconstruction of the surface, but also luminescence and exoelectronic emission of slow and fast electrons [3.181]. The mechanism of these phenomena is not yet completely understood. We do not know whether there is any connection between the centers of exoemission and A and B centers as observed by EPR and in the course of adsorption.

In experiments on the crushing of quartz and other solids, the EPR method, due to the reconstruction of the surface, apparently detects only a minor fraction of the surface radicals [3.182]. Measurements of chemisorption during crushing showed the concentration of active centers in some cases to be close to that of the surface atoms.

A relaxed surface is certainly quite nonuniform, since it is covered with steps, microcracks, dislocation exits, and other defects, and with domains of variously oriented superstructures. The electronic properties of such surfaces, as was pointed out in [3.173], are best investigated by methods used in the electronic theory of amorphous solids. The model proposed in [3.25] is one of the first steps in this direction. It involves Anderson localization, which is typical of disordered systems. The EPR signal in this case is influenced by the g-factor distribution, the hyperfine contact interaction, the hopping motion of the carriers, and the exchange interaction. For fairly amorphous surfaces it would be helpful to use the general principles of topology, which have been successfully applied to entirely amorphous and glasslike materials [3.183].

We have discussed the basic EPR data on dangling bonds and the hybridization of atoms on surfaces of certain semiconductors and dielectrics. The EPR technique has but limited application in investigations of atomically clean surfaces of metallic powders, because of their high electric conductivity.

3.1.5 The Genesis of a Real Surface

Chemisorption on an atomically clean surface leads to saturation of dangling bonds and changes in the hybridization of the surface atoms (for details see Chap.4). With some semiconductors and metals the rapid stage of chemisorption is followed by a slow stage in which a new surface phase forms (for instance, oxidation of metals and monatomic semiconductors—Si, Ge, etc.). The formation of the new phase is accompanied by complex processes of diffusion of both solid-state atoms and atoms or molecules of environment. The structure, chemical composition, and thickness of the surface phase depend on the temperature, the composition of the gaseous environment, the rate of synthesis, and a number of other factors. Often the structure of the newly formed three-dimensional phase differs greatly from the initial structure of the solid-state surface.

However, when chemisorption takes place at sufficiently low temperatures, the formation of a monomolecular layer of chemisorbed particles effectively prevents further chemisorption. In such cases, the concentration and topography of the chemical groups that spring up on the surface can be quantitatively estimated on the basis of data about the initial structure of an atomically clean surface. Such information has been obtained on the hydration of cleavage surfaces of certain metallic oxides.

Let us consider the hydration of different cleavage surfaces of α-quartz (3.10-14). Upon contact with water, the surface elements (3.10) and (3.11) on the rhombohedral cleavage face undergo hydration in the following manner:

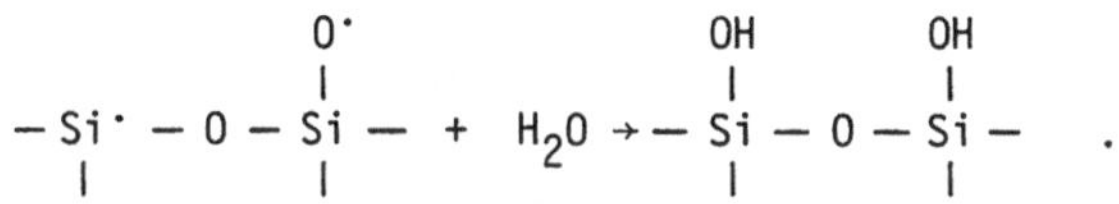

Scheme 3.1

The hydration of elements (3.12-14) will be accompanied by the release of hydrogen or oxygen. Hyperstrained bridges (3.17) are hydrated according to the following scheme:

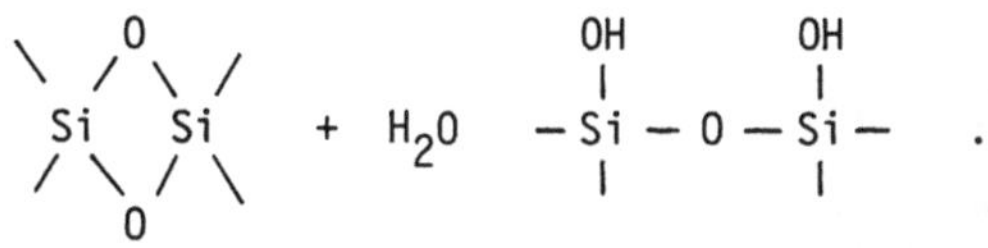

Scheme 3.2

For models of amorphous SiO_2 (silica gel, aerosil), one usually chooses planar cross sections of β-cristobalite and β-tridymite, since these crystal modifications of SiO_2 have about the same density as amorphous silica [3.184-186]. The short-range order is assumed to be basically retained in the transition from a crystallic surface to an amorphous one. The results of calculations of OH-group concentration (i.e., the degree of hydration of the surface, α_{OH}, in μmol • m^{-2}) are shown in Table 3.3 for different crystallic modifications of SiO_2.

An analysis of the cross-sections of crystallic modifications of SiO_2, e.g., cristobalite and tridymite (Fig.3.17) reveals that OH groups are arranged in one of two patterns: single hydroxyls on <111> faces of β-cristobalite and <0001> faces of β-tridymite, and pairs of hydroxyls belonging to the same Si

Table 3.3. Calculated concentrations of hydroxyl groups α_{OH} on the surfaces of certain oxides

Oxide	Modification	Cleavage surface	α_{OH} [$\mu mol \cdot m^{-2}$]	Ref.
SiO_2	α-quartz	<0001>	16.0	[3.164]
		<1011>	8.2	[3.164]
		Mean value	11.2	[3.165]
	β-tridymite	<0001>	7.6	[3.184]
	β-cristobalite	<111>	7.6	[3.185,186]
		<100>	13.0	[3.185]
Al_2O_3	γ-modification	<0001>	23.4	[3.187]
	Corundum	<0001>	25.4	[3.187]
	Hydrargillite	<001>	22.8	[3.187]
GeO_2	Tetragonal	<100>	26.4	[3.188]
		<110>	9.4	[3.188]
		Mean value	22.6	[3.188]
	Hexagonal	Mean value	10.6	[3.188]
TiO_2	Anatase	<111>	24-25	[3.189]
		<001>	11.6	[3.189]
		<110>	14.3	[3.189]
		<010>	9.2	[3.189]
		Mean value	20	[3.189]
	Rutile	<110>	17.4	[3.4]
		<100>	24.0	[3.4]
ZnO	Wurtzite	<0001>	18.2	[3.190]
		<1010>	19.8	[3.190]

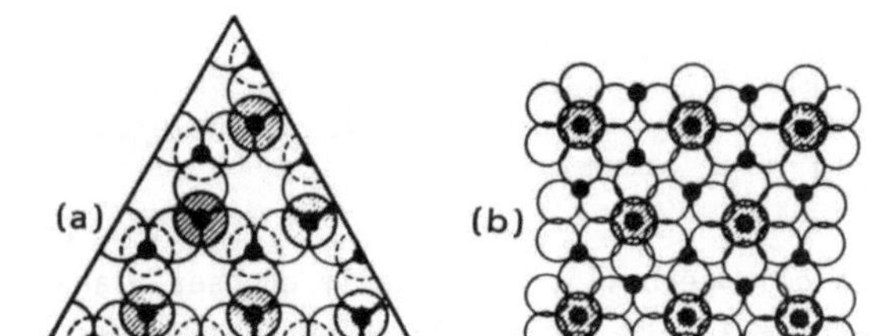

Fig.3.17 a-d. Completely hydrated surfaces of various planes of crystallic modifications of SiO_2 and Al_2O_3. (a) <0001> β-tridymite. (b,c) <111> and <100> β-cristobalite respectively. (d) <0001> γ-Al_2O_3. (*1*) OH groups. (*2,3*) Si and O atoms. (*4,5*) OH groups corresponding to Schemes 3. XIIf and 3.XIIb respectively

atom on <100> faces of β-cristobalite. Consequently, α_{OH} differs by nearly a factor of two for these arrangements (Table 3.3).

EGOROV [3.187] calculated the degree of hydration for ideal cross sections of certain modifications of Al_2O_3. The hydrated <0001> face of corundum ($\alpha-Al_2O_3$), for instance, can be presented as follows:

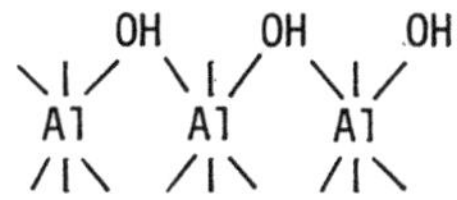 . (3.20)

Of special interest are calculations of α_{OH} for $\gamma-Al_2O_3$, which is a most valuable catalyst. The structure of $\gamma-Al_2O_3$ is that of a defect spinel. On the octahedral cleavage plane <0001> one may expect the formation of hydrate groups such as:

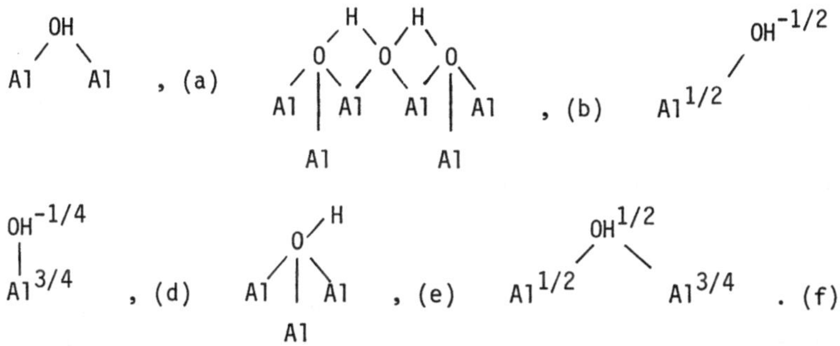

(3.21)

The fractional valence numbers denote the fraction of the electron density which is allocated to the bond. EGOROV [3.187] suggested that in the course of hydration the hydrogen atoms can penetrate the quaternary interstitial sites beneath the surface layer of anions, where there is a shortage of cations because of the deficient structure of $\gamma-Al_2O_3$. A similar possibility was put forward by TUNG and MCINISH [3.191]. Thus the surface of $\gamma-Al_2O_3$ contains about 23.4 $\mu mol \cdot m^{-2}$ of OH groups and 0.5 $\mu mol \cdot m^{-2}$ of H_2.

EGOROV, BOEHM and coworkers [3.187-190] calculated α_{OH} for different cross sections of oxides of Ti, Ge, and Zn. Their data are presented in Table 3.3. Although these calculations were based on a highly idealized concept of the hydrate cover of the oxide, the values obtained for α_{OH} agree well with experimental values from adsorption and spectroscopic investigations, which will be discussed in the following section.

So far we have been considering the cleavage surface, which is hydrated to saturation. PERI and coworkers [3.192,193] attempted a theoretical approach to dehydration of the surface of oxides, which for SiO_2 follows the scheme:

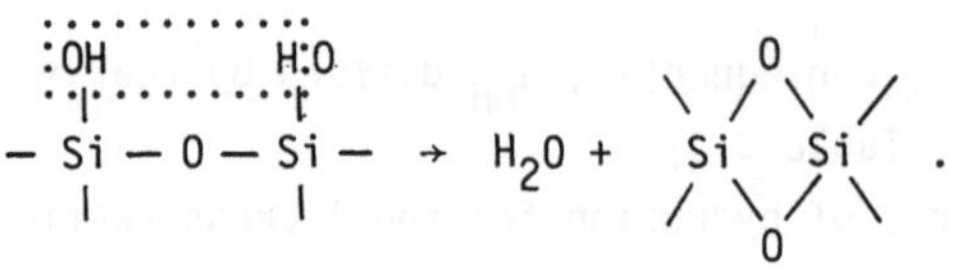

Scheme 3.3

They assumed this to be a random process, and employed the Monte Carlo method to analyze the probability of occurrence of different arrangements of radicals on the surface. For example, in the case of β-cristobalite [3.193], at temperatures below the temperature of dehydration ($T_{cal} < 900$ K) the surface ought to be occupied only by paired hydroxyls, and at $T_{cal} > 900$ K only by single hydroxyls. The steady-state value of α_{OH} for a hydrated <100> surface of cristobalite is, according to PERI and coworkers, about 7.6 $\mu mol \cdot m^{-2}$, or half the saturation value for this face (Table 3.3), and the surface is populated mostly by paired hydroxyls.

A quantum-chemical analysis of hydroxyl groups in the cluster approximation was performed in a number of works by LYGIN (see [3.194] and the references cited therein). With the aid of the CNDO/BW method, the authors of [3.195] investigated the mechanism of H-D substitution in hydroxyl groups on silica gel. PERI [3.196], on the basis of an analysis of P and R branches of vibrational and rotational spectra of the hydroxyl cover of SiO_2, estimated the Si-OH angle to be $\varphi = 113^{\circ}$. A similar value of φ was obtained in [3.197] with the aid of atom-atom potential.

PERI [3.192] carried out a similar calculation for the surface of γ-Al_2O_3. A highly idealized model of the surface was considered in which all the Al atoms are arranged in a hexadic coordination. In the course of partial dehydration of the surface there remain five types of OH-groups with a different number of bonds between the proton and the oxygen atom in the lattice; these are matched to the corresponding absorption lines in the IR spectrum of the oxide. This amounts to assuming the existence of the surface complexes (3.21 a-f) proposed earlier by Egorov. The model of the surface adopted in [3.192] does not reflect the real structure of γ-Al_2O_3, where besides Al atoms in an octahedral surrounding there are Al atoms in the centers of tetrahedrons. Direct proof of this was given by IR spectroscopy [3.198], in which oscillation bands with frequencies of 745 cm^{-1} (hexadic coordination) and 835 cm^{-1} (tetradic coordination) were detected in the absorption spectrum for the Al - O bond. DABROWSKI et al. [3.199] improved this model of Monte Carlo calculation by taking into account both possible coordinations of the surface atoms. It must be noted that the reliability of the theoretical model which was used as the basis for calculations via the Monte Carlo method in [3.193, 199] has not been put to the test. Applying a purely statistical method to calculate the probability of the removal of water from surface centers is scarcely justified. During dehydration the surface undergoes reconstruction involving a large number of atoms. The possibility of such reconstruction is not accounted for by the method employed in [3.192]. Presumably, interaction between a large number of atoms of the surface and the subsurface layer must be taken into consideration here.

3.2 The Real Surface of Semiconductors and Dielectrics

3.2.1 Hydrate Coverage of Oxides

Because oxides have quite a simple spectrum of surface chemical compounds, they have been most thoroughly investigated. The composition of all oxides includes structural water, which can be present in the lattice in the form either of OH groups or of molecules (hydrates and oxyhydrates in GLEMSER's classification [3.200]). Even oxides that are not encountered in crystallohydrate forms release a certain amount of structural water when heated to sufficiently high temperatures. In most cases this water comes from the surface. It was shown in Sect.3.1.5 that the cleavage surface of oxides undergoes hydration if brought in contact with water, and the dangling chemical bonds are saturated with formation of OH groups. The presence of hydroxyl groups on the surface of oxides is confirmed by analysis of the chemical reactions used to produce high-dispersion oxides. For example, the most highly dispersed silicon oxide, silica gel, is the product of condensation of orthosilicic acid, which is accompanied by the release of water [3.184]:

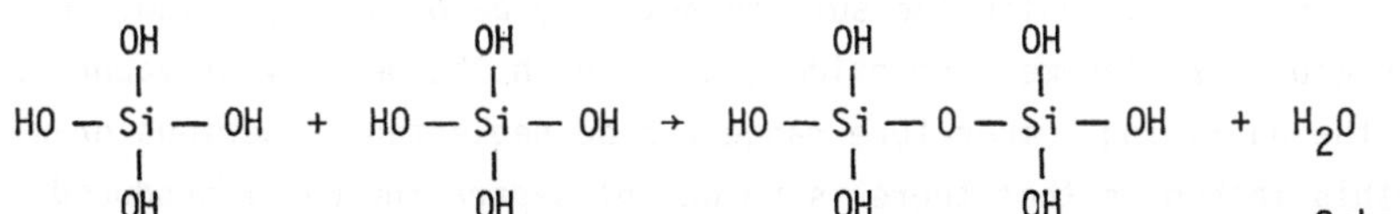

Scheme 3.4

When silica gel is dried, some of the OH groups remain on its surface up to quite high temperatures. As early as the 1930s, HOFMANN [3.201] and later RIDEAL [3.202], A.V. KISELEV [3.203], and CARMAN [3.204] showed the surface of silicate particles to be covered by OH groups under normal conditions.

Structural water is built into the composition of oxide films, which are always present on the real surfaces of monatomic and many binary semiconductors. For instance, single crystals of germanium and silicon are found, after etching, washing, and drying, to be covered by a dioxide film about 20 - 30 Å thick. Some researchers assume that the monoxides GeO and SiO are present in surface films [3.205-207]. The hybrid orbitals of Ge and Si in monoxide molecules correspond to π and σ bondings, as in a CO molecule. The formation of a more or less stable three-dimensional monoxide structure is impossible. What is usually called a monoxide is more likely a mixture of GeO_2(or SiO_2) with excess germanium (or silicon) [3.208]. The same conclusion is espoused by PHILIPP [3.209]. The data obtained by mass spectroscopy [3.205], thermochemical analysis [3.206], and IR spectroscopy [3.210] which are usually offered as proof of the existence of solid monoxide GeO (or SiO) do not in fact give direct confirmation of the existence of a pure monoxide phase. For instance, in [3.211] germanium monoxide observed in mass spectra of vaporized GeO_2 was shown to be a product of the decomposition of GeO_2. ESCA investigations have also failed to provide conclusive evidence of the existence of monoxide [3.212].

Mass spectrometry has been applied to the products of thermal desorption from real surfaces of germanium and silicon, and the release of large amounts of water and its fragments has been detected; this points to a high degree of hydration of the oxide film [3.205,213,214].[3] However, the small surface area of oxide films, and the fact that they are contaminated by diverse substances in the course of etching, have so far prevented a reliable identification of the forms of binding between water and the surface. For this reason, model systems represented by high-dispersion pure oxides have been receiving much attention [3.4].

The general content of structural water on the surface of an oxide, α_{H_2O} (in $\mu mol \cdot m^{-2}$), can be determined by measuring the amount of water released when the oxide is heated from a given temperature to a higher temperature (1300 K and higher) at which the surface actually becomes dehydrated. This method is good for high-melting oxides, with which the density of vaporized oxide in the operating temperature range can be neglected. A serious drawback of this method is that there is no way of separating water produced during dehydration of the surface from water coming from the bulk of the specimen, where it had been present either in the form of crystallohydrate water (if the oxide has crystallohydrate modifications) or inclusions which fill the defects of the crystal during growth. For instance, high-dispersion porous oxides (e.g., silica gel) are known to contain a certain amount of water after being handled under hydrothermal conditions.

---

3 Investigations of the thermal desorption of H_2O from atomically clean surfaces of Si were reported in [3.215,216].

To make it possible to measure the concentration of only those hydroxyl groups and water molecules that are present on the surface, attempts have been made to substitute these groups with other functional groups, so as to analyze later the products of the surface reaction. To that end, reactions with diborane [3.217,218], thionyl chloride [3.219,220], and diazomethane [3.221, 222] have been tried, as well as Zerewitinoff and ion exchange reactions [3.223]. However, FRIPIAT and co-workers [3.224], GLEMSER [3.200], and NEUMARK and CHERTOV [3.225] have shown that the exchange reactions are not always complete; moreover, small pores and defects make part of the surface inaccessible to large molecules of reagents. The method of isotopic exchange proposed by HALDMAN and EMMETT [3.226] does not have any of these drawbacks; in it, OH groups are replaced by OD groups. ZHURAVLEV and A.V. KISELEV [3.227] consider this method to be the most accurate one, since it makes it possible to separate the water on the surface from the water in the bulk.

Investigations carried out by BOEHM [3.223] have shown the isotopic method to yield higher values of α_{H_2O} than the method of thermal dehydration. This may be connected with the fact that the equilibrium constants for the exchange of H and D on a heterogeneous surface differ from those adopted for the homogeneous reaction in the bulk. BLIZNAKOV and co-workers [3.228] proposed measuring the concentration of OH groups by means of a reaction with potassium vapors. BAKARDJEV et al. [3.229] employed hydrolysis to determine the distribution of OH groups on the surface of SiO_2, using $TiCl_4$ as the reactant. This reaction is widely employed in the method of molecular layer deposition [3.230].

In cases where the oxides are water-free in the bulk, the method of thermal dehydration has provided direct reliable data on the general content of water on the surface [3.4].

In order to obtain quantitative information on the concentration of structural water on the surface, powders with rather high dispersion are necessary, and to compare data obtained with specimens having varying degrees of dispersion, one needs a reliable way of determining the specific surface of a specimen, s. The method commonly employed is the Brunauer-Emmett-Teller method (BET) [3.231].

Recall that to determine s, the experimental isotherm of adsorption a(h) (where a is adsorption in $\mu mol \cdot g^{-1}$ and $h = p / p_S$ is relative pressure of the adsorbate) is plotted in the coordinates of the BET equation of polymolecular adsorption:

$$\frac{h}{a(1 - h)} = \frac{1}{a_m C} + \frac{C - 1}{a_m C} h \quad . \tag{3.22}$$

Here a_m is the capacity of the monomolecular layer, and $C = C_0 \exp[(q_1 - q_L)/RT]$ is an energy constant, where q_1 is the mean molar adsorption heat in the first layer, and q_L is the condensation heat. Since (3.22) is linear with respect to h [3.231] in a wide range of relative pressures, the constants a_m and C can easily be derived from it. The specific surface is $s = a_m \omega N_A$, where ω is the area occupied by one molecule in a monolayer, and N_A is the Avogadro number. Often C_0 is set equal to 1, and the mean molar adsorption heat is derived from the value of C. To determine the specific surface some authors employ the adsorption isotherms for vapors of water, benzene, or methyl alcohol, which is methodologically convenient. However, adsorption of these substances is very sensitive to the conditions on the surface; in Chap.4 we shall show that the values of ω which enter the formula for s depend strongly on the extent of hydration of the surface. The employment of these substances in determining the specific surface can lead to results and conclusions which are grossly incorrect.

Reliable results are obtained by measuring low-temperature adsorption of nitrogen or inert gases, e.g., argon. This is accepted as the standard technique for determining the specific surface. Indeed, in [3.232,233] the specific adsorptive properties (with regard to nitrogen) were shown to be quite similar for the surfaces of such different substances as silica gel, alumosilicates, and oxidized and graphitized blacks. In [3.234] it was shown that specific isotherms of the adsorption of nitrogen by various samples of silica gel, aluminogel, and aluminosilicates, dehydrated in a wide temperature range, lie 5 - 7% from a common isotherm. The commonly accepted values for the adsorption of N_2 and Ar are $\omega_{N_2} = 16.2$ Å^2 and $\omega_{Ar} = 18.3$ Å^2 [3.235,236].

The only drawback in using nitrogen and argon as standard substances is the rather large size of the molecules. When dealing with highly porous specimens there is the danger of obtaining faulty results due either to disturbances caused by the pores, or simply to the inaccessibility of part of the surface to the molecule. A similar phenomenon was discussed in connection with the aggregation effect (Sect.3.1.1). According to [3.237], this effect becomes real with pores less than 20 or 30 Å in diameter (for N_2 and Ar). Therefore, in discussing the degree of hydration of a surface we will rely only on the data for nonporous disaggregated particles or macroporous oxides.

The hydrate cover of silicon dioxide has been most thoroughly investigated by the method of thermal dehydration: from macroporous silica gel and aerosil [3.185,238 - 241] to roughly disperse specimens of quartz or quartz glass [3.242, 243], with specific surface areas ranging from several square meters to several hundred square meters per gram. Figure 3.18 shows the dependence on calcination

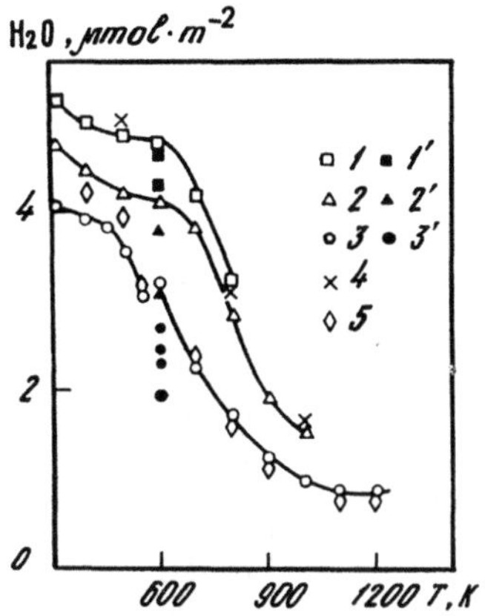

Fig.3.18 Structural water content in silica gel specimens KSK-1 (□), $s = 275\ m^2 \cdot g^{-1}$; KSK-2 (△), $s = 400\ m^2 \cdot g^{-1}$; K-2 (○), $s = 695\ m^2 \cdot g^{-1}$; K-3 (x), $s = 200\ m^2 \cdot g^{-1}$ [3.238]. ◇ – data from [3.240]. ■,▲,● – rehydrated specimens vacuumized at 600 K

temperature T_{cal} of the total structural water content α_{H_2O} on the surface of several samples of silica gel. In the temperature range $T_{cal} = 500 - 700$ K, the curves $\alpha_{H_2O}(T_{cal})$ show inflection. Similar anomalies were observed with thermogravimetric curves [3.244]. The maxima of the curves depicting the dependence of the heat of wetting by water, q_{wet}, upon T_{cal} are in the same range [3.245-247]. This suggests that water released below 500 - 600 K may have a different origin than water released at higher temperatures. Presumably, what is removed at temperatures below 500 - 600 K are mainly molecules of water in an amount approximately equal to the value of irreversible adsorption of water on specimens preheated at this temperature T_{cal} [3.4]. At higher temperatures the release of water is connected with the destruction of the hydroxyl cover of the oxide.

Similarly shaped curves of $\alpha_{H_2O}(T_{cal})$ have been observed for other oxides, which have no crystallohydrate forms, e.g., TiO_2 [3.248], ZnO [3.190], GeO_2 [3.188], and for oxide film on Ge [3.249]. Table 3.4 presents some typical values for the total content of structural water on the surface of these oxides in the temperature range $T_{cal} = 500 - 600$ K.

Serious difficulties are encountered in determining the extent of hydration of the surface of oxides which contain hydrates in the bulk. According to [3.253,254], the structural water can be separated into surface and bulk components if not only the water content, but also the adsorptive properties of the specimen (e.g., its wetting heat) are studied. Since the kinetics of the surface and the bulk hydration processes differ greatly, and the adsorptive properties pertain only to the surface, it is often possible to determine α_{H_2O} for the surface of the oxide. Thus the values of α_{H_2O} were obtained for aluminium oxides whose phase composition had been monitored by X-ray analysis [3.253] (Table 3.4). Similar values of α_{H_2O} were obtained for γ-Al_2O_3 in [3.256].

Table 3.4. Extent of hydration of the surface of certain oxides (experimental values)

Oxide	Specimen	Temp. anneal.[K]	s $[m^2 \cdot g^{-1}]$	d^a [mμ]	α_{H_2O} $[\mu mol \cdot m^{-2}]$	α_{ir} $[\mu mol \cdot m^{-2}]$	Ref.
SiO_2	α-quartz[b]	500	102 cm^2[c]			5.3	[3.250]
	Qu-6[d]	500	2.4	960	12.4	4.3	[3.242,243]
	Qu-5	500	31.7	72	9.9		[3.242,243]
	Qu-3	500	50.8	44	5.8	0.73	[3.242,243]
	KSK-1[e]	600	275	9.5	4.78	0.38	[3.238,239]
	KSK-2	600	400	6.6	4.12	0.32	[3.238,239]
	K-2	600	695	3.8	3.20	0.29	[3.238,239]
TiO_2	Anatase	560	90	18	3.4	7.7	[3.248]
	Rutile		4.5		5.0	4.6	[3.251]
ZnO	Wurtzite	500	28	40	16	2.5	[3.190]
GeO_2	Tetragonal	500	56	17	7	6	[3.4,188]
	Hexagonal	600	92		3.9		[3.252]
Al_2O_3	γ-form	500	322	5.6	9.4		[3.253,254]
	Corundum	500	158	9.1	11.1		[3.253,254]
$nAl_2O_3\ mSiO_2$	15% Al_2O_3	600	275	9	3.6	0.6	[3.255]
	30% Al_2O_3	600	496		7.3	0.6	[3.255]

[a] Average particle size
[b] Monocrystal
[c] Roughness of the surface taken into account
[d] Qu-3, Qu-5, and Qu-6 are disaggregated quartz powders
[e] KSK-1, KSK-2, and K-2 are silica gels

From Fig.3.18 and Table 3.4 it is evident that even for one and the same oxide (SiO_2), the values of α_{H_2O} depend not only on T_{cal}, but also on the magnitude of the specific surface area, i.e., on the particle size of the oxide or the size of xerogel globules. After having been subjected to the same treatment, roughly disperse particles of SiO_2 display greater hydration than high-dispersion particles. The magnitude of irreversible adsorption (α_{ir}) also decreases with a decrease in the size of the particles. This is also observed with other oxides, e.g., GeO_2 [3.4]. All these data disprove the commonly held opinion (see, e.g., [3.237,257]) that all silica specimens hydrated to saturation have a common "absolute" value of α_{H_2O}, regardless of their nature and dispersion That value, 6.3 $\mu mol \cdot m^{-2}$, was taken at first to be a kind of fundamental constant of the surface of SiO_2. It was later reduced by 30% and is now assumed to be 4 - 4.5 $\mu mol \cdot m^{-2}$ [3.227]. Further on we will discuss possible reasons why α_{H_2O} is dependent on the dispersion of oxides.

Neglect of the dependence of α_{H_2O} on s is one reason for discrepancies among the values proposed by different researchers for the hydration of oxides. Another reason is connected with neglect of the kinetics of dehydration: the time of heat treatment in air and in a vacuum varies from 0.5 to 24 hours with different researchers. On the basis of kinetic measurements of α_{H2O}, the standard heating time for the specimens in Table 3.4 was chosen to be 24 hours [3.238,239].

So far we have mainly been discussing the total content of structural water on the surface of oxides. But measurements of α_{H2O} and of adsorptive properties provide only qualitative information on the forms of binding between water and the surface. To learn more about the structure of the hydrate cover, we must turn to spectroscopic investigations.

3.2.2 The Structure of the Hydrate Cover of Oxides According to IR Spectroscopy and NMR

The analysis of IR spectra yields valuable information about the structure of the hydrate cover of oxides. Pioneering work on the IR spectra of surface phases was done by TERENIN and his school. They were the first to obtain important results on regularities in the behavior of the absorption bands of valence oscillations of OH groups on the surface of silica in both fundamental and harmonical spectral regions during hydration and dehydration of specimens [3.258 - 261]. The number of works dealing with investigations of the IR spectra of the hydrate cover of oxides now amounts to several hundred (see the reviews in [3.262-267]). Considerable progress in this direction was made with the help of simultaneous measurements of IR spectra and thermogravimetric adsorption analysis, and by employing multiple internal reflection spectroscopy (MIRS) to study the surface properties of monocrystals [3.267].

In spite of substantial achievements, there is no consensus regarding the interpretation of certain absorption bands in the IR spectrum of the hydrate cover, or the topography and the proton-donating properties of hydroxyl groups. Unfortunately, the majority of papers and reviews on IR spectroscopy of the surface do not consider results obtained by another promising method of investigating the hydrate cover, namely nuclear magnetic resonance (NMR). The very first works on NMR of protons on silica gels and aluminosilicates confirmed the presence of hydroxyl groups on the surface of these adsorbents [3.268,269]. The hydroxyl cover of the SiO_2 surface was shown in [3.269] to be nonuniform, and to contain tightly bound water molecules at temperatures up to $T_{cal} = 450 - 500$ K.

Let us consider, for example, the data on the IR and NMR spectra of the hydrate cover of Si, Ge, Al, and Ti oxides. This will be helpful in our discussion of many other surface characteristics of semiconductors and dielectrics.

SiO_2

The hydrate cover most thoroughly investigated by the methods of IR spectroscopy and NMR is that of amorphous silica. Figure 3.19 shows the behavior of the IR spectra of a nonporous specimen of SiO_2(aerosil) in the course of dehydration and adsorption of water vapor [3.270,271]. When the adsorbed water is present on the surface, the low-frequency region of the spectrum of valence oscillations of OH groups (3 μm) displays an intense wide band with a maximum at about 3400 - 3450 cm^{-1}, which is undoubtedly attributable to the oscillations of OH groups and water molecules bound to each other with hydrogen bonds [3.260,272]. Evacuation kills this band; it completely disappears in a vaccum at T_{cal} = 350 - 450 K. In the same region there is an increase in the width ΔH of the NMR signal from protons of the hydrate cover of SiO_2, indicating that the water molecules having the greatest mobility are removed from the surface (see [3.269,273] and references given in these works). With the increase in T_{cal} as the water is being removed three bands with maxima ν'_{OH} = 3750 cm^{-1}, ν''_{OH} = 3660 cm^{-1}, and ν_k = 3550 cm^{-1} can be distinguished in the asymmetrical band of valence oscillations of OH groups in the low frequency region. According to [3.270], they correspond to the three components of a spectrum of Lorentz form. The high-frequency band is attributed with good reason to isolated OH groups, with an average separation $\gtrsim$ 3 Å. As follows from Fig.3.17a,b, these groups are typical of the <111> face of β-cristobalite and the <0001> face of β-tridymite. The blurred maxima in the region of 3660 cm^{-1} are attributed by most researchers (e.g., [3.262,266]) to hydroxyl groups which are separated by less than 3 Å (this is typical of the <100>

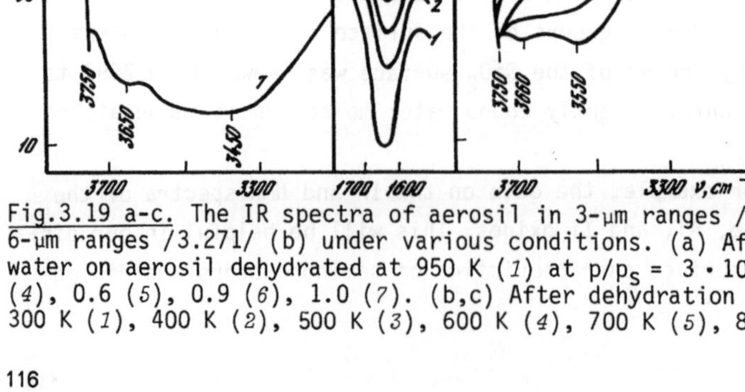

Fig.3.19 a-c. The IR spectra of aerosil in 3-μm ranges /3.270/ (a,c) and 6-μm ranges /3.271/ (b) under various conditions. (a) After adsorption of water on aerosil dehydrated at 950 K (*1*) at $p/p_s = 3 \cdot 10^2$ (*2*), 0.1 (*3*), 0.3 (*4*), 0.6 (*5*), 0.9 (*6*), 1.0 (*7*). (b,c) After dehydration of the surface at 300 K (*1*), 400 K (*2*), 500 K (*3*), 600 K (*4*), 700 K (*5*), 800 K (*6*), 950 K (*7*)

face of β-cristobalite; see Fig.3.17c) and are bound to one another by hydrogen bonds. Two alternative types of such bonds have been suggested: between OH groups belonging to a single Si atom [3.274], and between OH groups belonging to two adjacent atoms of Si [3.266]:

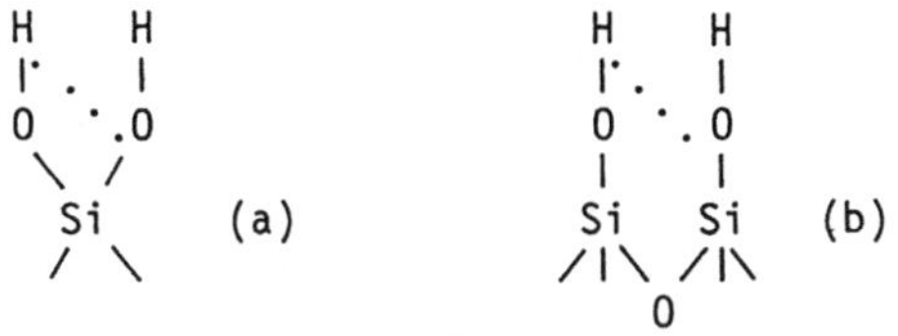

(3.23)

However, the formation of H bridges in structures of type (3.23a) is prevented by the small angle with the direction of the hydrogen bond. In a structure of type (3.23b), which is not typical of these cleavage faces (see Fig. 3.17a,b), the hydrogen bond will also be rather weak: even though the distance H ... O in the H bridge is 2.3 Å, the angle O - H - O is much smaller than 180°. The small frequency shift of these bands in relation to ν'_{OH} ($\Delta\nu = 60 - 200$ cm^{-1}) is also an indication of the low energy of the hydrogen bond. All this does not fit well with the high thermal stability of the low-frequency bands ν''_{OH} and ν_k, which disappear only at $T_{cal} = 900 - 1000$ K (ν''_{OH}) and $T_{cal} = 500 - 600$ K (ν_k) [3.264,270]. In high-temperature measurements of IR spectra [3.275] no increase in the intensity of ν'_{OH} due to the rupture of hydrogen bonds between adjacent OH groups could be detected; this does not confirm schemes (3.23) either.

In the region of valence oscillations (ν=3μm) one cannot strictly separate the bands corresponding to the valence oscillations of the water molecules from those corresponding to the structural hydroxyls. Especially interesting in this connection are investigations of the deformation vibrations $\nu_d = 1610 - 1640$ cm^{-1}, which are characteristic only of the molecular adsorption of water [3.276-278]. It was shown that the appearance and disappearance of the band ν_d during adsorption and dehydration are closely followed by a band in the region of the valence oscillations of OH groups: $\nu_k = 3550$ cm^{-1} (Fig.3.19). This gave grounds for attributing the band ν_k to tightly adsorbed water molecules [3.277]. In the case of SiO_2 observation of the band ν_d is hindered due to partial overlapping with a harmonic band of valence oscillations of Si - O bonds (1620 cm^{-1}) (Fig.3.19b). This obstacle was successfully overcome in [3.279], where deformation vibrations of HOD molecules were studied (ν_d = 1420 cm^{-1}). Both our own [3.264,277] and other researchers' data [3.276,278] indicate that the bands ν_k and ν_d remain in the IR spectrum up to T_{cal} = 500 - 600 K. The NMR spectrum up to T_{cal} = 500 K retains a component characteristic of water molecules firmly localized on the surface [3.269,280]. In order to explain such strong adsorption of water on the SiO_2 surface, the authors of [3.269] suggested that coordination bonds might form between water molecules and hydrated Si atoms of the silica surface. This question is discussed in greater detail in Chap.4 in connection with the adsorption mechanism.

According to NMR measurements [3.269,280], the concentration of coordination-bound water on the surface, n_c, is $(1-2)\cdot 10^{13}$ molecules $\cdot$ cm^{-2}. A similar value, $n_c = 7\cdot 10^{13}$ molecules $\cdot$ cm^{-2}, was later obtained in experiments on nuclear relaxation [3.281]. For pure aerosil, the NMR data [3.282] give $n_c = (1.5-3)\cdot 10^{13}$ molecules $\cdot$ cm^{-2}. Adsorption spectroscopy of acetyl acetone [3.283], a typical complexing agent for Si [3.284], showed the number of coordination-unsaturated centers on the surface of aerosil to be about $(0.8-1.5)\times 10^{13}$ cm^{-2}, which is close to the value of n_c. Thus the value of n_c is about 1/100 of the number of OH groups on the surface, and is close to the value of irreversible adsorption of water, α_{ir} (on SiO_2 specimens evacuated at $T_{cal} = 500-600$ K; see Table 3.4). From Table 3.4 one learns that α_{ir} increases with a decrease in the specific surface of SiO_2 symbatically with an increase in the extent of hydration α_{H_2O}. In a monocrystal almost every atom of Si on the surface develops a coordination bond with a molecule of H_2O.

Valuable information on the structure of the hydroxyl cover of the surface of silica was obtained in low-temperature studies of NMR spectra first performed by KVLIVIDZE [3.280]. At 90 K the mobility of protons is low, and the parameters of the NMR signal are wholly determined by the topography of resonating nuclei. An analysis of the shape of the NMR line (Fig.3.20a) by the method of linear anamorphoses revealed that in a first approximation the line can be presented as composed of three Gaussian components. The values of second moments S_2 and integral intensities were calculated for these components.

According to the Van Vleck formula [3.285],

$$S_2 = \frac{6}{5}\,\frac{I(I+1)}{N}\,g^2\beta^2 \sum_{j>k} r_{jk}^{-6} + \frac{4}{15}\,\frac{\beta^2}{N} \times \sum_{j>f} I_f(I_f+1)g_f^2 r_{jf}^{-6} \quad , \qquad (3.24)$$

where β is the nuclear magneton; g is the factor of nuclear spectroscopic splitting; I is the spin of the nucleus; f are the numbers of nuclei not taking part in the resonance; r_{jk} are the internuclear distances; and N is the number of resonating nuclei in the cell. Since S_2 rapidly decreases with distance, one usually takes only the nearest neighbors into consideration. The contribution from nuclei more than 5 Å away is accounted for by integration.

Since according to Sect.3.1.5 the short-range order varies little in silica, for an approximate calculation of S_2^{theor} one can employ the models of crystal-

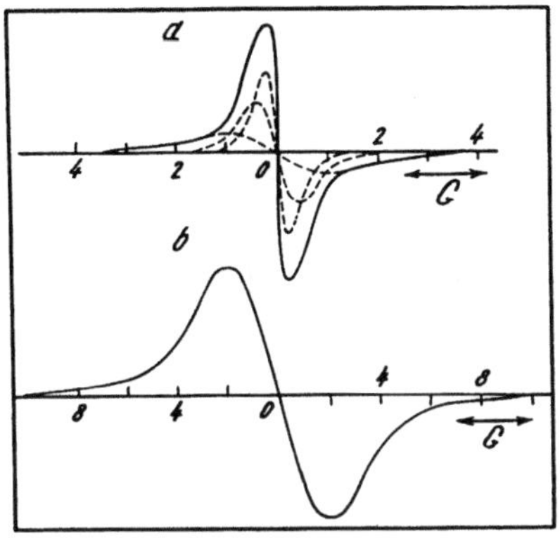

Fig.3.20 The derivatives of NMR signals from protons for hydrated silica gel specimens (a) (T_{cal} = 500 K) and γ-Al_2O_3 (b) (T_{cal} = 600 K). The dashed lines represent Gaussian curves singled out from the signal

line surfaces of β-tridymite and β-cristobalite (Fig.3.17). A comparison between calculated values S_2^{theor} and experimental values S_2^{expt} suggests the presence on the hydrated silica surface of at least three types of silanol patches, which differ from one another in the distance between the OH groups [3.286, 287]:

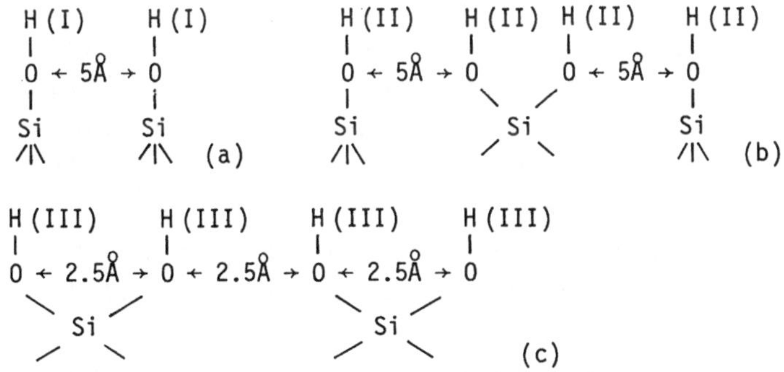

(3.25)

The hydroxyl groups of type I constitute 25% of the total number of hydroxyl groups on the surface, while those of type II constitute 30%, and those of type III 45%.

Further investigations of NMR spectra of partially dehydrated silica gels [3.287] showed the solitary hydroxyls to possess the greatest thermal stability: their concentration did not change in the investigated temperature range T_{cal} = 500 - 700 K. In the IR spectrum these hydroxyls are presumably represented by the band ν'_{OH}, whose intensity begins to decrease only at T_{cal} > 1000 K [3.272]. The dehydration of the surface proceeds mainly at the expense of hydroxyl groups at types III that decrease most abruptly. At the same T_{cal}, the intensity of the band ν''_{OH} in the IR spectrum decreases most noticeably. All this does not support the model of a hydrated surface proposed by Peri

(Sect.3.1.5), according to which the solitary OH groups of type I can exist on the surface only at $T_{cal} > 900K$.

The above interpretation of IR and NMR data agrees well with commonly accepted mechanisms of surface dehydration (see [3.4] and scheme 3.3, p.108). According to this scheme, in the temperature range $T_{cal} = 500 - 1000$ K the dehydration is primarily connected with the rotation of adjacent surface tetrahedrons, which leads to the formation of overstrained siloxane bonds

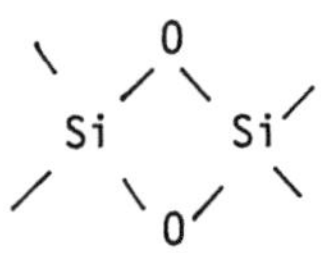

and to the release of H_2O molecules. Due to the rather high mobility of the surface tetrahedrons, this mechanism of dehydration can operate at relatively low T_{cal} [3.4]. The high mobility of the lattice elements in silicas is substantiated by the low values of phase transition heats: for instance, the heat of a $\alpha \rightleftarrows \beta$ transition for tridymite at 417 K is 0.16 $kJ \cdot mol^{-1}$, and for cristobalite at 550 K it is 0.8 $kJ \cdot mol^{-1}$.

To meet the needs of rapidly developing microelectronics technology, thin films of SiO_2 formed on the surface of silicon during oxidation have been studied, primarily with the aid of the MIRS technique [3.288]. Depending on the content of H_2O in the oxidizing agent, the concentration of OH groups in thermally grown films of SiO_2 varies from $5 \cdot 10^{18}$ to $20 \cdot 10^{18}$ cm^{-3} [3.289]. The estimates were based on the strength of the band: $\nu \sim 3600$ cm^{-1}. The smallness of the surface prevents effective separation of the ν'_{OH}, ν''_{OH}, and ν_k bands. Quite intense bands of hydride Si-H groups were also observed which are not typical of pure SiO_2 oxides.

The method of spin labels [3.290] has opened up new prospects in the investigation of the hydroxyl cover of oxides, including thin films of SiO_2. The authors of [3.290] employed stable 4-hydroxy and 4-amino-2,2,6,6-tetramethylpiperedin oxyl radicals as such labels. On the basis of the EPR response from these radicals interacting with silanol groups, certain conclusions have been made about the structure of the hydroxyl cover of high-dispersion SiO_2 (aerosil). It must be noted that this method has certain drawbacks connected with the separation of OH groups and coordination-bound water molecules.

<u>GeO_2</u>

Spectroscopic investigations of the hydrate cover of hexagonal germanium dioxide were first carried out by LOW and co-workers [3.252]. In general the behavior of the IR spectrum of this dioxide during dehydration and water adsorption is the same as with SiO_2. This is quite natural, since the structure of both oxides is built of $GeO_4(SiO_4)$ tetrahedrons. At $T_{cal} = 300$ K the spectrum displays an intensive band: $\nu_H = 3450\ cm^{-1}$. With an increase in T_{cal} the strength of ν_H drops, and the maximum shifts towards $\nu_k = 3550\ cm^{-1}$. This band has been attributed to H-bound OH groups; it would be more reasonable, however, to attribute ν_k to coordination-bound water [3.291], by analogy with SiO_2. This conclusion is supported by the presence of the band of deformation vibrations of water, ν_d, in the spectrum up to $T_{cal} = 500$ K. At higher T_{cal} the spectrum clearly displays the band $\nu'_{OH} = 3673\ cm^{-1}$, which belongs to the free hydroxyls. A detailed study of the hydration mechanism of deposited Ge films containing an oxide phase was recently carried out by METCALFE and SHANKAR [3.292].

Al_2O_3

Most spectroscopic investigations of Al_2O_3 have been concerned with its γ form, which is a very important catalyst in chemical reactions. IR spectroscopic data reveals that in the range $T_{cal} = 300 - 600$ K, the spectrum displays a band of deformation vibrations of water molecules $\nu_d = 1640 - 1615\ cm^{-1}$ [3.264, 293]. In the same temperature range a wide maximum, $\nu_H = 3470\ cm^{-1}$, which belongs to the water molecules linked by hydrogen bridges to the surface, drops abruptly, and at $T_{cal} > 500$ K the maximum shifts towards $\nu_k = 3550\ cm^{-1}$ [3.294,295]. DE BOER and co-workers [3.256] also suggested that at these T_{cal} a large number of water molecules are present on the surface of Al_2O_3, and are firmly bound to the oxygen in the lattice by H bonds. However, H bonds can hardly explain the high thermal stability of adsorbed water, so it would be more reasonable to attribute ν_k and ν_d to the coordination-bound water molecules.

The band of valence oscillations of OH groups on γ-Al_2O_3 specimens dehydrated at $T_{cal} = 650 - 1000$ K is much more complex than in the case of SiO_2. Various authors [3.192,264,266] have resolved three to five bands in this spectral region with frequencies of $3600 - 3800\ cm^{-1}$ and varying thermal stability. The appearance of low-frequency lines in the region $3600 - 3700\ cm^{-1}$ is attributed either to H bonds between adjacent OH groups [3.296], which as in case of SiO_2 seems improbable, or to varying coordination of the Al atoms on the surface of γ-Al_2O_3 [3.192]. It must be noted that at $T_{cal} > 1000$ K it is difficult to interpret the spectra because the surface of γ-Al_2O_3 at these temperatures is noticeably reconstructed into the high-temperature modification, α-Al_2O_3.

In order to obtain additional information on the structure of the hydroxyl cover of γ-Al_2O_3, the NMR spectra of protons were investigated at 80 K with pure specimens of γ-Al_2O_3 dehydrated in vacuo at $T_{cal} = 600 - 800$ K [3.297,298]. Analysis of the spectra (Fig.3.20b) revealed that at these T_{cal} the wide component from H_2O is absent, and the NMR response is due exclusively to the protons of hydroxyls. In this case the second moment of the signal is determined not only by proton-proton interaction, but also by interaction between protons and aluminuum nuclei.

A comparison between the values of the second moment for the octahedral face of γ-Al_2O_3, calculated theoretically according to (3.21) (Fig.3.17d), and the experimental value S_2^{expt}, suggests that the following proton configurations of hydroxyl groups are present on the surface at $T_{cal} = 600$ K:

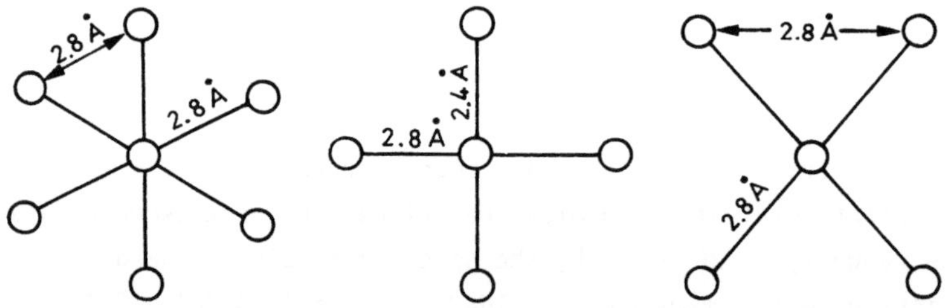

(3.27)

With an increase of T_{cal} to 800 K the value of S_2 decreases, and the configuration (3.27a and b) to (3.27c) corresponds to the removal of about 30% of the OH groups, which agrees well with direct measurements of the extent of hydration of the surface at these T_{cal} [3.254].

$nAl_2O_3 \times mSiO_2$

The IR spectra of man-made aluminosilicates with a low content of Al_2O_3 (up to 25%) differ little from those of pure silicon oxides [3.277,299]. As in the case of SiO_2, in the region of valence oscillations of OH groups one can observe an asymmetrical solitary band in the low-frequency region with a maximum at $\nu_{OH} = 3750$ cm^{-1}, which is usually attributed to the vibrations of Si-OH groups. As yet there is no reliable data on Al-OH groups in the literature. The first investigations of NMR from protons of aluminosilicates [3.300,301] also failed to reveal any peculiarities in comparison with SiO_2. In [3.302] the NMR response from protons of aluminosilicate dehydrated at 800 K was shown, like that for SiO_2, to be presentable as composed of several components; this indicates that several types of OH groups are present on the surface. At low temperatures of dehydration, up to $T_{cal} = 500 - 600$ K, the IR spectrum displays a band of deformation vibrations of water molecules, ν_d, and the band ν_k [3.277].

TiO_2

The IR spectra of the rutile modification of TiO_2 have been most thoroughly investigated [3.265]. Figure 3.21 presents typical spectrum of the hydrate cover of rutile, based on data from [3.303]. It is evident from that figure that, as with other oxides, heat treatment at T_{cal} > 400 K leads to the complete removal of H-bound water (ν_H bands). The fact that the bands $\nu_k = 3540\ cm^{-1}$ and $\nu_d = 1610\ cm^{-1}$ disappear during dehydration and appear during adsorption makes it possible to attribute them to coordination-bound water, which is removed at about 700 K. With the aid of the NMR technique, the presence of water on the surface of rutile at $T_{cal} = 650$ K was discovered [3.304]. The concentration of this strongly bound water is about $10^{14}\ cm^{-2}$ [3.248,251].

In the rutile spectrum at $T_{cal} > 400$ K, three bands are clearly distinguishable: $\nu'_{OH} = 3720\ cm^{-1}$, $\nu''_{OH} = 3656\ cm^{-1}$, and $\nu'''_{OH} = 3630\ cm^{-1}$, which belong to the structural hydroxyls.

It should be noted that other investigations have found one to four bands in this spectral range [3.251,265, 305-310]. The high-frequency band ν'_{OH} is attributed in [3.309] to contamination by SiO_2, which is hard to accept if one notes its shift during oxidation of the specimen [3.265]. The low-frequency

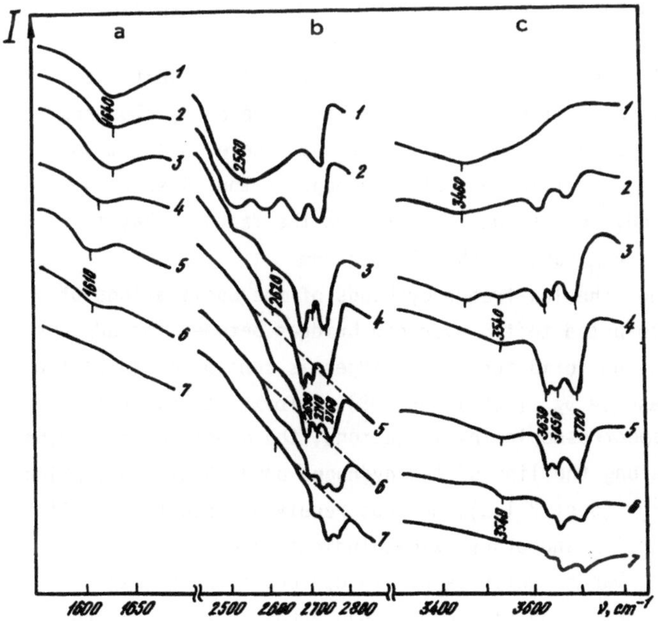

Fig.3.21 a-c. The IR spectra of rutile. (a,c) Initial hydrated and (b) deuterized rutile specimens after dehydration at various temperatures

bands ν_{OH} are attributed either to the existence of H bridges between adjacent hydroxyls [3.306,308], or to peculiarities in the crystal chemistry of the different faces of rutile and anatase [3.251,308,309]. This will be discussed in greater detail later. In the spectra of anatase, two or three typical bands of valence oscillations of OH groups can be observed in the range 3725 - 3650 cm^{-1} [3.305,309].

The number of bands and their position in the spectrum can be strongly dependent on the conditions of oxidation and reduction of the TiO_2 samples [3.310]. First, the reduction of the specimen sharply increases the absorption on the free charge carriers [3.311,312]. Second, the formation of oxygen vacancies (Sect.3.2.3) must necessarily tell upon the frequency of oscillations of the adjacent hydroxyl groups. Figure 3.21 shows the IR spectra of partially reduced and hydrated rutile specimens.

ZnO, MgO

Detailed investigations of the IR spectra of ZnO were carried out by GARTNER and MOLLWO [3.313], who analyzed the influence which doping of ZnO has upon OH bands ($\sim 3500\ cm^{-1}$) and OD bands ($\sim 2500\ cm^{-1}$). The hydroxyls on the surface of MgO were studied in [3.314-316], where two bands of OH groups($\nu'_{OH} = 3710\ cm^{-1}$ and $\nu''_{OH} = 3550\ cm^{-1}$) and the band of coordination-bound water molecules ($\nu_k = 3500\ cm^{-1}$) were observed.

From the above short discussion of IR and NMR data it follows that the hydroxyl cover of oxides is nonuniform. The surface is occupied by clusters consisting of hydroxyl groups in various configurations. There is still some controversy as to how valence oscillations of OH groups in the IR spectra of oxides should be interpreted. Let us dwell briefly on the factors that influence the position of the ν_{OH} band in the IR spectrum.

As was already mentioned, the low-frequency bands of the oscillations of OH groups are commonly attributed to the hydrogen bridges between the adjacent groups. For certain oxides the formation of H bridges is considered to be the only reason for the decrease in ν_{OH} [3.251,266,308]. Unfortunately, in the overwhelming majority of spectroscopic investigations, the question of lengths and especially of angles along the line of the hydrogen bond -O-H-...O-, which determine the binding energy (Sect.2.1.2), has not received the attention it deserves. Experiments with SiO_2 and other oxides indicate a high thermal stability of low-frequency ν_{OH} bands, which is not in conformity with small frequency shifts and thus low energy of the hydrogen bonds in these bridges.

EGOROV has shown that on ideal cleavage surfaces of crystallic oxides of Al [3.187] and Zn [3.190], the oxygen of hydroxyl can develop bonds with one, two, or three metallic atoms. Consider, for example, the models of hydroxyl clusters on the surface of ν-Al_2O_3 [see the surface elements (3.21)]. The constants of the bonds in these clusters, as well as the constants of bonds between metal and oxygen, will differ depending upon the coordination number for metal and for oxygen. By introducing fractional valence numbers, EGOROV [3.317] emphasizes the fact that the valence and the coordination number of Al in a defective spinel structure do not coincide. The difference in the frequencies ν_{OH} (and consequently of the bond constants) in the spectra of γ-Al_2O_3 [3.192] and rutile [3.309] are also attributed to the different coordination numbers of the metal atoms on the surface, to which the hydroxyls are bound.

An approach similar to the analysis of the IR spectra of a large group of oxides was employed by TSYGANENKO and FILIMONOV [3.318,319], who maintained with good reason that the frequency ν_{OH} is determined by the number of bonds between the oxygen of the hydroxyl and the metal atoms. Consequently, they assumed that there are three basic types of OH groups:

```
H        H          H
|        |          |
O        O          O
|       / \        /|\                       (3.28)
M      M   M      M M M
a        b          c
```

with a decreasing frequency of oscillations from (3.28a to c). However, the three suggested types of coordination of OH groups with respect to oxygen cannot explain the whole range of low-frequency ν_{OH} bands in the IR spectra of oxides. In cases where the number of bands in the spectrum is greater than that possible according to model (3.28), TSYGANENKO and FILIMONOV [3.318,319] again sought the answer in the concept of hydrogen bridges, which seems to us to be unnecessary. The range of dynamic constants and the frequencies ν_{OH} will be greater than that predicted by (3.28) if one takes into account the different coordination numbers of metal atoms on the surface. For instance, for γ-Al_2O_3 the surface elements (3.21a and f) or (3.21c and d), which according to the model (3.28) have an equal number of M-O bonds, will display different frequencies because of their different fractional valence bonds.

So far we have been considering the morphology of ideal faces of various crystal lattices. In reality, however, as we have repeatedly pointed out, the surface layer of a crystal is always essentially deformed. In [3.291] we showed that in many cases the existence of several types of OH oscillations on the surface, or the asymmetry of the ν_{OH} band in the low-frequency region, can be explained by the nonuniformity of the surface, without speculative concepts

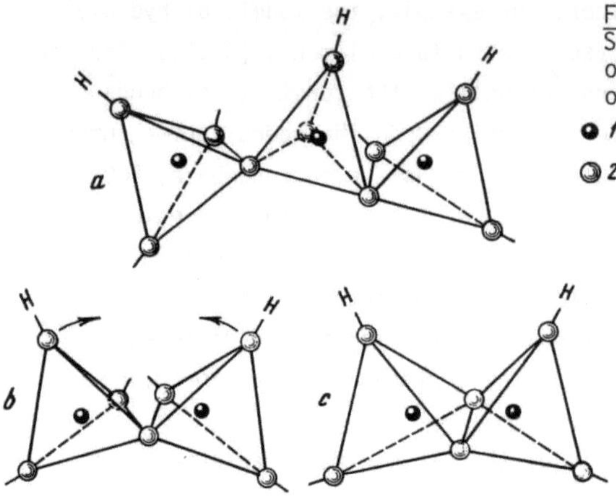

Fig.3.22 a-d. Configurations of SiO_4 tetrahedrons on the surface of SiO_2. (*1*) Si atoms. (*2*) O atoms

about hydrogen bridges. As an example, let us consider the amorphous structure of SiO_2 (Fig.3.22).

According to PAULING [3.320], structures built of SiO_4 tetrahedrons are stable when the adjacent tetrahedrons only have common vertices (Fig.3.22a). The mean equilibrium angle in Si-O-Si bonds in the glasslike modifications of SiO_2 is $143 \pm 17^\circ$ [3.321,322]. Since the 3d orbitals of silicon are rather low in position, one can assume the $d_\pi - p_\pi$ interaction to play an important role in the structure of SiO_2 [3.322-324]. If the angle between the tetrahedrons is decreased (Fig.3.22b), the tetrahedrons will undergo deformation due to an increase in the repulsive forces between the Si atoms. If adjacent tetrahedrons come to have common edges (Fig.3.22c) or faces (Fig.3.22d), which correspond to the scheme of dehydration of SiO_2, then the length of the common edges of the adjacent tetrahedrons will be reduced [3.320].

In a strained and amorphous structure of the hydrated surface of SiO_2, the angles between the edges of adjacent tetrahedrons can vary greatly, their values being quite different from the equilibrium values. This is directly indicated by a detailed analysis of the IR spectra from oxide films of SiO_2 [3.288], in which the dependence of the length and angles of Si-O-Si valence bonds upon the size of SiO_2 globules [3.224] was discovered.

It is known [3.322,325] that in crystal modifications of SiO_2 the Si-O-Si angles, and consequently the lengths of Si-O bonds can vary within wide limits. For instance, the Si-O-Si angle is about 180° for β-cristobalite and about 144° for α-quartz, and the lengths of the bonds are about 1.55 Å and 1.61 Å, respectively. The energy spectrum of amorphous SiO_2 is determined primarily by variations in the angles, and not by topological disorder [3.325]. For that reason NUCHO and MADHUKAR [3.326] attribute variations in the density of localized states in the energy spectrum of SiO_2 to variations in both Si-O-Si and O-Si-O angles. The electronic structure of SiO_2 is also greatly influenced by O-O interaction. We suggested earlier [3.291] that the deformation of surface hydrated tetrahedrons, due to alterations in angles and lengths of bonds, leads to variations in the frequencies of oscillations of OH groups in the vertices of tetrahedrons (Fig.3.22b). If, for instance, the uppermost frequency ν'_{OH} belongs to the OH groups of undistorted tetrahedrons (3750 cm^{-1}), the low-frequency maximum ν''_{OH} = 3660 cm^{-1} (Fig.3.19) can characterize the maximum of the distribution of fractional OH groups for the distorted tetrahedrons.

Deformation and hydration of the surface tetrahedrons leads to a change in the effective charges q_{eff} of silicon with respect to the mean value of q_{eff} for the bulk, The silicon atoms in these tetrahedrons which possess the greatest q_{eff} presumably serve as centers of water adsorption in accordance with the coordination mechanism. When water enters the coordination bond, the coordination number increases from 4 to 5. As was mentioned in Sect.2.1.5, experiments have confirmed that a specimen of SiO_2 expands during adsorption of water according to the coordination mechanism [3.4].

Direct proof of a change in q_{eff} in the grid of Si-O-Si bonds can be obtained by analyzing ESCA spectra. For instance, GRUNTHAMER and MESERJIAN [3.327] attributed variations observed in chemical shifts of ΔE peaks of Si(2p) in the ESCA spectra from thermally grown SiO_2 films to variations in the rings of silicon-oxygen tetrahedrons (and consequently in the q_{eff} of Si atoms). They had observed a noticeable variation in chemical shift during hydration of SiO_2, which was attributed to a variation in the Si-O-Si angles.

YATSIMIRSKY and co-workers [3.328] used ESCA to detect an increase in positive charge on Si during desorption of firmly bound water from the surface. All these facts verify our proposed coordination mechanism of adsorption of water on the surface of SiO_2 [3.291,329].

It was found [3.272,330] that the intensity of the low-frequency bands ν''_{OH} and ν_k in the spectrum of aerosil (Fig.3.19) increases and the intensity of the ν'_{OH} band decreases when a pellet of SiO_2 is compressed. This is attributed to a shortening of the distance between the hydroxyls of the aerosil, which facilitates the formation of hydrogen bonds. According to [3.270], the average distance r between the OH groups of aerosil is $\sim 5-6$ Å. To permit the formation of H bonds, r must become smaller than 3 Å, which is hardly possible at the employed pressures of 10^3 atm. However, the point contact between globules being taken into account, this pressure is sufficient to bring about changes in the angles between some tetrahedrons in the contact region, and to cause their deformation. Similar pressures exerted on a bulk sample of quartz glass gave rise to noticeable frequency shifts of siloxane bonds in the surface region of the sample [3.331]. An increase in the concentration of firmly bound water on an aerosil tablet subjected to compression was also noticed [3.278].

A similar situation presumably takes place with other oxides as well, for instance TiO_2 [3.291,303]. Due to an alteration in the coordination of surface atoms of titanium, hydrated titanium-oxygen octahedrons TiO_mH_n are always distorted. The resulting redistribution of the electron density within the octahedron leads to changes in the dynamic constants of Ti-O bonds. It is interesting to note that fluorination of the rutile surface, which does not change the relationship between the crystal faces in the sample, leads to a noticeable decrease in the strength of the band of OH groups [3.332], to a shift in the low-frequency band ν'''_{OH} by 20 cm^{-1}, and to a weakening of the thermal stability of OH groups [3.291]. The penetration of the coordination sphere of the surface atoms of titanium by fluorine atoms results in a change in the interatomic distances and the dynamic bond constants in the titanium-oxygen octahedrons.

All of this makes it possible for the complex structure of the IR spectrum of the hydroxyl cover to be consistently interpreted in terms of the distorted coordination sphere of the metal atoms on the amorphous surface of oxides. The variations in the oscillation frequencies of OH groups are due not only to the different coordination numbers of the oxygen in the hydroxyl, but also to significant variations in the valence angles and lengths of bonds in the distorted surface region of the solid. As will be shown later, an analysis of the phonon spectrum of the surface region supports this conclusion.

In addition to optical methods, low-angle X-ray scattering methods have also been employed lately to investigate amorphous structures. The EXAFS method (Sect.3.1.3) has also been used, for example, to study GeO_2 structure [3.333]. However, there is still some controversy with regard to the distribution of the bond lengths derived with this technique [3.334].

The proton-containing clusters on the surface, as was indicated in Sect. 2.2.3, play an important role as Brönstedian or proton-donor centers in catalysis. The proton acidity of the surface centers determines the adsorptive and catalytic properties of the oxide surface with respect to numerous molecules whose adsorption follows the hydrogen-bond mechanism. In this section we have discussed the two main types of proton-containing clusters: OH groups and water molecules bound firmly to the surface by coordination bonds. Which of these two dominates as a proton donor in adsorption and catalysis is discussed in Chap.4 in connection with a general discussion of adsorption mechanisms and the reactivity of adsorbed molecules.

3.2.3 Electron-Acceptor and Electron-Donor Centers on the Surfaces of Oxides

The role of electron-acceptor (EA) and electron-donor (ED) active surface centers is no less important in many catalytic reactions than that of proton-containing clusters. In isolated atoms and ions the ability to accept or to give off an electron is measured by the electron affinity A_0 or by the ionization potential J (Sect.2.1.2). Needless to say, the values of A_0 and J for atoms of a solid-state surface will be quite different than those for isolated atoms. The electron-acceptor and electron-donor properties of atoms of a solid-state surface are determined not only by the structure of the atoms themselves, but also by the ionicity of their bonds with neighboring atoms and with the chemical compounds of the surface. The polarity of chemical bonds is determined by the effective charges q_{eff} of the interacting atoms. For a qualitative description of the ionicity of bonds and q_{eff} of the atoms in the bulk the concept of electronegativity is often employed [3.335]. In the case of surface atoms, such an approach is complicated by distortion of the coordination sphere and by the presence of surface complexes, often of an unknown nature.

A critical analysis of investigations of adsorption and catalysis reveals that it is defect, and not the regular atoms of the lattice, that play the dominant role as EA and ED. Even in the absence of impurities, at $T = 0\,K$ the bulk of the solid always displays a small concentration of biographical defects. They can be Frenkel defects (interstitial Me atoms and vacancies in the lattice sites where the Me atoms used to be) or Schottky defects (anion and cation vacancies, whose concentration obeys stoichiometric conditions). Of especial significance in adsorption and catalysis are nonstoichiometric crystals characterized by an excess or deficit of components as compared with an ideal stoichiometric crystal. The surface, which is the most defect-ridden region of the lattice, is the starting point for defects which form when the crystal is heated or result from interaction with the environment.

As was mentioned in the preceding section, the point at which the lattice elements become mobile, which predetermines the commencement of self-diffusion, is sometimes roughly described by the Tamman temperature, $T_T = 0.5\ T_{melt}$. Schottky and Frenkel defects are commonly assumed to arise at $0.3\,T_{melt} < T_T < 0.5\,T_{melt}$. However, since T_T decreases with an increase in dispersity (e.g., with a decrease in particle size; see Sect.3.1.3), one can anticipate a noticeable concentration of surface vacancies at relatively low calcination temperatures of oxides in a vacuum or a reductive (oxidative) atmosphere. The concentration and distribution of vacancy-type defects depend to a large extent not only on the conditions of heating, but also of cooling. An effective way of obtaining reproducible concentrations of vacancy defects on the surface of certain oxides is to quench specimens by quickly cooling them after reduction. Evidently, highly mobile defects are partially annihilated at a slower cooling rate.

In a solid body, defects can play the role of electron and hole traps, and thus determine what type of conductivity the crystal has. As early as the 1930s, before solid-state zone theory was developed, Frenkel, Schottky, Wagner and some other researchers [3.336,337] showed conclusively that the type of conductivity exhibited by pure ionic crystals is determined by the departure from stoichiometry. In the oxides of 3d metals the size of the O^{2-} anion ($r_{O^{2-}} = 1.36$ Å) is greater than the size of the cation. Consequently, the excess oxygen dissolved in an MeO oxide supplements the lattice with the formation of a Schottky defect: a cation vacancy □ and hole p

$$1/2\ O_2 \rightleftarrows O^{2-} + \square + 2p \quad .$$

The holes may contribute to the hole-type conductivity of a crystal or be captured by the neighboring metal ions, which will enter a higher oxidation state Me^{n+1}.

A classic example of a p-type semiconductor is Cu_2O. The movement of holes takes place here through electron exchange between Cu^{2+} and Cu^{+}. The adsorption of oxygen will naturally change the electroconductivity because of the change in the number of vacancy defects in the crystal. According to Wagner, this reaction can be written in the form

$$O_2(\text{gas}) \rightleftarrows 2\ Cu_2O + 4\ \square'\ Cu + 4p \quad ,$$

where □' Cu is a vacancy in the sublattice of Cu^{+}, and p is a hole. Applying the law of mass action to this reaction, we get the equilibrium constant K:

$$K(p,T) = \frac{[\square'\ Cu]^4 [p]^4}{[O_2(\text{gas})]} \quad .$$

Since $[\square'\ Cu] \simeq [p] \simeq K[O_2(\text{gas})]$, then, assuming the concentration of the oxygen molecules to be proportional to the pressure, we find the conductivity to be

$$\sigma \sim [p] \simeq \text{const}\ P_{O_2}^{1/8} \quad .$$

A similar exponent, $\sigma = \text{const}\ P_{O_2}^{1/7}$, is obtained from experiment.

When there is a deficit of oxygen, Frenkel defects usually form, i.e., cations implanted into the interstitial sites M^{2+}, and electrons:

$$M^{2+} + O^{2-} \rightleftarrows 1/2\ O_2 + 2e \quad .$$

The electrons can be excited into the conduction zone or trapped by a metal atom in the vicinity of the defect (including the ion M^{2+}) with formation of a lower oxidation state $M^{(n-1)+}$.

An example of a thoroughly studied n semiconductor is TiO_2. At moderate temperatures (900-1000 K) the main source of disorder in TiO_2 crystals are Ti^{3+} ions in the interstices ΔTi. The equation of reaction with oxygen is:

$$TiO_2 \rightleftarrows Ti\Delta Ti + e + O_2(gas) \quad .$$

Employing the law of mass action as with Cu_2O, we get $\sigma = \text{const}\ P_{O_2}^{-1/2}$, which is also in good agreement with experiment. With an increase in temperature, oxygen vacancies become the more probable defects, and $\sigma = \text{const}\ P_{O_2}^{-1/4}$.

The law of mass action is widely employed in solid-state physical chemistry to estimate the equilibrium concentration of defects in a crystal. When calculating the thermodynamic parameters of defects of a crystal which is in equilibrium with its own vapor, one must take into account the equilibrium between the defects and the crystal, including the interaction between defects and the formation of associations. For that purpose the quasi-chemical approximation which we encountered in our discussion of the tangential interactions of adsorbed atoms (Sect.2.1.4) is usually employed. For instance, the reaction of interaction between defects A and B is written in the form:

$$nA + mB \rightleftarrows (A_nB_m) + \Delta H,$$

where ΔH is the change in enthalpy resulting from the formation of an association. Then, by the law of mass action,

$$\frac{[A_nB_m]}{[A]^n[B]^m} = K_A = \exp(-\Delta F / kT) \quad ,$$

where ΔF is the change in the Gibbs free energy. Correspondingly, the change in the configuration entropy will be $\Delta\eta_c = k \ln (G/z)$, where G is the number of ways in which the association can be formed [determined as in (2.40)], and z is the number of neighbors. If the defects are electrically active, the equilibrium between the defects and the electron subsystem of the solid must be taken into account. Interaction between defects can sometimes be accompanied by autocompensation of both equilibrium and nonequilibrium conductivity of the crystal [3.338,339]. Then the value of σ will exhibit a more complicated dependence on the vapor pressure of one of the lattice components and on the concentration of defects. These questions are discussed in greater detail in [3.340].

With high-dispersion oxides the defect type depends strongly on the process by which the specimen is prepared. For instance, TiO_2 fired in a vacuum at about 700 K is a typical n semiconductor. However, when subjected to reoxidation and subsequent reduction at a lower T_{cal}, it can become a p-type semiconductor, as indicated by TEICHNER [3.341]. Mass spectroscopic investigations have demonstrated that depending on the value of T_{cal}, either titanium or oxygen is removed from the surface of TiO_2 [3.342].

There now exists a well-developed statistical theory of equilibrium and nonequilibrium bulk defects in crystals (the Onsager theory) [3.336-338,343, 344]. However, it has but limited application to surface defects. This is primarily due to the general disorder of the surface phase, the great distortion of the coordination sphere of the surface atoms, and the presence on the surface of two- and three-dimensional chemical compounds which remain there up to quite high T_{cal}. For that reason the values of the bond energies, the enthalpy of defect formation, the coefficients of self-diffusion, and the electronic parameters of the defects will differ greatly from those in the bulk. Moreover, with the solid-state high-dispersion particles employed in adsorption and catalytic measurements, we are often dealing with nonequilibrium systems in which the equilibrium both between the defects and with the gaseous phase is violated. Thermodynamic methods cannot be rigorously applied to surface defects; at best they give only a qualitative idea of the nature of defect formation on the surface of disperse particles.

The two main groups of experimental techniques developed for the study of bulk defects, namely radioactive tracer and electrophysical methods have little or no applications to disperse particles. Investigations of the temperature dependence of the Hall constant and of electroconductivity, which yield information on the concentration and energetic parameters of a crystal, become inaccurate when used to study high-dispersion specimens. Great complications arise with spectral methods because of spectral line broadening and the impossibility of drawing a line between surface and bulk defects. In certain cases, however, these complications can be overcome, and optical and radio spectroscopic methods provide valuable information on the nature and parameters of surface defects.

A widely used method in the investigation of electron-acceptor and electron-donor surface defects is that of spectral "sondes." They are essentially based on an analysis of changes in the spectral properties of donor (D) or acceptor (A) molecules, which have respectively a low ionization potential or a high electron affinity, during interaction between these molecules and EA and ED centers on the surface. All the intermediate cases can then be observed (Sect.2.1.2), from complexes of type $D^{+\delta} \cdot M^{-\delta}$ or $A^{-\delta} \cdot M^{+\delta}$ (where δ is smaller than the electron charge) to complexes with transfer of charge (CTC), where $\delta \to q$. The transfer of charge is accompanied by changes in the IR and UV spectra of the adsorbed molecules, and in the case of CTC an EPR response can be observed from the anion or cation radicals which form [3.345,346].

Let us consider first the result obtained by this method in investigations of ED centers on the surface of an oxide. Pioneering work conducted by TERENIN and his colleagues [3.347,348] demonstrated that adsorption of such strong electron acceptors as molecules of parabenzoquinone, fluoroanil, chloranil, and tetracynaoethylene on partly reduced oxides of Zn, Mg, and Ti (anatase) gives rise to anion radicals ($A^- \cdot Me^+$); these were detected in electronic absorption spectra and by EPR. The EPR signal from the fluoroanil anion radical on ZnO and MgO was found to possess an ultrafine structure (UFS) associated with the interaction of an unpaired electron with fluorine nuclei [3.348]. This proves conclusively that the electron is completely transferred from the ED center of the surface to the orbitals of the adsorbed molecules, i.e., that a UFS is formed [3.348,349]. A similar EPR response was later observed during adsorption of p-Bq on rutile [3.350] and on real surfaces of Ge and GeO_2 [3.351], CdS [3.352], and PbS [3.353]. The concentration of spin centers N_S ranged from 10^{10} spin $\cdot$ cm^{-2} (GeO_2) to 10^{11} spin $\cdot$ cm^{-2} (TiO_2).

The formation mechanisms of ion radicals and the parameters of the EPR signals proved to be approximately the same for narrow-zone (Pbs, Ge) and wide-zone (CdS, ZnO, TiO_2) semiconductors and for typical dielectrics (GeO_2, MgO). Evidently, with dielectrics the formation of radicals is due not to the free charge carriers of the oxide, but to the electrons localized in the surface traps. The role of such traps can be played by oxygen vacancies (Fig.3.23a,

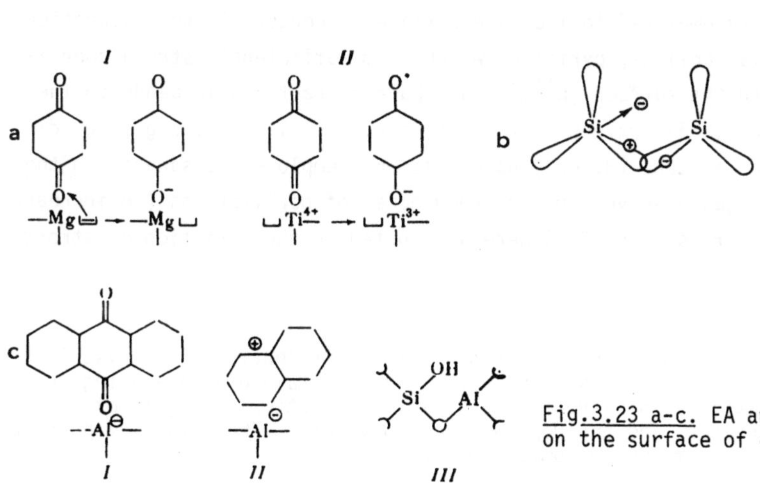

Fig.3.23 a-c. EA and ED centers on the surface of oxides

model I). In [3.354-357] EPR signals from electrons localized in oxygen vacancies on MgO [3.354,355] and on excess Zn atoms in ZnO [3.356,357] were detected. Detailed studies of vacancy defects on ZnO [3.357] indicate the existence of three types of paramagnetic defects: an oxygen vacancy with a trapped electron; an O^- atom in the interstice, and a charged oxygen vacancy interacting with an O^{2-} ion in the interstice. EPR signals from surface vacancy-type defects were also found with SnO_2 [3.358], and the ELS method (Sect.3.1.3) has supplied valuable information on vacancy defects on monocrystals of TiO_2, Ti_2O_3, and V_2O_3. HEINRICH et al. [3.359] observed the formation of Ti^{2+} and Ti^{3+} ions, as well as electron transitions in ions of vanadium, following ionic bombardment.

If an atom M next to an oxygen vacancy has a low redox potential, then its charge state can be changed due to the formation of a CTC [3.360]. There have been numerous discussions on the question of where the electron is localized: in a complex of type $\square Ti^{4+}$, on the orbitals of Ti which goes over into Ti^{3+}, or in vacancy (Fig.3.23a, model II). Calculations of the parameters of EPR signals from Ti^{3+} centers in the bulk [3.361] point to a significant delocalization of the unpaired electron. This agrees well with data obtained in studies of positron annihilation [3.362], which indicate a high degree of covalence of Ti-O bonds.[4] The extent of delocalization depends strongly on the presence, in the neighborhood of the defect, of hydroxyl groups or other chemical complexes. Consequently, any further probing into the details of the structure of the ED center is hardly possible now. We shall deal with ED centers again later on in connection with our discussion of the radical chemisorption of oxygen and other molecules (Chap.4); we now turn our attention to the nature of EA centers.

A method widely employed to indicate surface EA centers is the adsorption of bases (ammonia, aniline, pyridine) which form sufficiently strong coordination bonds with the surface ($D^{+\delta}M^{-\delta}$) and have characteristic bands in the IR spectrum [3.265,266]. Oxygen-containing molecules—ligands, e.g., NO, dimethyl ether, and acetone [3.265], have also been employed as sondes. In the latter case, changes observed in the frequencies of oscillations (an increase for -C-O and a decrease for =C=O) were attributed to the formation of strong coordination bonds [3.265].

---

[4] Valuable information on the electron density distribution in a crystal, and particularly on the properties of defects, can be obtained by analyzing the rate of annihilation of positrons in a solid, the scattering angles of the γ-quanta thus created, and the Doppler frequency shift. The method of positron annihilation has become quite common in surface science [3.363].

More recently, UV spectral investigations of molecules that have entered the donor-acceptor bond [3.364] and EPR methods for cases of CTC formation have gained wide acceptance. Convenient adsorbates for these purposes are acenes (naphthalene, anthracene, perilene) and aromatic amines (diphenylamine, triphenylamine, N-methyl diphenylamine, etc.) [3.346]. Also employed as indicators are dyes, which change color upon transition into ionized form [3.345]. And one last promising sonde for EA centers must be mentioned, namely the EPR spectra of adsorbed stable iminoxyl radicals [3.290,365].

Let us now briefly discuss some aspects of EA centers on the surfaces of certain oxides which will be important in our subsequent discussion.

SiO_2

Mass spectroscopic investigations [3.366] reveal that the main product of dehydration of pure SiO_2 is water. This means that out of two possible courses a and b

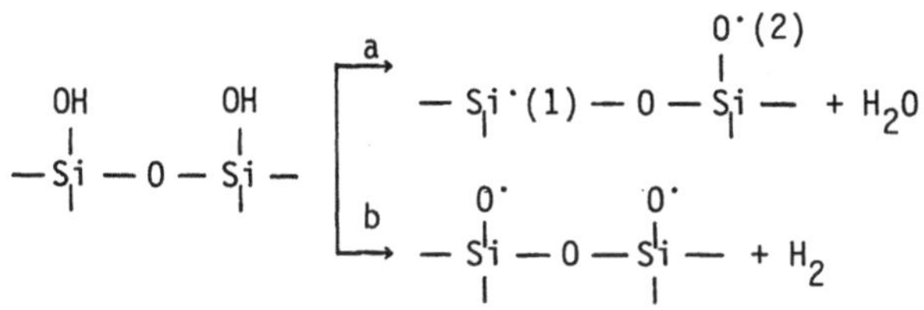

Scheme 3.5 a,b

dehydration mainly follows route a.

The surface resulting from Schemes 3.5a,b ought to exhibit high adsorptive activity. This, however, is not supported by experimental measurements of the adsorption of water. The adsorptive activity drops with dehydration of the surface (Chap.4). Evidently, the majority of broken bonds are closed up according to Scheme 3.3 (p.108), with the formation of overstrained siloxane bridges. Only a few dangling bonds arising from the destruction of solitary hydroxyls remain on the surface. The surface of an extremely pure SiO_2 specimen dehydrated at $T_{cal} = 700$ K starts to adsorb noticeable amounts of oxygen (Fig.3.24) above all on centers of type 1 (Scheme 3.5a).

It is readily seen from Fig.3.24 that specimens with excess oxygen content (i.e., higher than stoichiometric) display a pronounced oxidative ability which increases with an increase in the temperature of heat treatment in a vacuum symbatically with an increase in chemisorption.

FREUND and GENTSCH [3.367] used an omegatron (miniature mass spectrograph) to detect the release of a very small quantity of H_2 during heating of silica gel. The first maximum of H_2 was observed at $T_{cal} = 540$ K and the second at $T_{cal} = 900$ K. The concentration of H_2 released (about 10^{-4} $\mu mol \times m^{-2}$) was close

in order of magnitude to the concentration of oxidative centers discovered in our experiments (Fig.3.24). FREUND suggests two mechanisms of dehydration of the oxide. The first involves tunneling of a proton from one hydroxyl group to another, and the formation of a water molecule. This mechanism, however, is not valid for solitary OH groups. FREUND suggests two mechanisms of dehy-
is not valid for solitary OH groups. Freund [3.368] then assumes that exoelectronic emission takes place in the course of heating. The exoelectron hits upon the OH groups, causing a hydrogen atom to be produced:

$$e^- + OH^- \rightarrow H + O^{2-} \quad .$$

Then a certain number of unsaturated oxygen atoms appear on the surface, which agrees well with our data on the oxidative ability of the surface. However, the number of exoelectrons observed experimentally is always several orders of magnitude smaller than the number of oxidative centers.

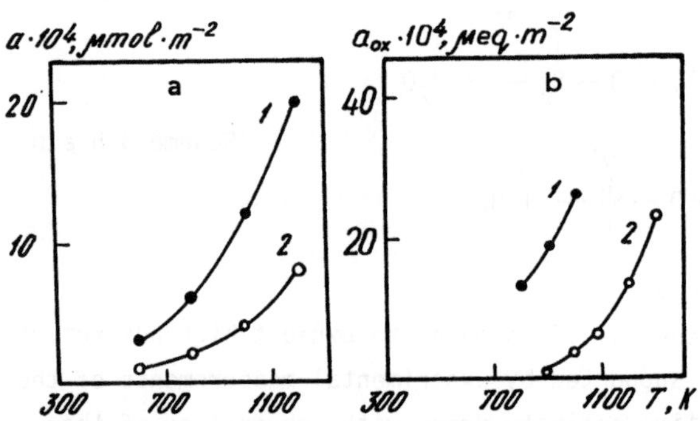

Fig.3.24 a,b. Dependence of the adsorption of oxygen (a) and the oxidizing power (b) of silica gels KSK-2 (*1*) and K-2 (*2*) upon the heat treatment temperature T_{cal}

It must be noted that in many investigations of the electronic and EPR spectra of molecules of amines and acenes adsorbed on pure SiO_2 [3.369] cation radicals have not been observed. This is an indication that the Si atom has poor electron acceptor properties with respect to these molecules. It was shown in [3.370] that if a pure SiO_2 specimen (aerosil), dehydrated at T_{cal} = 800 K with adsorbed molecules of dimethylamine on its surface, is irradiated by UV light, a strong EPR signal arises from cation radicals, and a corresponding characteristic absorption band appears in the electronic spectrum. According to [3.370], the concentration of EA centers is about $4 \cdot 10^{11}$ per cm^2, which differs little from the number of centers of chemisorption of oxygen on thermally dehydrated silica gels (Fig.3.24). BEZNOZDREV and KAZANSKY [3.370] presume the centers of chemisorption of amines to be the positively charged Si atoms in the tricoordination environment. NISHI [3.371] observed EPR signals from vacancy defects in SiO_2 layers on thermally oxidized silicon, which he attributed to trivalent silicon. CAPLAN and coworkers [3.372] have conducted thorough investigations of paramagnetic vacancy defects on the Si-SiO_2 boundary, both by EPR and by NMR relaxation of liquids contacting the surface.

When SiO_2 is irradiated by UV or γ-quanta, a strong dehydration of the surface can be observed [3.373]. For instance, after irradiation by γ-quanta at a dose of about 10^6 R, the amount of water lost by aerosil was equivalent to thermal dehydration at $T_{cal} = 1100$ K. When the water released was completely evacuated in the course of irradiation, the specimen afterwards became noticeably active in the chemisorption of oxygen. Evidently, the radiation removes the isolated hydroxyl groups and thus helps to conserve centers of type 1 in sufficient concentration (Scheme 3.5a). In the EPR spectra from pure SiO_2 specimens irradiated at 77 K, one can observe a doublet from H atoms stabilized on the surface at such temperatures, and a complex central line resulting from radiation-induced defects [3.374,375]. The $H^{\cdot}$ atoms disappear when the temperature rises to 90 or 100 K. The authors of [3.376] detected an abrupt drop in the signal from $H^{\cdot}$ atoms with specimens preheated at 800 or 1000 K; this is connected with dehydration of the surface of SiO_2.

FEIGL et al. [3.377] propose, as the most probable model of a vacancy defect E' formed in SiO_2 under the influence of radiation, two adjacent Si atoms with slightly overlapping sp^3 orbitals. Such a defect is electrically neutral and diamagnetic. Having captured nonequilibrium holes produced by the effect of radiation (or having lost an electron), this defect becomes positively charged and paramagnetic. By analyzing various possible combinations of reconstructed orbitals of three neighboring Si atoms, the whole range of EPR signals from radiation-induced quartz defects could be explained [3.377,378]. This model of a vacancy center recently received further confirmation in experiments reported in [3.379]. Photoinjection of holes from germanium into the surface oxide layer was observed to cause positive charging of the oxide (due to capture of the holes in traps) and the appearance of the corresponding EPR signal. The parameters of the signal were satisfactorily described within the framework of the vacancy defect model illustrated in Fig.3.23b. Similar EPR signals were observed with SiO_2 and GeO_2 specimens exposed to hard radiation [3.380,381]. It is thus clear that the formation of relatively active EA centers on SiO_2 and GeO_2 is caused by the excitation of nonequilibrium holes, which are subsequently captured in the p_z orbitals of the "defects" Si (or Ge) atoms that form the basis of these centers. The concentration of these centers at an irradiation dose of about $2 \cdot 10^8$ rad (with Co^{60}) ranges from 10^{17} cm^{-3} (Si-SiO_2 structures) to 10^{16} cm^{-3} (pure SiO_2) [3.382]. According to [3.382], nonstoichiometric oxides show a lower resistance to radiation than stoichiometric oxides. Similar E' defects are formed when Si-SiO_2 structures are penetrated by ions [3.383], when SiO_2 is irradiated with neutrons [3.384], or when the SiO_2 surface is treated with plasma [3.385]. Very interesting experiments have shown the g factor of the EPR signal to be only slightly dependent on the surroundings of the active center, while the spin-lattice relaxation times are quite sensitive to adsorption (O_2, H_2, p-Bq) [3.386]. The mechanism of interaction between tricoordinated silicon (centers 1 in Scheme 3.5a) and oxygen has long been a subject of debate. FRIEBELE et al. [3.387] were able to observe the EPR line from the peroxide Si-O-O˙ radical on a very pure irradiated SiO_2 specimen enriched by O^{17}. This confirms the mechanism of O_2 adsorption proposed in [3.366]. A similar signal was detected by RADŻIG [3.167] in investigations on the interaction of O_2 with quartz during crushing.

$\underline{Al_2O_3}$

The presence of strong EA centers on the surface of γ-Al_2O_3 dehydrated at sufficiently high T_{cal} was proved in a large-scale investigation using all of the above-mentioned spectral sondes. Data on IR spectra from adsorbed ammonia [3.192,388], pyridine [3.294,389-391], and certain other bases point to the existence of at least three types of EA centers of varying activity. This is explained by STOLZ and KNÖZINGER [3.389] within the framework of PERI's model of the surface [3.192]. It ought to be noted that the existence of several types of coordination of the Al atom on a partially dehydrated surface of detective spinel directly follows from the surface model 3.21 (p. 107) proposed by EGOROV [3.187].

For a long time there was no general agreement regarding the possibility that cation radicals form on the surface of dehydrated γ-Al_2O_3. The ion radicals detected in some experiments were attributed to interaction between adsorbed molecules and excess oxygen (higher than stoichiometric), that is, to the formation of anion radicals. Very interesting results in this regard were obtained in [3.392]. Characteristic EPR signals with ultrafine structure (UFS) from the Al nuclei (Fig.3.23c, model I) were observed during adsorption of quinones on dehydrated γ-Al_2O_3. The parameters of these signals were quite different from those of EPR signals from anion radicals formed as a result of interaction between quinones and ED centers (see above). This confirms again that the predominant role in the formation of CTCs on the surface of γ-Al_2O_3 vacuumized at 800 K is played by EA centers, and not by excess oxygen. Analysis of the IR spectra of anthraquinone adsorbed on EA centers confirms the presence of cation radicals [3.393]. Experiments on the adsorption of iminoxyl radicals lead to the same conclusion. Analysis of the multiplet EPR spectrum from these radicals reveals that the unpaired electron interacts with the Al nucleus (spin 5/2) [3.365]. On the basis of these experiments the authors of [3.365] proposed that the role of EA centers is played by the tricoordinated Al atoms.

Estimates of the concentration of EA centers on the surface of γ-Al_2O_3 dehydrated at 800 K, based on simultaneous gravimetric and spectroscopic investigation of the adsorption of bases (NH_3 and pyridine), yield the value $n_c \sim 10^{12} - 10^{13}$ centers $\cdot$ cm^{-2} [3.389]. Naturally, the concentration of the centers and their activity will depend on the coordination of the active Al atoms, and consequently on the process by which the specimen is prepared.

nAl_2O_3 $mSiO_2$

As in the case of pure aluminum oxide, the adsorption of ammonia, pyridine, and other bases on dehydrated surfaces of aluminosilicate is accompanied by the appearance of typical absorption bands of coordination complexes of these molecules in the IR spectrum [3.266]. However, in contrast to pure Al or Si oxides, the adsorption of amines and acenes always leads to the formation of cation radicals with their characteristic absorption bands in the electronic and EPR spectra [3.346]. The UFS in the EPR spectra is due to the interaction of an unpaired electron with the Al nucleus, e.g., when a π electron of naphthalene moves to the p orbital of Al [3.369] (Fig.3.23c, model II).

FRIPIAT and co-workers [3.394] assume that in aluminosilicates with a high (up to 80%) aluminum content the atoms of Al are found mostly in the octahedral surrounding. Upon dehydration, part of these complexes become tetrahedrons. Aluminosilicates with an aluminum content of less than 30% are assumed in [3.394] to display tetrahedral coordination. Certain authors assume that the role of EA centers is played by tricoordinated Al atoms. A similar view (see below) is expoused in [3.395,396] with regard to decationized zeolites (Fig. 3.23c, model III). PERI [3.397] points to the existence of three types of EA centers based on tricoordinated atoms in overstrained ordinary and double aluminum-oxygen bridges ($Al{<}^{O}_{O}{>}Al$). The concentration of the most overstrained bridges (the strongest EA centers) is about $6.6 \cdot 10^{12}$ cm^{-2}, which differs little from the experimental data.

TiO_2

The existence of EA centers on the surface of TiO_2 dehydrated at 800 K has been irrefutably proved by investigations of the IR spectra of adsorbed ammonia and pyridine molecules [3.308,398,399]. Two types of EA centers have been observed: the most active ones, which are formed when solitary OH groups are removed from the surface, and the least active ones, which arise when coordination-bound water molecules are removed. Indeed, a noticeable distortion of the coordination sphere of Ti^{3+} atoms on a partially reduced surface of rutile was detected by EPR during adsorption of H_2O and NH_3 [3.360]. Whether it is possible for cation radicals to form on the surface of partially reduced TiO_2 during adsorption of donor molecules is not yet clear. Observations of EPR signals from cation radicals of perilene adsorbed on anatase are reported in [3.308,400]. On the other hand, PIMENOV [3.349] points out that a necessary condition for the formation of cation radicals on the surface of anatase is the presence of chemisorbed oxygen.

From this brief discussion of the data on EA and ED centers on the surface it follows that the basis for such centers is provided by the "defect" atoms on the surface.

The effective charges of these atoms, q_{eff}, and their coordination sphere differ from those of the atoms in the bulk. These parameters determine the value and sign of the transferable charge δ in the complexes $A^{-\delta}(D^{+\delta})\cdot M^{q_{eff}\pm\delta}$. The method of spectral sondes described above gives only a qualitative and, as will be shown in Sect.4.2, not always indisputable idea of the parameters of EA and ED centers of the surface. To gain a more profound understanding of the nature of these centers, direct information is needed on the coordination and character of the orbitals of the "defect" atoms which form the basis of the centers. With paramagnetic centers, radio spectroscopic methods have proved to be most informative. EPR, and especially electronic nuclear magnetic resonance (ENMR) [3.401-404] methods yield valuable data on the coordination sphere of the defect atom and its deformation, on the distribution of defects in the solid-state body, and on their mobility. Also of interest are investigations of EPR and ENMR spectra in the electric field, which provide information on the symmetry of the paramagnetic center [3.403-406]. These methods have been successfully used to investigate bulk F centers in halide crystals, and donor and acceptor impurities in silicon. In cases where the basis of EA and ED centers is formed by atoms with Mössbauer nuclei, equally valuable information on their coordination is obtained from analysis of nuclear gamma resonance (NGR).

Unfortunately, the applicability of all these methods to the study of surface defects is still limited because of the smallness and nonuniformity of the surface phase. The EPR spectra, and in certain systems the NGR spectra of transition-metal ions on the surface, have been most thoroughly investigated thus far.

3.2.4 The Structure of Zeolites

Zeolites are constructed of silicon-oxygen and aluminum-oxygen tetrahedrons. Silicon can be isomorphously replaced by aluminum in the skeleton, so in all zeolites [Si + Al] : [O] = 1 : 2, and [Al] $\leq$ [Si]. The bonds -Si-O-Si- and -Si-O-Al- are possible in zeolites, while -Al-O-Al- bonds are forbidden because [AlO_4] tetrahedrons are prevented from coming together by electric repulsion forces (the Levenstein rule). The excess negative charge of the aluminum in the tetrahedral skeleton structure must be compensated by the introduction of cations, so that the number of positive charges equals the number of [AlO_4] tetrahedrons. Sometimes a proton can play the role of such a cation [3.407,408].

The skeletal structure of zeolites is characterized by a regular system of channels and open spaces whose size is comparable to the size of molecules. Figure 3.25 shows the open structure of cuboctahedrons found in the most abundant zeolites of type A and faujasite (type X or Y). Each cuboctahedron, or so-called sodalite cell, contains Al or Si atoms in the vertices and an O atom near the center of each face. The distance Al-O (1.73 Å) is 0.02 Å greater than the distance Si-O. The sodalite cells are joined together by oxygen bridges. In zeolites of type A the cuboctahedrons are linked by tetrahedrons and organized into a simple cubic NaCl-type lattice, with a sodalite cell playing the role of the lattice ion (Fig.3.25a). In zeolites of type X or Y the cuboctahedrons are joined together by hexagonal prisms and form a diamond-type lattice. The ratio SiO_2/Al_2O_3 is $\sim$2 in zeolites of type A, $\sim$2.2 - 3.0 in zeolites of type X, and 3.2 - 5.0 in zeolites of type Y.

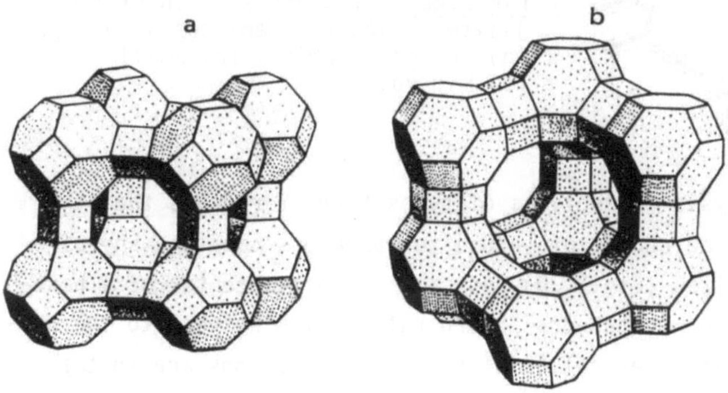

Fig.3.25 a,b. Spatial structure of cuboctahedrons in zeolites of type A (a) and types X or Y (b)

The space formed between the cuboctahedrons in such a skeleton (in the center of the cell in Fig.3.25) is called the big cavity. In zeolites of types A, X, and Y it has a diameter of about 12 Å and a volume of about 800 $Å^3$. Big cavities are interconnected by passage about 9 Å in diameter in zeolites of types X and Y and about 4.2 Å in diameter in zeolites of type A. The space inside a sodalite cell is called the small cavity; it has a diameter of about 6.5 Å and a volume of about 150 $Å^3$. The six-sided oxygen passages between the big and the small cavities have a length of about 2 Å [407-411].

The passages, cavities, and pores ensure selective absorption of small-size molecules by zeolites, which in this case have the properties of a molecular sieve. Naturally, the molecular-sieve properties of zeolites are determined not only by the geometry and dimensions of the skeleton, but also by their chemical properties and the potential within the spaces. It is easy to calculate that the surface of the big cavities amounts to $10^3 - 10^4\ m^2 \cdot g^{-1}$. However, the concept of surface loses its meaning when applied to zeolites [3.412]. The curvature of the cavities in zeolites is of about the same order as the radius of action of molecular forces, so that the absorption of a gas by a zeolite is better considered to be the formation of a solid solution, and called sorption [3.413] (see also Sect.2.1.3).

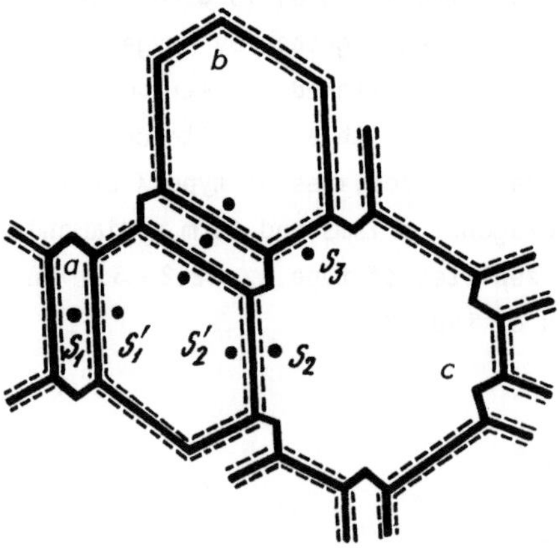

Fig.3.26 a-c. Cross section of faujasite structure (zeolites X and Y). (a) Hexagonal prism. (b) Small void. (c) Big void

X-ray diffraction analysis of zeolites of types X and Y reveals that the cations which compensate the charge of [AlO_4] tetrahedrons can take any of three main positions or sites (Fig.3.26). At site S_1 in the center of a hexagonal prism, the cations show octahedral coordination with respect to the oxygen in the skeleton (16 sites per cell); at site S_2 the cations are in the center of six-sided oxygen passages connecting big cavities to small ones (32 sites per cell); and at site S_3 they are in the center of four-sided oxygen rings which connect the big cavities (48 sites per cell). A more subtle differentiation would take into account the position of the cation with respect to the plane of the rings. If the position in the center of the six-sided oxygen ring is shifted towards the big cavity, such a site is labeled S_2; if it is shifted towards the small cavity the center is denoted S_2'. Similarly, the position in the center of the passage between sodalite cell and hexagonal prism, when shifted towards the small void, is labeled S_1'. In zeolites of type A there are no S_1 centers.

The bonding between cations occupying these sites, and the Al-O-Si skeleton, is similar to an ionic-type bond. MINACHEV et al. [3.62] used X-ray electron spectroscopy to study the effective charge of Na^+, K^+, and Ca^{2+} cations in zeolites of type Y, and found it to be close to the maximal charge of cations in salts. For instance, in the zeolite NaY, the effective charge of Na takes a value between that in NaCl($q_{eff} = 0.66$) and that in NaF($q_{eff} = 0.8$), though closer to the latter.

DEMPSEY [3.410] adopted the model of a purely ionic bond, and calculated the Madelung potential for the sites S_1, S_2, S_1', and S_3 in zeolites of types X and Y: $\varphi = \sum_i (e_i / r_i)$, and where e_i is the charge of the ion i, and r_i is its radius. It turned out that the potential φ for bivalent and trivalent cations in S_1 centers is always lower than for those in S_2, S_1', and S_3 centers regardless of the degree to which these centers are occupied. The value of φ decreases with an increase in the ion's radius. On this basis it was concluded that bivalent and trivalent cations first occupy the S_1 sites and then, when all of these are occupied, the remaining ions take S_2 sites (when the ion radius of M^{2+} and M^{3+} is lower than 1.3 Å) or S_1' sites (when $r > 1.3$ Å).

A more complicated situation is encountered with univalent cations. For a small cation radius ($r < 1.0 - 1.3$ Å) the curves depicting the dependence of the Madelung potential upon the radius, $\varphi(r_1)$, for centers S_2 lie below those for centers S_1. Consequently, these cations (Li^+, Na^+, etc.) should occupy S_2 sites first. With univalent cations the energy for S_1' centers is very close to that for S_1 centers, and thus when the centers of type S_2 have all been occupied, the S_1' and S_1 centers begin simultaneously to be occupied (of course, two adjacent S_1' and S_1 centers cannot both be occupied at the same time). Univalent cations with greater radii occupy centers of type S_1 first. The Madelung potential is always greater for S_3 centers than for the other centers, so that they are the last to be occupied.

X-ray analysis in general confirms these conclusions. Bivalent cations occupy centers of type S_1 first. Some investigations indicate that in CaX zeolites the Ca^{2+} cations are divided between the S_1 and S_2' centers. Dempsey assumes that this is due to a large residual water content. The above calculations were carried out under the assumption of complete dehydration of the zeolites.

The presence or absence of water in the voids of zeolites has a great effect on their properties. A large volume of experimental data now exists on the IR spectra of zeolites. Several rather narrow bands in the range of $3500-3750\ cm^{-1}$ can usually be observed in the region of valence oscillations of OH groups in IR spectra. The position of these bands in the spectrum is dependent upon the nature of the exchange cation of zeolite. However, opinions on the origin of the various bands diverge. The interpretation of the spectra is further complicated by the fact that the position of lines in zeolites with the same exchange cations depends on the conditions under which the zeolites are synthesized and prepared [3.414].

In [3.415], four characteristic regions are given for the frequencies of valence oscillations of OH groups in zeolites: (1) $3744-3750\ cm^{-1}$, (2) $3673-3715\ cm^{-1}$, (3) $3630-3660\ cm^{-1}$, (4) $3540-3605\ cm^{-1}$. Apparently there is reason to attribute the first of these to oscillations of Si-OH groups. The other bands have not yet been conclusively explained. Some authors attribute them to the structural hydroxyls belonging to the Al atoms, while others relate them to the existence of decationized sites. One group of researchers clings to the traditional explanation that the frequency ν_{OH} is lowered through the formation of OH bonds between adjacent hydroxyls (as in the case of pure oxides). Another group of researchers attributes the low-frequency bands to the oscillations of OH groups in the adsorbed water molecules. The authors of [3.266] assume that in zeolites with alkaline cations the structural OH groups are absent, and the respective bands in the IR spectrum belong to the water molecules, which are removed at $T_{cal} \sim 300$ K. They assume that structural hydroxyls are found only in zeolites with rare-earth and alkaline-earth exchange cations. On the other hand, from [3.415-419] it follows that in zeolites with polyvalent cations (Mg^{2+}, Ca^{2+}, Sr^{2+}, Cr^{3+}, La^{3+}), the bands of deformation vibrations of water molecules remain in the spectrum up to $T_{cal} \sim 800$ K.

As was already indicated, the NMR method is a very promising technique of separating structural water into hydroxyl groups and water proper. NMR investigations of the zeolites NaX, ScNaY, LaNaY dehydrated at $T_{cal} \sim 650-700$ K, and of decationized zeolite Y [3.420] have revealed OH groups of various types and water molecules which were thermally very stable. Hydroxonium ions were not detected at these T_{cal}. The presence of H_2O molecules with increased interproton distances was also detected in NMR spectra of zeolites with alkaline-earth cations [3.421]. The high thermal stability of these water molecules is attributed in [3.420,421] to coordination bonds, and in [3.422] this is also considered a possible explanation.

Based on the shape of the NMR lines of hydroxyls in zeolites, certain conclusions can be made regarding the distribution of these groups in different types of zeolites. While in the NaY zeolites these groups are totally removed at $T_{cal} = 700$ K, they remain in the spectra of zeolites with polyvalent cations and decationized zeolites up to $T_{cal} = 800$ K [3.420,421]. The NMR signal from OH groups of zeolites with polyvalent cations can be approximated by two Gaussian curves corresponding to two types of OH groups which have different minimal distances between the protons (3 - 5 Å and 2.4 - 2.6 Å).

The nature of EA (Lewis) centers in cationized zeolites is not yet completely clear. The catalytic activity of zeolites is often connected with the polarizing effect of cations [3.423]. However, the assumption that exchange cations play the predominant role as EA centers of zeolite fails to explain the whole range of experimental data, and in particular the dependence of the activity of zeolite upon the ratio SiO_2/Al_2O_3. RUBINSTEIN and co-workers [3.424] point out a correlation between the concentration of cation-free regions of zeolite structure which resemble the corresponding acidic centers of amorphous aluminosilicates, and the catalytic activity of zeolite. The activity of these metal-free patches can be quite high [3.425]. The influence of cations on catalytic activity apparently lies not in the direct capture of the electrons of adsorbed donor molecules, but in the deformation of the zeolite lattice and the activation of metal-free EA centers, e.g., tricoordinated Al atoms [3.426, 427].

A recent achievement of great import was the discovery of a new class of zeolites with a SiO_2/AlO_2 ratio as high as 20 or even 50. A typical representative of this new class of zeolites is the ZSM-5 zeolite. By using organic cations (templates) to produce these zeolites, instead of inorganic cations as formerly, one can obtain zeolites with calibrated pores penetrating the entire crystal structure. These new zeolites made it possible to conduct adsorption and catalysis experiments with a great degree of selectivity ("selectoforming"). For instance, the isomerization of xylenes gives a high yield of para-xylene (a)

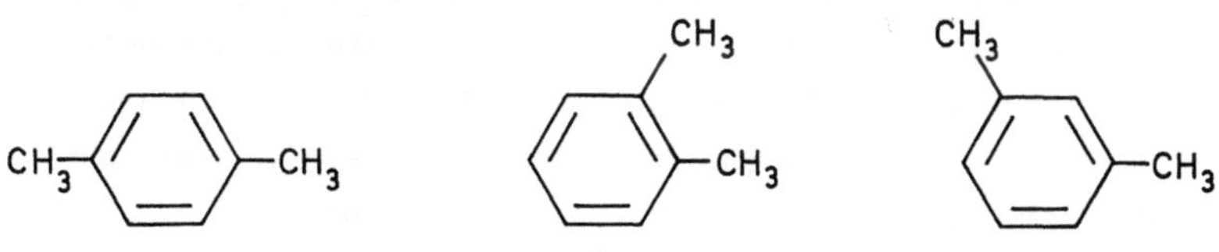

a b c

and yields no ortho-xylene (b) or meta-xylene (c), since only in the first case is the diameter of the molecule small enough to allow drift through the pores of the zeolite. Mobil Oil Company used the ZSM-5 zeolite in experiments on the synthesis of engine fuel from methanol with a high percentage of aromatic hydrocarbons, e.g., durene $C_6H_2(CH_3)_4$. This process is universally considered to be an alternative to the production of gasoline from crude petroleum [3.428,429].

The crystal structure of the ZSM-5 zeolite consists of two two-dimensional networks of intersecting channels [3.430] formed by a ring of ten oxygen atoms; the cross-sectional dimensions are about 5.4 Å by 5.6 Å. The total number of acidic centers in the zeolite corresponds to the number of aluminum-oxygen tetrahedrons and $[AlO_4]^-$ anions, which replace $[SiO_4]$. This number, and the aluminum content as well, are lower than in zeolites of types X, Y, and A, though the strength of the acidic centers is greater. Such centers are assumed [3.431] to be found at the intersections of the channels. When organic molecules interact with such a center, low-mobility carbonium ions are formed. At the same time the initial reactants and the products drift easily in the channels of zeolite.

3.2.5 Surface Energy. The Dependence of the Concentration of Active Adsorption Centers on Particle Size

The surface energy ε is one of the principal energetic parameters of a solid. The overwhelming majority of experimental techniques provide information on the energy not of the ideal solid-state surface, but of a certain reconstructed subsurface layer of finite thickness [3.2-4] (see also Sect.3.1.2). It follows from the analysis of adsorption thermodynamics (Sect.2.1.3) that adsorption leads to a decrease in ε.

Hydration of the surface and the presence of various vacancy-type defects also influence the value of ε. WEYL [3.1] pointed out quite early that the surface energy of a silicate decreases due to screening of the surface Si atoms by hydroxyl groups, but the number of experimental works on the effect which hydration of the surface of an oxide has on the surface energy is still quite limited.

BRUNAUER et al. calculated in surface energy of amorphous SiO_2 [3.432] on the basis of data on the heat of solution of partially hydrated silica gels in hydrofluoric acid. They assumed that the surface enthalpy of the silanol surface, ε_1, is equal to the energy of the dehydrated surface, ε_2, minus the energy of hydration q_h, i.e., $\varepsilon_1 = \varepsilon_2 - q_h$. Then the surface energy of a partially hydrated oxide is

$$\varepsilon = \varepsilon_2 - q_h \alpha_{H_2O} \quad . \tag{3.29}$$

Based on the assumption that the processes of hydration and dehydration are completely reversible, BRUNAUER and co-workers [3.432] arrived at the following values for the enthalpies of silanol (ε_1) and siloxane (ε_2) surfaces: $\varepsilon_1 = 129\ \mathrm{erg \cdot cm^{-2}}$; $\varepsilon_2 = 259\ \mathrm{erg \cdot cm^{-2}}$. On the basis of the experimentally measured hydration heat $q_h = 4.8\ \mathrm{kcal \cdot mol^{-1}}$ [3.433] and the degree of hydration of SiO_2 [3.238,239], the value of ε was estimated in [3.433,434] for various partially hydrated SiO_2 specimens. As was predicted in [3.1], the value of ε increased with dehydration of the surface of SiO_2.

Other conditions being equal, the surface properties of high-dispersion solid substances, in particular the specific surface energy, must depend on the size of the particles. Indeed, the process of dehydration is connected with reorientation of the surface tetrahedrons. As was indicated in Sect.3.1.3, the starting point for mobility of the lattice elements is described by the Tamman temperature $T_T = 0.3 - 0.5\ T_{melt}$. However, T_{melt} decreases with a decrease in particle size (Sect.3.1.3), so the rotation of the SiO_4 surface tetrahedrons, and thus the dehydration of the surface, will proceed at a lower temperature with small-size particles. Consequently, the value of α_{H_2O} will be lower for small particles than for bigger ones, which agrees with experiment. From Table 3.4 it is evident that among oxides treated in the same way, the value of α_{H_2O} is lower for fine-dispersion SiO_2 oxides than for coarsely ground specimens. Similar behavior of α_{H_2O} was observed with GeO_2 specimen [3.4].

Independent proof of the dependence of the adsorptive properties of SiO_2 upon the value of the specific surface was obtained with spectroscopic measurements [3.435]. An almost fourfold decrease in the intensity of the absorption band of single OH groups in the IR spectrum was observed when the specific surface of aerosil was reduced from 360 to 60 $\mathrm{m^2 \cdot g^{-1}}$.

The correlation between α_{H_2O} and s has a rather general character. Figure 3.27 shows results of measurements of α_{H_2O} for 35 samples of SiO_2 [3.243], grouped according to the conditions of heat treatment and rehydration. In all groups there is a clear tendency of α_{H_2O} to decrease with an increase in s, i.e., with a decrease in particle size. The curves for coarsely ground quartz specimens (small values of s) do not go smoothly into the curves for silica gels, which reflects the different structure of their surfaces. At the same time, it has been noted [3.243] that with close values of s, powders of crystal quartz and of quartz glass display similar values of α_{H_2O}. Apparently α_{H_2O}, and consequently the topography of the tetrahedrons in the surface layer of the SiO_2 particle, depend not only on the structure of the amorphous or

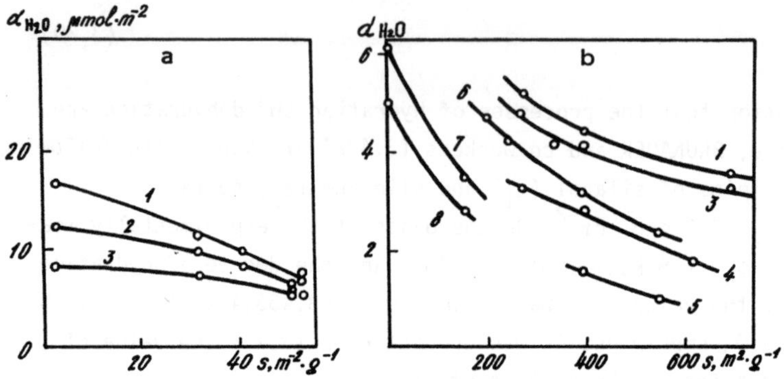

Fig.3.27 a,b. Dependence of the extent of hydration αH_2O of the surface of quartz powders (a) and silica gels (b) upon specific surface area s. The specimens were dehydrated in vacuo at (*1*) 300 K, (*2*) 500 K, (*3*) 600 K, (*4*) 700 K, (*5*) 800 K, (*6*) 1000 K, (*7,8*) 1200 K. Specimens (*6-8*) were rehydrated in liquid water and vacuumized at 600 K, 300 K, and 500 K respectively, after being subjected to heat treatment

crystalline nucleus of the particles, but also on its size. In this connection we again recall [3.224], where changes in the valence oscillations of Si-O-Si bonds were found to depend upon the change in size of silica gel globules.

The dependence $\alpha_{H_2O}(s)$ reported in [3.239,243] was independently proved in [3.436], where the adsorptive properties were compared with the curvature of the adsorbent surface. The vapor pressure p/p_S over the concave meniscus of adsorbate in the capillary of a porous body of radius r is described by the Kelvin equation, which for an ideally wetting surface has the form

$$\ln(p / p_S) = -2V\gamma / r \quad , \tag{3.30}$$

where γ is the surface tension of the adsorbate. In accordance with (3.30), BAKARDJEV [3.436] took the quantity $\ln(p/p_S)$ as the measure of the surface curvature and introduced the distribution function of the surface over the curvature in the following form:

$$\left(\frac{da}{d\ \ln(p / p_S)}\right)_T \ln \frac{p}{p_S} \quad , \tag{3.31}$$

where da is the isothermal adsorption in the interval $d\ln(p/p_S)$ of the relative pressure of the adsorbate vapor. With the aid of these functions it was shown that the concentration of OH groups on the surface decreases with an increase in the surface curvature, i.e., with a decrease in particle size (Fig.3.28).

The dependence $\alpha_{H_2O}(s)$ implicitly reflects the connection between the deformation of the surface cells, variations in the interatomic distances, and

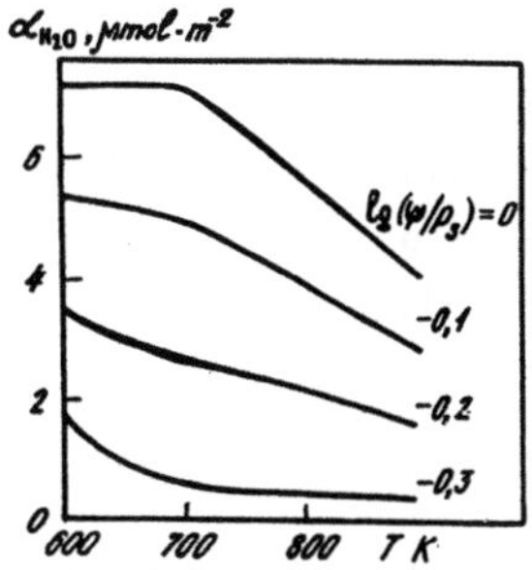

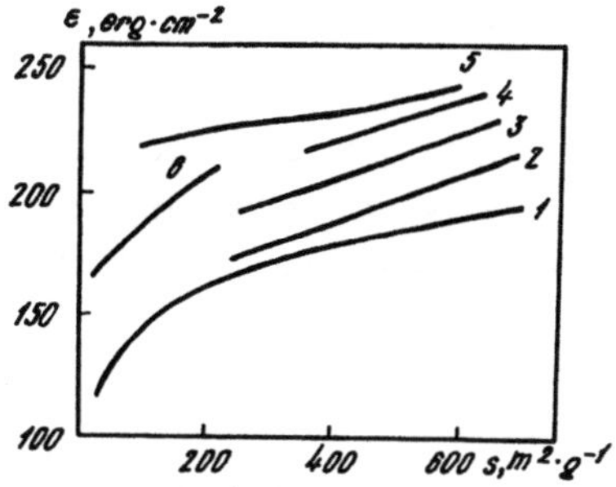

Fig.3.28 Dependence of the structural water content on the surface of an oxide upon the curvature of the surface /3.436/

Fig.3.29 Dependence of the surface energy (enthalpy) of SiO_2 specimens on the specific surface area s. The specimens were treated under the same conditions at (*1*) 500-600 K, (*2*) 700 K, (*3*) 800 K, (*4*) 900 K, (*5*) 1000 K. Specimen (*6*) was rehydrated and vacuumized at 500 K after being subjected to heat treatment at 1200 K

valence angles on the one hand, and the size of the adsorbent particles on the other. For this reason the concentration of EA centers on the surface must also be dependent upon the dispersity. BAKARDJEV and co-workers [3.437,438], with the aid of the distribution functions of adsorptive properties of the surface over curvature, see (3.31), demonstrated theoretically and proved experimentally the dependence of basic and acidic properties of the surface, as well as the catalytic activity of aluminum oxide and aluminosilicates, upon the dispersity of the catalyst.

Since according to (3.29) the surface enthalpy depends upon the concentration of structural water on the surface (α_{H_2O}), and the latter in turn varies with dispersity, it is interesting to investigate the dependence of ε upon the specific surface s. The results of such calculations are presented in Fig.3.29 [3.433,434]. In groups of specimens subjected to the same treatment, ε increases with an increase in the dispersity of specimens, e.g., with a decrease in particle size.

The bulk and surface properties of amorphous bodies and strongly reconstructed crystals (nonequilibrium structures) depend on the conditions of preparation and on chemical and thermal treatment. The comparison of data on ε given in Fig.3.29 is based only an uniform conditions of thermal treatment; all other factors were more or less uncontrollable. It should be noted that the nonuniformity of silanol and siloxane surfaces can greatly influence the absolute value of the surface energy. Consequently, one can only speak of qualitative relationships between α_{H_2O}, ε, and dispersity, and not of quantitative regularities. These results agree with WEYL's conclusions [3.1] regarding the increase in the surface energy with an increase in dispersity, which he confirmed by measuring the solubility of fine quartz powders.

The values for the heat of wetting of SiO_2 specimens by water were thoroughly investigated in [3.433], so it is worth estimating the values of interphase energies on the solid-vacuum boundary, (ε_{s-v}), and the solid-liquid boundary ($\varepsilon_{s-\ell}$):

$$q_{wet} = \varepsilon_{s-v} - \varepsilon_{s-\ell} \quad . \tag{3.32}$$

The result of calculations of the interphase energy on the SiO_2-water boundary are presented in Fig.3.30. It is evident that $\varepsilon_{s-\ell}$ decreases with an increase in the extent of hydration, and becomes negative for strongly hydrated quartz specimens. This is due to the strong interaction between water and the hydrated surface of the oxide. Extrapolating $\varepsilon_{s-\ell}$ to $\alpha_{H_2O} = 0$, which corresponds to a siloxane surface, yields $\varepsilon_{s-\ell} = 209\ erg \cdot cm^{-2}$, while $\varepsilon_2 = 259\ erg \cdot cm^{-2}$.

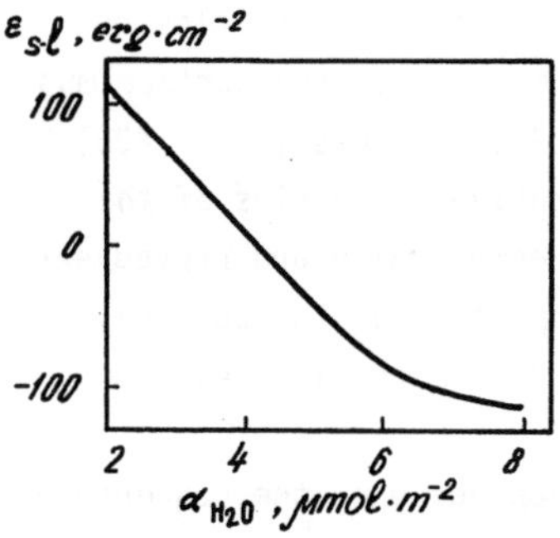

Fig.3.30 Dependence of the specific interphase energy on the solid-liquid interface, ε_{s-1}, upon the extent of hydration of the surface of SiO_2 specimens heat treated at 500-600 K, and rehydrated specimens

The surface energy decreases when the SiO_2 specimen is submerged in water. This is in agreement with the conclusions of REBINDER [3.439] and other researchers.

A number of works have discussed the connection between the specific adsorptive and catalytic properties of the surface and the properties of the adsorbed phase on the one hand, and the size of particles and pores of the adsorbent on the other. We remarked in Sect.2.2.3 that the activity of metallic catalysts deposited on an inert carrier depends upon the dimensions of the metallic phase. This dependence is found to be different for catalytic reactions of different types. The sensitivity of the chemisorptive activity to the size of adsorbent particles has also been discovered for pure oxides. The effects of dispersity on the temperature of ferroelectric phase transitions in $BaTiO_3$ powder were studied in [3.440]. The adsorption of donor molecules (CO) lowered the temperature of phase transition by several degrees, while the adsorption of acceptor molecules (O_2) raised it. A thorough analysis of catalytic processes which are sensitive to the particle size of the catalyst was carried out some time ago by SCHLOSSER [3.441], but even today the true nature of scale effects in adsorption and catalysis is not completely clear. Even in simple adsorption systems the specific activity is greatly influenced by surface defects, both macroscopic (dislocation exits, cracks, steps, etc.) and microscopic (point defects and congregations). The concentration of defects depends not only on the size of particles, but also on the entire history of the adsorbent. When dealing with semiconductors, one must take the distortion of the zone structure in particles of small size into account [3.340].

Equally interesting in the physics of disperse systems are investigations of scale effects in the adsorbed phase. Before the 1950s changes in the thermodynamic parameters of adsorbate (in particular, changes in the melting temperature T_{melt}) in a porous adsorbent were explained in terms of capillary condensation by the lowering of the vapor pressure over the meniscus. In 1959 it was shown [3.442] in the laboratory of one of the authors of this book that the decrease in the melting point of water in capillaries of porous solids displays a practically linear dependence upon the inverse radius of pores. It has also been demonstrated that the decrease in the temperature of phase transition bears no connection to the shape of the meniscus, but is determined solely by the dimensions of the disperse water phase and the value of the interphase energy of the water-solid boundary [3.4]. Later, with the aid of calorimetric and NMR measurements, a decrease in T_{melt} was detected in mono- and polymolecular water adsorption on nonporous disperse solids. The hydrophobization of the surface led to a rise in the phase transition temperature as compared with the hydrophilic surface [3.443-446]. In [3.447] a scale-dependent phase transition was discovered in liquid crystals adsorbed in pores of silica gel.

4. The Mechanism of Adsorption on the Semiconductors and Dielectrics and the Reactivity of Adsorbed Molecules

In this chapter we consider the mechanism of adsorption in two extreme situations: atomically clean surfaces and real surfaces of semiconductors and dielectrics. General conclusions about the multistep interactions of some simple atoms and molecules with surfaces are drawn from extensive experimental data on adsorption, heat of adsorption, IR, NMR and EPR spectroscopy.

4.1 Adsorption of Simple Molecules on Atomically Clean Semiconductor Surfaces

4.1.1 Adsorption on Graphite Surfaces

As we pointed out in Sect.3.1.1, graphite is a rewarding object for experiments on adsorption due to its high dispersity and lack of tendency to form agglomerates. At least three types of active centers of chemisorption are present on the reactive surface of prismatic faces of graphite which emerges in the course of crushing under a high vacuum: A_1 — dangling σ bonds of carbon in sp^3 hybridization, B_1 — atoms with paired antiparallel spins in s^2p^2 hybridization, and overstrained quasi-acetilene bonds, see Fig.3.14b and Models (3.19). If the crushing is carried out at room temperature (series K), then the atomically clean surface of prismatic faces is populated mainly by centers of type 1, see Model (3.19b), and types 1 and 2, see Model (3.19c). Specimens crushed at 77 K (series N) are characterized by the presence of centers of type 1 in low concentrations, see Model (3.19a), which determine the EPR response from these specimens.

The chemisorption of oxygen exhibits a pronounced irreversibility [4.1-3] (Fig.4.1, Curve 1). The magnitude of irreversible adsorption α_{ir} for a specimen in series K is $\sim 0.65\ \mu mol \cdot m^{-2}$. The concentration of oxygen atoms, reduced to the surface area of prismatic faces, n_0, is $\sim 1.4 \cdot 10^{15}\ cm^{-2}$, which is close to the number of carbon atoms on a square centimeter of such a face. The isotherm has an initial vertical segment which corresponds to the occupation of the

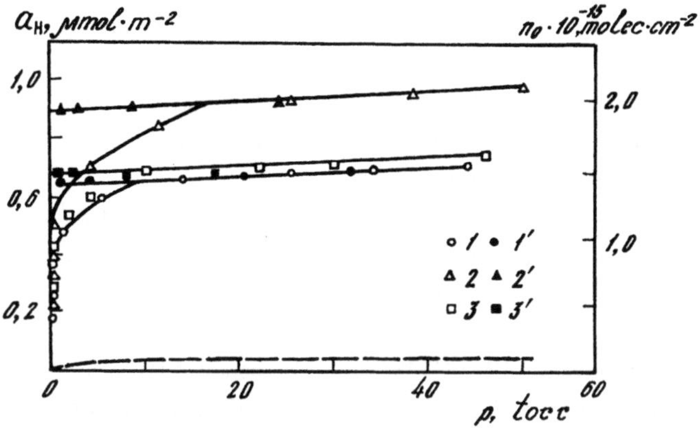

Fig. 4.1 Adsorption of oxygen on atomically clean surfaces of graphite of series K (*1*), series H (*2*), and series H after heat treatment in a vacuum at 800 K (*3*). (*1'*-*3'*) Desorption. The dashed line represents secondary adsorption

most active centers of the surface ($n_0 = 0.5 \cdot 10^{15}$ cm^{-2}). Here the values for the differential heat of adsorption are great: ~ 460 kJ $\cdot$ mol^{-1} (Fig.4.2). This segment of the isotherm is characterized by the rapid establishment of equilibrium pressure (in 30-40 minutes). With further injections of oxygen the tilt of the isotherm becomes lower, and the time in which adsorption equilibrium is established increases to 5 or 6 hours. The differential heat of adsorption decreases nearly twofold (Fig.4.2). At an oxygen pressure of about 10 torr the isotherm is almost horizontal (Fig.4.1), and it takes dozens of hours to establish equilibrium, making calorimetric measurement impossible. The insignificant absorption of oxygen in this region is due to its solution in the

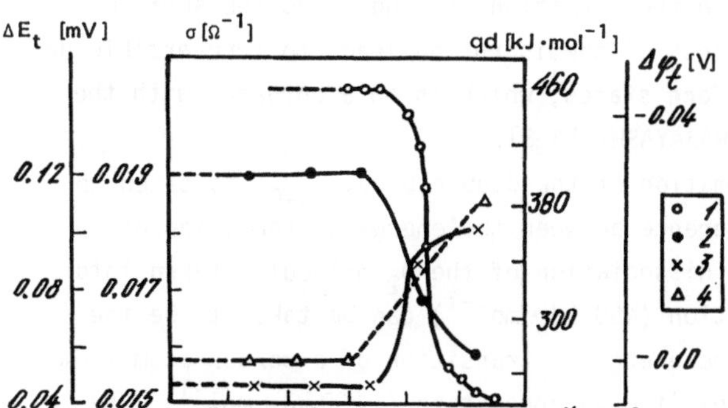

Fig. 4.2 Dependence of (*1*) the differential heat of adsorption, q_d, (*2*) conductivity σ, (*3*) thermal emf E_t, and (*4*) thermionic work function $\Delta\varphi_t = \varphi_{graphite} - \varphi_{gold}$ upon the quantity of oxygen adsorbed on an atomically clean surface of graphite of series K

graphite lattice, which is partially destroyed in the course of crushing. Analysis of the desorption shows that nearly all the adsorbed oxygen is irreversibly bound on the surface (Fig.4.1). Secondary adsorption of oxygen on oxidized graphite is negligibly small, and is primarily due to the diffusion of oxygen into the interfacial gaps of the graphite particles.

According to oxidation theory [4.4], the first stage of oxidation is the formation of peroxide radicals. However, investigations of the EPR spectra from graphite specimens of series K failed to reveal the presence of spin centers. On the basis of measurements of the static magnetic susceptibility of graphite during adsorption of oxygen, ZARIF'YANTS [4.5] showed that all the chemisorbed oxygen is in diamagnetic form, i.e., in the valence saturated state.

Irreversible chemical adsorption of oxygen up to an occupation n_0 of about 10^{13} cm^{-2} did not change the conductivity of graphite, σ, the thermionic work function $\Delta\varphi_t$, or thermal emf E_t [4.6] (Fig.4.2), and neither did the differential heat of adsorption, q_d, change in this region. This is an indication that in the initial stage of occupation, chemisorption does not cause any changes in the surface charge or the curvature of the energy bands. Presumably the initial surface of the prismatic faces is negatively charged with respect to the bulk, due to the localization of some π electrons on the Shockley states and the formation of singlet pairs (B centers in Fig.3.14b). When these centers interact with the oxygen, rehybridization of bonds occurs, which corresponds to the transition of the lateral carbon atoms from s^2p^2 into the sp^3 state and the formation of carbonyl groups $>C=O$. Since the adsorption proceeds directly on the charged surface states, the electrons localized on these states take part in the formation of bonds, and the surface charge does not change. The process of chemisorption leads to irreversible destruction of these acceptor surface states, which is in accordance with the ideas of MROZOWSKI [4.7] and HIRABAYASHI [4.8].

If the average energy of formation of the double bond, $E_{C=O}$, is taken to be 730 $kJ\cdot mol^{-1}$, then the difference between the energy of formation of $>C=O$ bonds (with the energy of dissociation of the O_2 molecules taken into account) and the heat of adsorption (460 $kJ\cdot mol^{-1}$) can be taken to be the qualitative characteristic of the energy of transition of a carbon atom from state 1, Model 3.19b, into the excited valence state ($\sim$290 $kJ\cdot mol^{-1}$). The energy of transition of a free carbon atom in the gaseous phase from the s^2p^2 state into the excited valence sp^3 state is $\sim$670 - 750 $kJ\cdot mol^{-1}$. Apparently, the energy of such a transition is much lower for an atom in a solid-state lattice with conjugated bonds.

For occupations exceeding 10^{13} cm^{-2} there is an abrupt drop in conductivity σ and an increase in the work function $\Delta\varphi_t$ of graphite (Fig.4.2), indicating the appearance of the charged form of chemisorption. As a rule, oxygen is chemisorbed in the acceptor form. An increase in $\Delta\varphi_t$ indicates the appearance of an additional negative charge on the surface which is due to the capture of a certain number of electrons from the π band by surface traps[1]. The change in surface charge and the twofold decrease in q_d are evidence that the predominant role in the process of adsorption is taken over by other types of centers, presumably centers of types 1 and 2 [Model (3.19c)]. On these centers the oxygen can be adsorbed with the formation of ketones and heterocyclic compounds. Radical forms of chemisorbed oxygen were not found in this stage of oxidation either.

Oxidation of an atomically clean graphite surface does not completely suppress the chemisorptive activity of that surface. When such a surface is brought in contact with water vapor, the surface oxides undergo hydration, and hydroxyl and hydroperoxide groups are formed. The initial value of water adsorption heat is $\sim 170\,kJ \cdot mol^{-1}$ [4.9]. Chemical functional analysis was carried out in [4.2] following oxidation and hydration of a graphite surface and carbonyl groups ($\sim 2.5 \cdot 10^{13}\,cm^{-2}$), carboxyl groups ($\sim 2 \cdot 10^{14}\,cm^{-2}$), acidiferous carboxyl groups ($\sim 6 \cdot 10^{13}\,cm^{-2}$), and hydroperoxide groups ($\sim 3 \cdot 10^{13}$ cm^{-2}) were detected. (All concentrations were calculated for the prismatic faces of graphite.) It is interesting to note that the concentration of the carbonyl groups is close to the value for the range of occupation (Fig.4.2) where σ, $\Delta\varphi_t$, and E_t do not change. This supports the proposed mechanism of chemisorption on centers of type 1, Model (3.19b).

Let us now consider the chemisorption of oxygen by specimens of series N containing centers of type 1, see Model (3.19a): unsaturated carbon atoms with a concentration of $\sim 10^{13}$ cm^{-2} (Table 3.2) [4.10]. The EPR signals from these centers retain their shape when the specimen is defrosted to a temperature of 300 K. The adsorption of oxygen at this temperature is also completely irreversible, and the magnitude of irreversible adsorption is greater than the corresponding value for a specimen of series K (Fig.4.1). This difference in the concentration of active centers cannot be attributed solely to the presence of free radicals. Apparently, the nature of the cleavage and the activity of other centers [Model (3.19b), center 1; Model (3.19c), centers 1 and 2] also depend on the temperature at which crushing has been carried out.

---

[1] With powders the data on changes in σ and E_t allow only very qualitative conclusions on changes in the form of adsorption.

If specimens of series N crushed at 77 K are calcinated in a high vacuum at 800 K, their adsorption isotherm lies close to that for specimens of series K (Fig.4.1, Curve 3).

The injection of oxygen at 77 K leads to a broadening of the EPR response from specimens of series N, while subsequent evacuation restores the shape of the line. Here we encounter a typical reversible oxygen effect. At 300 K the adsorption of oxygen leads to an irreversible distortion in the EPR signal. When oxygen was pumped out (in order to preclude the "oxygen effect"), a nearly symmetrical signal with $H = 20$ G and $g = 2.0036$ could be observed, whose intensity was close to the strength of the signal from the initial specimen. The signal from radical forms of oxygen is thermally stable, and can be destroyed only by heat treatment at 800 K. We see that in contrast to oxidation of a surface which contains no spin centers (specimens of series K), in this case the oxidation is accompanied by the appearance of the radical forms of chemisorption: $>CO^{\cdot}$ or $>C\text{-}O\text{-}O^{\cdot}$. It is hoped that further advances in X-ray photoelectronic spectral techniques will help to provide a detailed picture of the forms in which oxygen is bound to graphite surfaces. The high sensitivity of this technique has already been proved in pioneering investigations on the surface of graphite monocrystals [4.11,12].

Interesting results have been obtained in studies of the chemisorption of hydrogen [4.13]. Figure 4.3 shows that the adsorption of H_2 on graphite specimens of series K is totally irreversible, with $\alpha_{ir} = 0.5$ µmol · m^{-2}, or $n_{ir} = 1.1 \cdot 10^{14}$ H atoms per square centimeter, which is almost 13 times lower than the corresponding value for oxygen. The adsorption of oxygen was measured

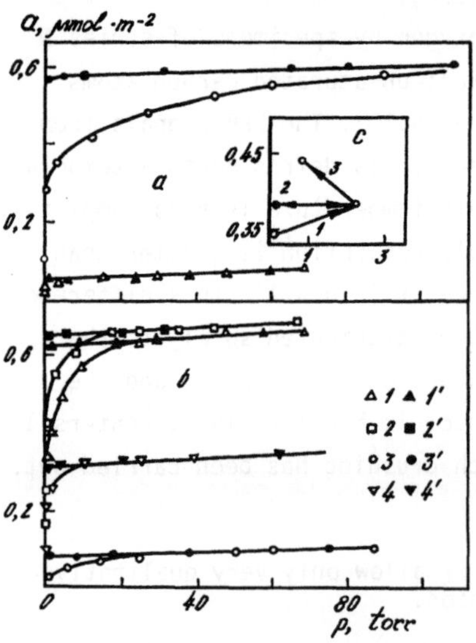

Fig. 4.3a-c. Adsorption of hydrogen (a) on an atomically clean surface of graphite of series K at 300 K (*1*) and 650 K (*3*) and oxygen (b) on hydrogenized specimens at 300 K (*1*) and 650 K (*3*) on the initial specimen (*2*) and on the specimen hydrogenized in glow discharge (*4*). (c) Initial part of the isotherm. (*1*'-*4*') Desorption

after prolonged treatment of the specimen in a hydrogen atmosphere and subsequent evacuation; as would be expected, it differed little from adsorption on an atomically clean surface (Fig.4.1). The sum $n_0 + n_{ir}$ coincides with n_0 for a freshly prepared surface. Most probably, adsorption of H_2 at 300 K proceeds primarily on centers of type 1, Model (3.19b) with the formation of groups

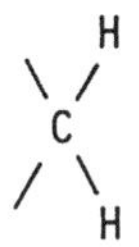
.

In low molecular aromatic hydrocarbons, the analogs of these compounds are perinaphthene ($C_{13}H_{10}$) and perinaphthenone ($C_{13}H_8O$). The concentration of these groups ($\sim 5 \cdot 10^{13}$ cm^{-2}) was close to the concentration of the carbonyl groups ($3 \cdot 10^{13}$ cm^{-2}) formed on centers of type 1, see Model (3.19b). As with oxygen (Fig.4.2), the adsorption of hydrogen on these centers does not change the conductivity of graphite.

The irreversible adsorption increased sharply when hydrogen atoms were adsorbed from hydrogen plasma created in glow discharge. In this case the value of n_0 reached a maximum of $1.3 \cdot 10^{15}$ cm^{-2}. The irreversible adsorption of oxygen on a specimen treated in glow discharge was 0.3 μmol $\cdot$ m^{-2} (or $n_0 = 0.6 \cdot 10^{15}$ cm^{-2}) (Fig.4.3). Thus the sum $n_{ir} + n_0$ turns out to be greater than n_0 in adsorption of oxygen on the initial specimen (Fig.4.1). Chemisorption of hydrogen atoms appears to lead to a certain increase in the number of adsorption centers on the prismatic faces.

As was noted in Sect.3.1.4, an EPR signal with $\Delta H = 11$ G and $N_S = 2 \cdot 10^{12}$ cm^{-2} was observed following adsorption of hydrogen. The parameters of that signal are characteristic of the interaction between an unpaired electron of a broken carbon bond, and a C-H bond. These paramagnetic centers may have emerged either as a result of interaction between hydrogen atoms and centers of type B (Fig. 3.14b), or due to the rupture of quasi-acetilene bonds [centers 1 and 2 of Model (3.19c)].

The adsorption of chlorine at 300 K on atomically clean surfaces of graphite specimens of series K (Fig.4.4) [4.14] proceeds similarly to the adsorption of oxygen (Fig.4.1). The magnitudes of irreversible adsorption are also close in both cases. Up to an occupation of about 10^{13} cm^{-2} the adsorption of chlorine does not change the conductivity of the powder. This suggests that Cl_2 interacts with centers of type B at this stage (Fig.3.14b), resulting in the formation of groups

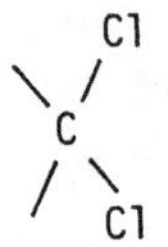

.

The adsorption of chlorine on specimens of series N causes the EPR signal from centers of type 1, Model (3.19a), to disappear completely which indicates that valence-saturated complexes of chlorine are formed.

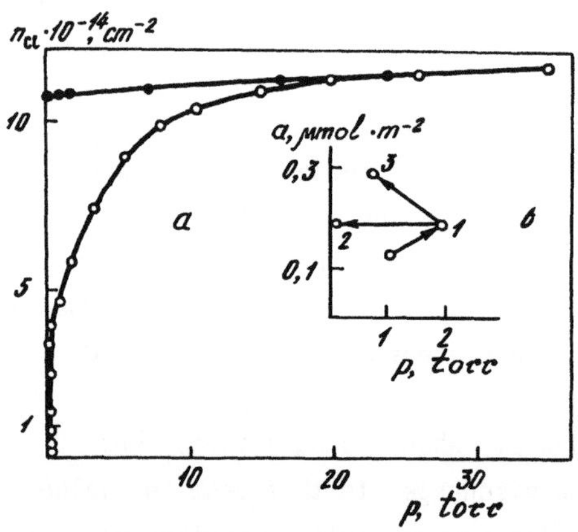

Fig. 4.4 (a) Adsorption of chlorine on an atomically clean surface of graphite of series K (reduced to the surface of the prismatic faces). (b) Initial part of the isotherm

An analysis of the shape of adsorption isotherms for oxygen and chlorine on atomically clean graphite surfaces reveals the nonelementary nature of this adsorption. Even at 300 K, adsorption of O_2 and Cl_2 is accompanied by a complex surface chemical reaction. Indeed, one might conclude from Figs.4.1 and 4.4 that because such adsorption is dependent upon pressure and involves an equilibrium gaseous phase, it ought to be at least partially reversible. However, experiment proves that at every degree of occupation the oxygen is irreversibly bound to the surface, and the desorption is quite low. The adsorption being completely irreversible, the isotherm ought to display a typical inverted-L shape. The dependence upon pressure cannot be explained by kinetic factors either. In some experiments the time of observation was increased to 5 or 6 days, but no increase in adsorption was noticed. The absence of nonadsorbable impurities in the oxygen or chlorine, which could be responsible for the gaseous phase over the adsorbent, was proved by mass spectrometry. One can only conclude that the gaseous phase is created in the act of adsorption itself, and inhibits the process of oxidation. In order to check this hypothesis, additional experiments were carried out in which the equilibrium gaseous phase was replaced by pure gases. This was always found to lead to an increase in adsorption. For instance, in adsorption of oxygen, periodic replacement of the gaseous phase by pure oxygen yielded a maximal value of $\alpha_{ir} = 0.65\ \mu mol \cdot m^{-2}$ at a pressure of 0.1 torr, and an isotherm having a nearly true inverted-L shape.

Similar observations were made with Cl_2 (see Fig.4.4b; the designations are the same as in Fig.4.3c). If, after a specimen has been kept for a prolonged period at a pressure of 2 torr (point 1), evacuation is conducted, the desorption is equal to zero (point 2). If the system is then filled with fresh chlorine at the same pressure (2 torr), further adsorption takes place, reducing the pressure to 0.7 torr (point 3).

Chemical adsorption of H_2 on graphite at 300 K actually proceeds at almost zero pressure (Fig.4.3c). The formation of gaseous hydrocarbons is hardly probable at 300 K. The isotherm has a typical inverted-L shape. An increase in the temperature to 650 K stimulates chemisorption of H_2 (Fig.4.3a, Curve3), but not increase in the concentration of the active centers is observed at these temperatures. As can be seen from Fig.4.3a, the isotherm declines from the inverted-L shape. A gaseous phase equilibrium with the adsorbent exists in a wide range of pressures. The gaseous products of the reaction between hydrogen and graphite inhibit the process of chemisorption of H_2. The replacement of this gaseous phase by pure hydrogen leads, as with oxygen and chlorine, to a sharp increase in adsorption (Fig.4.3c).

Mass spectroscopic analysis of the equilibrium gaseous phase during adsorption of O_2, H_2, and Cl_2 revealed small quantities of the reaction products CO and methane, and fragments of chlorine-carbon compounds respectively [4.2,13, 14]. It is interesting to note that in adsorption of O_2, the introduction of small quantities of CO into the gaseous phase sharply decreased the magnitude of chemisorption [4.9].

The products of oxidation are reported to hamper or completely stop oxidation in some cases. For instance, Semenov and Ryabinin (in [4.4]) discovered that SO_2 has an inhibiting effect in the oxidation of sulfur and a similar phenomenon was observed in the oxidation of phosphorus. Semenov attributed this to a breakage in the chain reaction. It is possible that in our case the role of inhibitors is played by the intermediate products of oxidation. The phenomenon of inhibition is not limited to oxidation of atomically clean surfaces; we observed similar effects during chemisorption of hydrogen and chlorine. Thus, when investigating chemisorption on atomically clean surfaces, one must sometimes take into account complications resulting from the inhibiting action of the products of the surface reaction.

Also of interest here is the grafting of high-molecular compounds to the freshly prepared surface of a solid. In this connection, adsorption and the

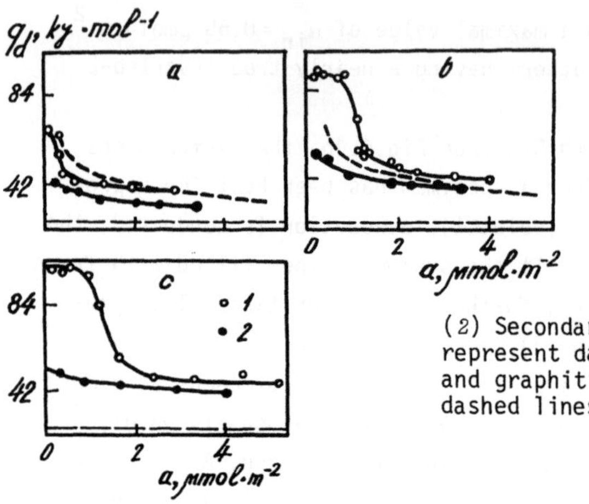

Fig. 4.5a-c. Differential heat of adsorption of butane (a) on atomically clean (*1*) and oxidized (*2*) graphite surfaces. (b) Butene-1. (c) isoprene. (b,c) - (*1*) Primary adsorption on an atomically clean surface. (*2*) Secondary adsorption. The dashed curves represent data from [4.16] for channel (a) and graphitized (b) black; the horizontal dashed lines mark the heat of condensation

differential heats of adsorption on atomically clean surfaces of graphite specimens of series K for butane, butene-1 (ethylethylene), and isoprene were studied in [4.2,15]. The adsorption of saturated hydrocarbon (butane) is practically reversible; after adsorptive measurements and subsequent evacuation the specimen was again active in the chemisorption of oxygen. The differential heats of adsorption for butane in the initial stage (Fig.4.5a) exceed the corresponding values for the oxidized carbon, and approach those for the channel black [4.16]. In both cases the initial values of q_d are much greater than those for the passive graphitized black. The probability of chemical binding between molecules of saturated hydrocarbon and centers of types 1 and 2, Models 3.19b and c, which would demand the rupture of σ bonds in the molecules, is quite low, and does not lead to any significant energy gain. One cannot exclude the possibility that unpaired π electrons of radicals on the surface of the black are partially drawn onto the orbitals of butane. Such a mechanism of reversible quasi-chemical interaction with such centers was discussed in [4.9] for water adsorption, and in [4.17] for adsorption of oxygen at decreased temperatures.

In contrast to the case of butane, the adsorption of unsaturated hydrocarbons is characterized by a marked degree of irreversibility. From Fig.4.5 one sees that the initial values of the adsorption heat are 92 kJ · mol^{-1} for butene-1 (Fig.4.5b) and 100 kJ · mol^{-1} for isoprene (Fig.4.5c), which is close to the heat of polymerization of unsaturated hydrocarbons. The region of high q_d values in Fig.4.5 approximately corresponds to the magnitude of irreversible adsorption. In the initial stage of adsorption the $\pi\sigma$ bond of the molecule is dissociated and a simple σ bond is established with the lateral carbon atoms of graphite. The molecule then retains one unsaturated π bond:

```
    H   |   H   H
C - C - C - C - C - H   .
    H   H   H   H
```

Indeed, after adsorption of butene-1, a signal was detected in the EPR spectra with a g factor close to the g factor of a free electron. The wide component of the signal, with a linewidth of about 63 G and an intensity $N_S = 10^{14}$ spin $\cdot$ cm^{-2}, which is close to the magnitude of irreversible adsorption of butene-1, is apparently due to interaction between the unpaired π electrons of the molecule and its C-H bonds. The low-intensity narrow component ($\Delta H = 4.2$ G) is most probably connected with the π electrons of graphite which are partially localized in the vicinity of the adsorptive complex [4.2].

We see that in contrast to oxygen, the interaction of unsaturated hydrocarbons with an atomically clean surface of graphite containing no spin centers follows the radical mechanism. It is interesting to mention here the results of an investigation of the NMR of polyethylene loaded with freshly ground graphite powder of series K as a filler [4.18]. It turned out that the emergence of spin centers in the adsorbed molecules leads to a retardation of the end groups of macromolecules which is stronger than in the case of polyethylene filled with oxidized graphite.

4.1.2 Adsorption on Semiconductor Surfaces

The most thoroughly investigated type of adsorption is that of oxygen, hydrogen, and some other simple atoms and molecules on atomically clean surfaces of germanium and silicon. Such investigations have employed the volume, weighting, and mass spectroscopic techniques of adsorption measurement, as well as ELEED, Auger spectroscopy, electron and ion emission spectroscopy, refractometry, and ellipsometry (see [4.19-26] and references therein).

The first investigations of the kinetics of oxidation of silicon and germanium surfaces were conducted in the 1960s, and established the existence of two stages in this process: a fast stage (formation of a monolayer) and a slow stage (growth of a three-dimensional oxide phase). The second stage obeys the logarithmic Zel'dovich-Roginsky law (2.67) [4.19]. The constants B and C differ with various researchers, apparently because of differences in the conditions of oxidation and preparation (cleaning) of the surface.

Under the clean conditions of an ultrahigh vacuum the fast stage, up to the formation of a monolayer, takes place at an exposure to oxygen of about 10^{-6} torr $\cdot$ s; the slow stage proceeds at exposures greater than 10^{-3} torr $\cdot$ s [4.27]. The superstructure then disappears, and the density of the surface states abrubtly changes [4.23-25]. Here we shall be chiefly concerned with the initial stage of chemisorption and the nature of the surface complexes which form.

The characteristic feature of chemisorption on atomically clean semiconductor surfaces is the relative smallness of the sticking coefficients k_S. Even for so active a gas as oxygen, the initial values of the sticking coefficient k_S turn out to be much lower than 1. For instance, according to [4.23, 28-30], the values of k_S are $10^{-3}-10^{-4}$ for Ge; $10^{-1}-10^{-3}$ for Si; 10^{-5} for GaSb and InSb; 10^{-5} for GaAs and InAs; $3\cdot 10^{-6}$ for CdS; $4\cdot 10^{-5}$ for GaP; and less than 10^{-10} for <1010> ZnO. MARK [4.30] pointed out that the sticking coefficients for ordered nonpolar surfaces of compound semiconductors are always lower than those for ordered surfaces of monatomic semiconductors; this is due to the different bond ionicity. Chemisorptive activity is considerably greater for disordered surfaces than for ordered ones. Adsorption alters the stability of vicinal surfaces, and at the same time the steps on these surfaces influence the kinetics of adsorption [4.26]. With an increase in the degree of occupation θ of the surface by adsorbed molecules, the value of k_S drops abruptly.

It should be noted that both the absolute value of k_S and the dependence $k_S(\theta)$ vary from researcher to researcher. There are several reasons for this. One is that in investigations of oxygen adsorption quite large errors may result from the presence of up to 20% of CO in the ultrahigh-vacuum apparatus, due to the interaction of oxygen with carbon contaminants on the hot cathode [4.31-33]. In ELEED, AES, and ELS studies, possible interaction between the primary electron beam and the adsorbed molecules which can result in partial desorption or dissociation of the molecules [4.33-36] and subsequent diffusion of atoms into the lattice [4.37], has often been neglected. Furthermore, the newly formed atoms of oxygen will react with the surface in a more active way, since the value of k_S for elemental oxygen is much greater than for O_2(e.g., $k_S = 10^{-1}$ for adsorption of O on Ge [4.38]). FRANTSUZOV [4.28] attributes the differences in the reported values of k_S to variations in the condition of the initial silicon surface. The purity of the surfaces studied and of the gaseous phase, which is in equilibrium with the adsorbent, is still a point of controversy. Only in isolated studies has the purity of the adsorbate gases recently been monitored by quadrupole mass spectrometer [4.39]. Investigations of the adsorption phase conducted with photoelectron methods employing hard enough ultraviolet (UPS) or X-ray (ESCA, sEXAFS) radiation should be viewed with caution; the resulting information on the adsorptive interactions will not always be correct, because of the effects of photodesorption and photodissociation of a number of adsorbed molecules.

Adsorption on atomically clean silicon and germanium surfaces has often been monitored by measuring the conductivity σ in the course of adsorption. In adsorption of O_2 on an atomically clean germanium surface the curves σ(p) or σ(t) display pronounced maxima. Some of these are presented in Fig.4.6 [4.40-42]. The shape of the curves is attributed either to the lowered mobility of holes with a simultaneous increase in concentration [4.43], or to variations in the form of chemisorption of O_2 at different stages in the occupation of the surface [4.44]. The maximum on the curve σ(p) is so typical of adsorption on atomically clean germanium surfaces that certain researchers have proposed employing this dependence as the criterion for purity of the surface [4.45,46]. However, it has been shown elsewhere [4.32] that this maximum may be due to the presence of CO in the vacuum apparatus.

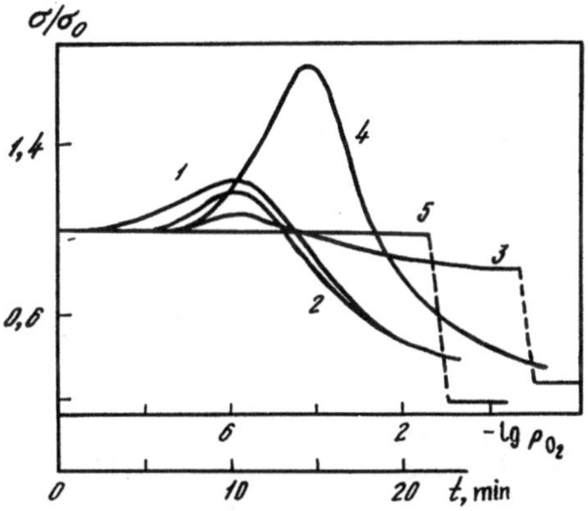

Fig. 4.6 Dependence of the relative conductivity σ/σ_0 of germanium specimens with atomically clean surfaces upon pressure (*1,4,5*) and exposure time at 10^{-2} torr (*2*) and 10^{-5} torr (*3*) according to [4.40] (*1*), [4.41] (*2*), [3.42] (*3*), and [4.49]. (*4*) With conventionally dried oxygen. (*5*) With superdry oxygen

As we pointed out in Sect.3.1.1, the purity of experiments with atomically clean surfaces can be described by the contamination coefficient ϰ, which is defined as the ratio of residual gas molecules in the measuring cell to the total surface area of the specimen under investigation. An analysis of the published data reveals a tendency of k_S to become smaller with a decrease in ϰ. For instance, the greatest values of k_S have been obtained in experiments with autoelectronic emission, where ϰ is rather large due to the minuteness of the surface area of the needlepoint [4.47]. An exceedingly low value of k_S ($\sim 10^{-5}$) was recently obtained for the Ge - O_2 system by the "flash" technique [4.48]. This value decreased to zero when the occupation reached about 1% of a monolayer.

DEMIDOVICH [4.49] studied the influence of oxygen adsorption on the conductivity of germanium powders with an extremely low value of ϰ: $\sim 6 \cdot 10^5$ molecules $\cdot$ cm^{-2}. This was done by using powders with a very high dispersion

($s \sim 5 \cdot 10^4$ $cm^2 \cdot g^{-1}$). An interesting new phenomenon was discovered in this work, namely that when "superdry"[2] oxygen was let into a container with freshly prepared germanium powder within a wide range of pressures ($10^{-8} - 10^{-2}$ torr), no change in conductivity was observed (Fig.4.6). At the same time oxygen dried in the conventional manner (by means of low-temperature desiccators with adsorbing agents) produced maxima on the $\sigma(p)$ or $\sigma(t)$ curves (Fig.4.6). These results led to the conclusion that "superdry" oxygen does not oxidate atomically clean germanium surfaces at all.

This important conclusion was later confirmed by direct adsorption measurements [4.50]. The injection of "superdry" oxygen at pressures of $10^{-4} - 10^{-1}$ torr into the cell did not result in any noticeable chemisorption (with an accuracy of $3 \cdot 10^{-2}$%) or produce any adsorption heat (within an accuracy of $5 \cdot 10^{-5}$ J). The injection of conventionally dried oxygen resulted in somewhat lowered chemisorption. In a region of occupation of about 5 percent, the differential heat of adsorption was $\sim 480 \pm 41$ kJ $\cdot$ mol^{-1}, which is in good agreement with earlier measurements of the values of the heat of adsorption of oxygen on vacuum-deposited germanium films (Fig.4.7) [4.51].

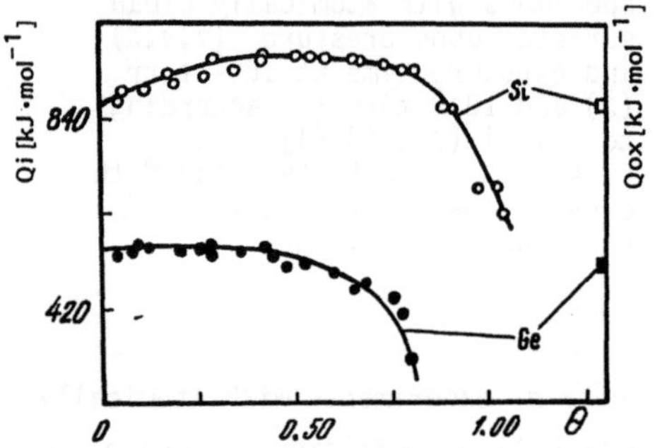

Fig. 4.7 Integral heat of adsorption of oxygen, Q_i, on atomically clean surfaces of germanium and silicon [4.51]. The small squares on the right mark the table values of heat of oxidation Q_{ox}

It follows that "superdry" oxygen is not chemisorbed on atomically clean surfaces of Ge, and that the sticking coefficient k_S is almost zero. Apparently, in order for the surface to be oxidated, the gaseous phase must contain some promoters, perhaps a microscopic quantity of water. This is a well-known feature of the corrosion of metals: even so active metal as potassium retains its glossy surface in an atmosphere of absolutely dry oxygen. However, in the presence of only traces of water, rapid oxidation and even self-ignition of the potassium occurs. The strong promotive action of a small admixture of

2 The oxygen used was produced from CuO calcinated with the whole vacuum installation for 40 hours at 700 K.

water in the oxygen may be an indication of the nonelementary mechanism (or perhaps chain mechanism [4.49]) of oxidation of germanium. It is possible that the adsorption of small amounts of water initiates reconstruction of the surface (see below), which accelerates the oxidation rate of germanium. These results, as well as the data obtained in [4.32], indicate that the purity of an adsorption system cannot be judged by the presence or absence of maxima on the $\sigma(p)$ curves.

In contrast to the case of germanium, experiments with silicon under the same conditions have shown [4.52] that the first injection of "superdry" oxygen at about 10^{-4} torr is accompanied by the release of a rather large amount of adsorption heat (875 ± 125 kJ $\cdot$ mol^{-1}, which is close to the data in [4.51]), see Fig.4.7. The reasons for such different behavior of Ge and Si in the course of oxidation merit further discussion. Quite dissimilar objects, namely powders [4.50,52] and films [4.51], produced close values of adsorption heat, which indicates that their oxidation mechanisms have a common nature. These values are close to the values for the heat of formation of bulk phases for these oxides. For comparison, the initial heat of chemisorption of O_2 on an atomically clean surface of PbS is $\sim 330 \pm 83$ kJ $\cdot$ mol^{-1} [4.52].

So far we have been discussing the results of investigations of oxygen adsorption at room temperature. With a rise in temperature the sticking coefficients for oxygen adsorption on atomically clean surfaces of Ge [4.28,48,53], GaAs [4.29], and Si [4.38,48] are greatly increased, which formally indicates an activization of adsorption. At increased temperatures an important role is played by the desorption of products of the reaction between oxygen and the surface. For example, in the case of Ge, mass spectroscopic analysis detected the release of GeO, beginning with temperatures of 700 or 800 K [4.28,48,54]. In addition, diffusion processes accompanied by the formation of new phases (GeO_2 and SiO_2) gain in importance at increased temperatures. Studies of the temperature dependence of sticking coefficients [4.28] have not taken into account diffusion of impurities from the bulk semiconductor towards the surface. At higher temperatures this may lead to misinterpretations of experimental data. In view of all of this, one can hardly ascribe any physical meaning to sticking coefficients and adsorption activation energies measured at increased temperatures in investigations of the kinetic of chemisorption on Ge and Si.

One very controversial question has to do with the nature of the surface complexes formed during oxygen adsorption. GREEN and LEE [4.55], on the basis of a qualitative thermochemical analysis of reactions between molecular oxygen and ideal polyradical germanium surfaces, concluded that it is possible that the following oxygen complexes form on the Ge<111> face:

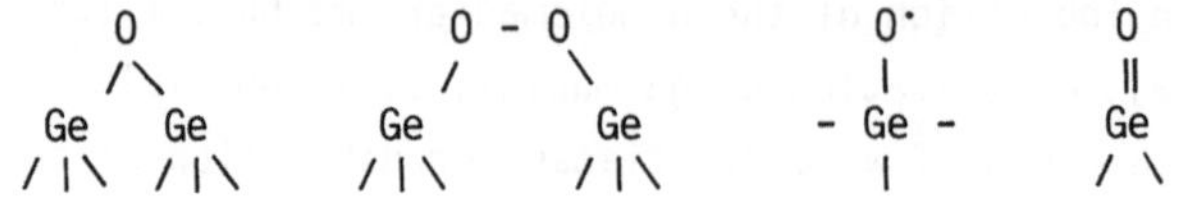

(a) (b) (c) (d) (4.1)

Complex (4.1d) is the most advantageous from an energetic viewpoint, but GREEN and LEE gave preference to the peroxide structure (4.1b), since double bonds are not typical of oxygen-containing germanium compounds.

In recent years, electron spectroscopy methods have been widely employed to investigate the nature of oxygen complexes (AES, LES, UPS, XPS; see footnotes in [4.56-58]). However, the results of such investigations are not conclusive. Some authors have supported a nondissociative (molecular) form of oxygen adsorption in the early stages of oxidation of Si ([4.33,59-61] and references cited therein). GARNER and coworkers [4.61] pointed out an interesting feature of the early stage of oxygen adsorption on Si. The exposure of a <111> (2×1) Si surface to oxygen at 10^{-8} torr caused peaks of the surface states to disappear from the electronic spectra; at the same time, the adsorption caused no change in the chemical shift of 2p orbitals of Si. The authors attributed this to the formation of an intermediate state of adsorbed oxygen. This assumption that the adsorption of oxygen has molecular form is hardly compatible with the high initial values of adsorption heat of O_2 on silicon [4.52]. Earlier [4.17] we had observed quasi-chemical bonding of molecular oxygen with atomically clean graphite surfaces at low temperatures. The adsorption heat ($165\ kJ \cdot mol^{-1}$) was much greater than the heat of Van der Waals adsorption ($29 - 33\ kJ \cdot mol^{-1}$), though much lower than the heat of irreversible chemical adsorption ($\sim 400\ kJ \cdot mol^{-1}$). At room temperature only irreversible chemical adsorption was observed.

Another group of researchers has advanced quite convincing arguments in favor of dissociative adsorption [4.56-58]. An analysis of data obtained by sEXAFS lead STÖHR et al. [4.57] to the conclusion that the formation of bridge structures of types (4.1a,b) and peroxide structures - Si - O - O˙ is most probable. VON WIENSKOWSKI and FROTZHEIM [4.62] also assume that peroxide structures are formed during oxidation of germanium.

None of the works cited above made use of EPR data, which could have provided valuable additional information on the nature of the groups formed during chemisorption. As was shown in Sect.3.1.4, the cleavage surface of germanium displays two types of chemisorptive centers: unoccupied Shockley states represented by free germanium valence bonds (Fig.3.14b, centers A), and occupied Shockley states (centers B in Fig.3.14b). States of the first type were present only on specimens dispersed at low temperatures (Ge_{77}). On specimens dispersed at room temperatures (Ge_{300}) the prevailing states were of type B. The adsorption of oxygen at 300 K on Ge_{300} specimens did not produce an EPR signal at either 77 K or 300 K.

We assumed [4.50,63] that as in the case of graphite [4.3], the initial doses of oxygen are adsorbed on the occupied Shockley states B on Ge, and structures of type (4.1d) are formed. The $\pi\sigma$ rehybridization in the surface structures of germanium is also considered possible in the Lander-Morrison surface model (Fig.3.10). The formation of a double bond is estimated to be more advantageous energetically than the formation of single σ bonds. The energy of formation of complexes of type (4.1d) [4.55] does not differ greatly from the experimentally observed initial adsorption heat (Fig.4.7). The EPR data do not confirm the formation of complexes of type (4.1c). Complexes of type (4.1d) may provide the basis for the formation of GeO molecules, which, as has already been mentioned, are sharply resolved in the mass spectra of thermal desorption from the oxidized germanium surface.

A similar mechanism of formation of double bonds apparently functions in the case of silicon. Mass analysis of SiO molecules is complicated because their mass coincides with that of CO_2 molecules ($M = 44$), which are always present in the vacuum system.

LUDEKO and KOMA [4.64], on the basis of an analysis of the absorption spectra of low-energy electrons convincingly demonstrated the possibility that $>$Ge = O complexes form during low-temperature oxidation of atomically clean germanium surfaces. This is in complete accord with the schemes constructed earlier on the basis of EPR data (Fig.3.14b) [4.50,63]. The $>$Si = O model of the center is also adopted in [4.58,65] on the basis of data obtained by AES, XPS, and ELEED. This configuration does not contradict the results of quantum-chemical calculations carried out in cluster approximation [4.66].

The most stunning results were obtained in studies of the EPR spectra of oxygen adsorbed on the surface of Ge_{77} specimens with a rather high concentration of dangling Ge$^{\cdot}$ bonds ($\sim 10^{13}$ spin $\cdot$ cm^{-2}). It was discovered that the adsorption of conventionally dried oxygen at 300 K does not alter the shape of the EPR line from dangling Ge bonds, nor does it change the concentration N_S of the spin centers. A similar result was also obtained for atomically clean Si surfaces [4.67,68].

CHUNG [4.69] noted small changes in the parameters of the EPR signal from dangling Si bonds during adsorption of O_2. He attributed this to the formation of O^- or O^{2-} ions. These changes, however, are too small to explain the high stability of the majority of Si$^{\cdot}$ radicals ($N_S \sim 10^{14}$ spin $\cdot$ cm^{-2}). From data obtained by electron spectroscopy the authors of [4.60] concluded that SiO_2^- groups form on the surface of silicon. STill unresolved is the question of the possible role played by the primary electron beam in the formation of these ions [4.34,35].

Thus, the EPR data [4.67,68] indicate that a large fraction of dangling Si and Ge bonds do not take part in the initial stage of chemisorption of oxygen. Prolonged exposure of Si and Ge specimens to oxygen led to the formation of the oxide phases SiO_2 and GeO_2, and the free radicals were buried under the oxide layer. The $Si^\cdot$ and $Ge^\cdot$ radicals could be made to disappear only by etching away the oxide over cover of the semiconductor [4.50]. The high stability of $Si^\cdot$ and $Ge^\cdot$ radicals observed in EPR experiments does not contradict the conclusion, based on an analysis of the overlapping integral of Si atomic orbitals in the two uppermost layers of the Si<111> plane, that a number of orbitals of the unpaired $Si^\cdot$ electron turn over towards the bulk of the solid [4.70] (see Sect.3.1.4). An alternative explanation is given by the Haneman surface model with microcracks (Sect.3.2.3).

In order to obtain additional information on this question we carried out EPR measurements on specimens of Si calcinated in a vacuum at 1000 K [4.71]. At these temperatures the majority of spin centers on the atomic surface of Si are annealed, and the value of n_S is $\sim 10^{-12}$ spin $\cdot$ cm^{-2}. The remaining paramagnetic centers interacted with acceptor molecules such as O_2, p-Bq, and tetracyanethylene. Simultaneously with the disappearance of the signal from Si at a very slow rate (of the order of hours) there appeared a signal from the radical forms of the adsorbed molecules. The parameters of the signal were the same for all three types of molecules. The characteristic feature of these signals was the presence of ultrafine structure (splitting) from Si^{29} nuclei, which is in agreement with the natural content of this isotope. This indicates that the major fraction of the electron density of the unpaired electrons of the adsorbed molecules is drawn to the Si atoms.

A most striking feature is the totally different behavior of Ge and Si cleavage surfaces towards "superdry" oxygen: complete passivity of Ge and normal chemisorption in the case of Si. The sticking probabilities for these two semiconductors are also quite different ($k_S \to 0$ for Ge; $k_S \sim 10^{-1} - 10^{-3}$ for Si [4.48]). One possible explanation proved, is the different activity of the hybrid orbitals of Si and Ge atoms. While for the bulk atoms of C and Si the s and p orbitals take about equal part in the chemical bonds, the 4s orbitals of Ge participate less actively, and 4s electrons of Ge are rather inert (the Sidgwiss effect) [4.66]. In principle, this peculiarity can affect the interaction between centers of type B and oxygen.

The promotive role of water molecules in the injected oxygen might be due to the formation of H_2O coordination bonds with Ge atoms, which stimulates the reorientation of dangling bonds. The reconstruction of the surface connected with chemisorption of O_2 on the occupied Shockley states alters the coordination surrounding of the paramagnetic centers (empty traps), which may explain the small changes in the EPR signal parameters observed during O_2 adsorption on silicon [4.68].

In contrast to oxygen, molecular hydrogen is not actually adsorbed on atomically clean surfaces of germanium and silicon [4.19,21,55]. The initial sticking coefficient of H_2 in a wide temperature range is lower than 10^{-8} for Ge and about $2 \cdot 10^{-3}$ for Si [4.21,72]. A drastic change is observed when it comes to the adsorption of atomic hydrogen. Atomic hydrogen is actively chemisorbed and forms a monolayer on atomically clean surfaces of Ge and Si [4.21,72-74]. The value of k_S for atomic hydrogen is $\sim 10^{-1}$ for Ge and ~ 1 for Si [4.21,48]. Depending on the plane of cleavage, thermodesorption from the silicon surface displays two or three peaks of hydrogen release, indicating that hydrogen forms with surface bonds of varying energy [4.23]. The results of UPS investigations [4.73] point to the existence of a peak in the adsorption of H atoms on Si; the corresponding energy is close to the ionization potential of the t_2 orbital in gaseous SiH_4. The estimated energy of the Si-H bond is also close to that of the hydride bond in an SiH_4 molecule [4.23].

The majority of researchers assume that the H atoms saturate the dangling bonds on the Si or Ge surface, forming hydride groups. It was from this point of view that IBACH and ROWE [4.73] interpreted the ELEED data, which indicated that silicon superstructures <111> Si(7×7) and <100> Si(2×1) are stable with regard to adsorption of atomic hydrogen. The lack of detailed information on the change in the EPR signal from dangling Si$^\cdot$ and Ge$^\cdot$ bonds during adsorption of atoms and molecules of hydrogen makes it impossible to draw final conclusions on the nature of hydride groups. LEMKE and HANEMAN [4.75] pointed out a noticeable change in the EPR signal from dangling Si bonds during adsorption of molecular H_2. It is not yet clear how this should be matched with the low sticking coefficients for H_2 [4.21,72]. The first experiments on the chemisorption of hydrogen on a Ge_{300} specimen whose surface only has occupied states of type B (Fig.3.14b) failed to yield any EPR signals in the course of absorption [4.50]. The hydrogen atoms are most likely adsorbed on states B forming groups

```
     H
 \  /
  Ge      .
 /  \
     H
```

This does not contradict the data obtained by emission spectroscopy [4.73,74]. It is worth noting here that IWASAKI and NAKAMURA [4.76] detected an increase in the EPR signal from dangling Si bonds during adsorption of O_2, H_2, and Cl_2. This result can be explained through unpairing of spins in states of type B, but not within the framework of the Haneman "cracks" model.

Of interest in this connection are data on the chemisorption of chlorine on atomically clean surfaces of silicon and germanium. MADIX and SCHWARZ [4.77] observed the release of $GeCl_2$ and $SiCl_2$ in the products of thermal desorption from Ge at 600 - 1000 K and Si at 800 - 1500 K. The basis of these molecules is most probably formed by the complexes

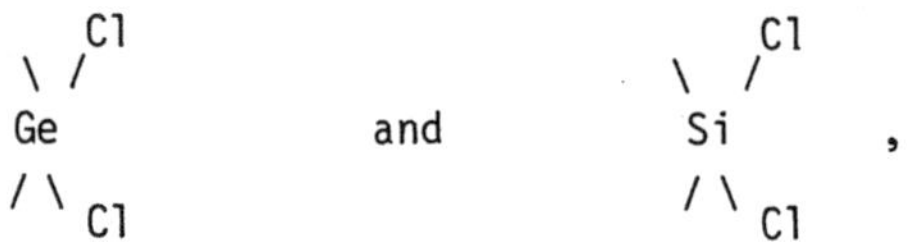

which are produced, as in the case of hydrogen, in the interaction between chlorine and states of type B (Fig.3.14b). Investigations using the UPSAR technique [4.78] reveal that chlorine forms simple covalent complexes on the surface of Si. The positions of the resonance peaks in the spectra are in excellent agreement with theoretical predictions (see references in [4.78]). Experiments on simultaneous adsorption of H_2 and O_2 can be very useful in studying the mechanism of interactions between molecules of these gases and atomically clean surfaces. Thermal desorption methods were employed in [4.21, 72] to detect the release of water molecules during adsorption of hydrogen on a germanium surface containing chemisorbed oxygen. A similar effect was observed when germanium monoxide instead of oxygen was adsorbed on the germanium surface. NISHIJIMA and MUROTAMA [4.79] also noted that successive adsorption of oxygen and hydrogen on an atomically clean silicon surface was accompanied by an increase in the partial pressure of water vapor in the vacuum cell. The peak of adsorbed hydrogen was then absent from the Auger-electron spectra. It is possible that oxygen atoms interacting with occupied Shockley states of type B (Fig.3.14b) were responsible for the surface reaction between hydrogen and oxygen.

There is a large volume of data on the adsorption of triatomic molecules (HCl, HBr, H_2S, H_2Se, NH_3, PH_3) on atomically clean surfaces of germanium and silicon [4.19,80-82]. Measurements of the kinetics of chemisorption have provided the basis for conclusions about the number of surface atoms per adsorbed molecule, or the order of the reaction [4.19]. The ELEED data [4.80-82] indicate that there is significant reconstruction of the surface during adsorption, and in some cases quite stable superstructures are formed.

Detailed investigations of the chemisorption of water on atomically clean germanium surfaces were carried out by HENZLER and TÖPLER [4.83] with the aid of ELEED and AES. Adsorption was shown to lead to significant reconstruction of the surface superstructure even in its earliest stage. There were two adsorption sites to each adsorbed molecule; HENZLER and TÖPLER [4.83] took this as an indication of the dissociative nature of adsorption. However, more recent investigations with ELS and UPS have shown that the adsorption of H_2O, NH_3, and H_2S on the <111> Si surface proceeds not in dissociative, but in molecular form [4.84,85].

Of particular interest are investigations of chemisorption of O_2 and H_2O on atomically clean Ge and Si surfaces, since oxidation of these substances with a minimum possible distortion of chemical bonds (and consequently with a minimum density of the surface states on the Si - SiO_2 boundary) is vital for the creation of the basic microelectronic elements—MOS (metal-oxide-semiconductor) structures. Numerous investigations of this interface with the aid of ESCA, AES, sEXAFS, SIMS, etc., have revealed a marked deviation from stoichiometry with respect to oxygen in the oxide layer adjacent to Si. This agrees well with the results of the above-discussed UPS, ESCA, and ELS investigations of atomically clean silicon surfaces, which also indicate that a growing surface layer of silicon oxide differs from the regular network of SiO_4 tetrahedrons. Such questions, however, are beyond the scope of this book, and the reader is referred to specialized reviews [4.86,87].

Although there are now several hundred works dealing with adsorption on atomically clean semiconductor surfaces, our knowledge of the elementary act of desorption is still quite limited. The existing data on graphite and on the Ge - O_2 system attest to the important role of impurities in the gaseous phase as inhibitors or promoters of the surface chemical reaction. The main experimental efforts have so far been aimed at studying structural changes in the surface with the aid of ELEED, and at determining the number of adsorbed molecules by Auger and mass spectroscopic techniques. In order to understand the adsorption mechanism, information obtained with these techniques must be supplemented with data on surface complexes provided by optical and radio spectroscopic methods.

4.2 Adsorption of Certain Simple Molecules on Hydrated Oxide Surfaces

4.2.1 General Aspects of Adsorption

The dehydration of an oxide surface has long been observed to reduce its adsorptive activity with regard to the vapors of water, alcohol, and benzene. At the same time there is little change in the low-temperature adsorption of inert gases and nitrogen or in the adsorption of saturated hydrocarbons. The adsorption of water is most sensitive to dehydration of the surface. In [4.88] it was demonstrated that the calcination of silica gel leads not only to a lowering of the adsorption isotherms of water, but also to changes in their shape. At $T_{cal} > 800^{\circ}$ the usually convex isotherm shows a concavity characteristic of adsorption on hydrophobic surfaces. Thus, we must first use the data on the hydrate cover of oxides presented in the previous chapter to consider the quantitative relations between the adsorptive properties of the surface and the extent of its hydration, α_{H_2O}. These relations were most thoroughly investigated for SiO_2 specimens of varying dispersity [4.89]. Figure 4.8 presents the initial sections of adsorption isotherms for adsorption of water vapor on three specimens of silica gel before and after thermal dehydration. Information on the extent of hydration of the surface is given in Fig.3.18 and

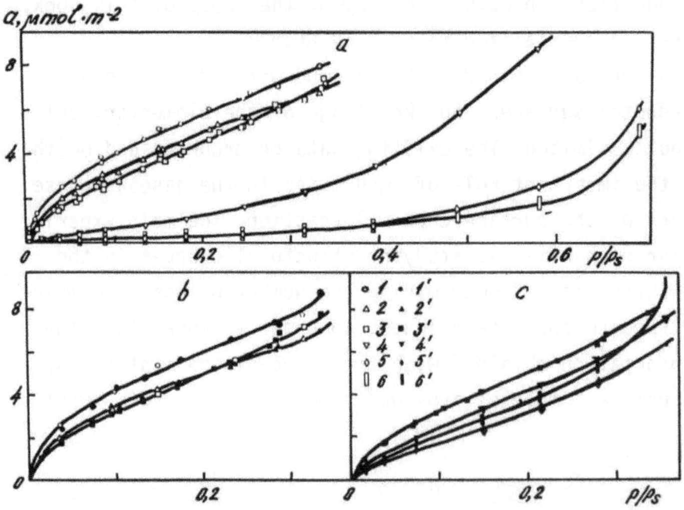

Fig. 4.8a-c. The initial sections of isotherms of primary adsorption (a, *1-6*); desorption (b, *1'-3'* and c, *3'-6'*); and secondary adsorption (b, *1-3*) of water vapor at 300 K on silica gel specimens KSK-1, heat treated at 600 K (*1,1'*); KSK-2, $T_{cal} = 600$ K (*2,2'*); K-2, $T_{cal} = 600$ K (*3,3'*); K-2, $T_{cal} = 800$ K (*4,4'*); K-2, $T_{cal} = 1000$ K (*5,5'*); K-2, $T_{cal} = 1200$ K (*6,6'*)

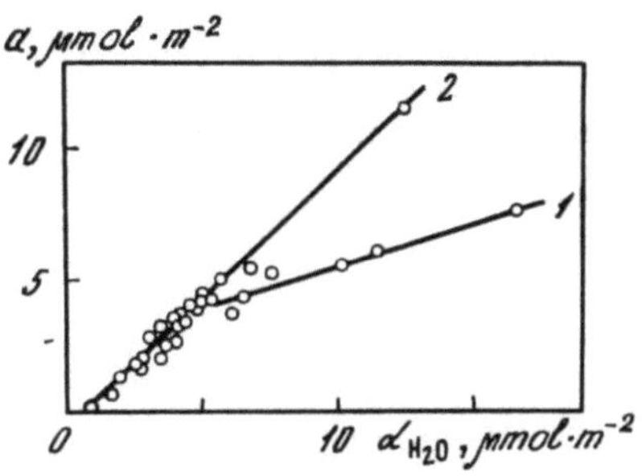

Fig. 4.9 Dependence of the specific adsorption of water vapor at $p/p_S = 0.1$ upon the extent of hydration of SiO_2 specimens vacuumized at 300 K (*1*) and 500 K (*2*)

Table 3.4. The specimens were initially evacuated at 600 K. The figure shows that the isotherms for the initial specimens follow the order of increasing α_{H_2O}. The isotherms for strongly dehydrated specimens are concave. After rehydration of all the specimens in saturated water vapor for 48 hours, the isotherms of adsorption were measured again; they are presented in Fig.4.8b for the initial specimens and in Fig.4.8c for the dehydrated specimens. For convenience of comparison all the isotherms have been reduced to a common zero point; that is, the value of irreversible adsorption, α_{ir}, has been subtracted from each of them. The values of α_{ir} are small for the initial specimens evacuated at 600 K (Table 3.4). The secondary adsorption isotherm coincides with the curve of desorption. If the initial specimens are evacuated at 550 K after desorption, the secondary isotherm coincides with the primary one. It follows that calcination at 550 K does not result in any significant dehydration of the surface. Figure 4.9 shows the dependence of the magnitude of adsorption of water vapor upon the extent of hydration of the surface at a certain fixed point $p/p_S = 0.1$ (close to the point of inflection of the isotherm). The values of adsorption a for all of the silica gels studied increase linearly with an increase in α_{H_2O} in the range of 1 to 5 or 6 $\mu mol \cdot m^{-2}$. A similar relationship between a and α_{H_2O} was reported in [4.90-92].

Similar dependences of the adsorptivity of SiO_2 on the extent of hydration of the surface and on the degree of dispersity were also obtained later for roughly dispersed disaggregated powders of quartz and quartz glass [4.93,94]. The high sensitivity of the piezoquartz balance[3] ($\sim 10^{-10}$g) made it possible to measure the adsorption of water vapor on two α-quartz monocrystals having different surface roughness coefficients (34 and 50) as determined via low-temperature adsorption of argon [4.95]. The adsorption data are presented in Fig.4.10, and the corresponding values for the extent of hydration are lis-

[3] This instrument is based on measurement of the eigenfrequency shift of an AT-cut quartz crystal resulting from adsorption. It was developed in.[4.96,97].

Table 4.1. Adsorption characteristics of the surface of certain oxides. All specimens (Qu—disaggregated quartz powders, KSK—macroporous silica gel, K-2—microporous silica gel) were vacuumized at 300 K (a_m is the capacity of the monolayer, ω is the "landing site" of the molecule in the monolayer, and C is the Brunauer constant)

Specimen	α_{H_2O} [$\mu mol \cdot m^{-2}$]	s [$m^2 \cdot g^{-1}$]	a_m [$\mu mol \cdot m^{-2}$]	ω [nm^2]	C	Reference
α-quartz (monocrystal)	--	--	12.7	0.13	-	[4.95]
Qu-6	16.7	2	10.0	0.17	25	[4.93,94]
Qu-5	11.5	32	9.0	0.18	13	[4.93,94]
Qu-3	6.5	51	6.9	0.24	12	[4.93,94]
KSK-1	5.1	275	6.3	0.27	14	[4.93,94]
KSK-2	4.4	400	5.2	0.32	13	[4.93,94]
K-2	3.5	695	5.1	0.33	10	[4.93,94]
TiO_2	12.0	83	7.9	0.21	15	[4.98]
ZnO	9.6	29	5.0	0.33	19	[4.99]
GeO_2	21.0	55	5.0	0.33	14	[4.100]

ted in Table 4.1. The characteristic feature of the quartz powders is the high value of irreversible adsorption on specimens evacuated at $T_{cal} = 500$ K (see Table 3.4). An extremely high value was observed with an α-quartz monocrystal.

As was already mentioned in Sect.3.2.1, the surface of SiO_2 specimens evacuated at 300 K has, in addition to hydroxyl groups, a small number of water molecules that are tightly coordination-bound to the surface. Evacuation at

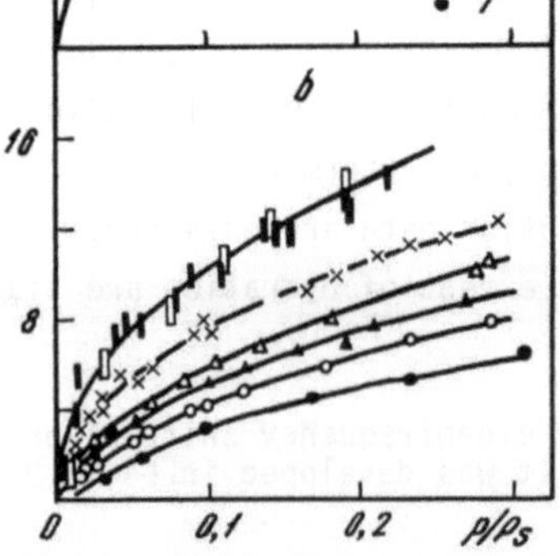

Fig. 4.10a,b. The specific adsorption isotherms for argon (a) at 77 K on silica gels (solid curve) and quartz monocrystals with surface areas of 102 cm² (1) and 150 cm² (2), and water vapor (b) at 300 K on quartz monocrystals with surface areas of 102 cm² (1) and 150 cm² (2) and on disaggregated quartz powders with a specific surfaces s equal to 2.4 $m^2 \cdot g^{-1}$ (3), 31.7 $m^2 \cdot g^{-1}$ (4), 41.1 $m^2 \cdot g^{-1}$ (5), 51 $m^2 \cdot g^{-1}$ (6) and on silica gel with s = 695 $m^2 \cdot g^{-1}$ (7)

500 K removes most of this water, resulting in an increase in the adsorptive activity. If, in accordance with what was said in Sect.3.2.2, we choose $T_{cal} = 500$ K as the lower temperature limit for the removal of water in molecular form, then the magnitude of irreversible adsorption on a surface hydrated by OH groups can be taken as a qualitative characteristic of the concentration of coordination-bound water, α_c. According to spectroscopic data, the total water content on the surface of a specimen at $T_{cal} < 500$ K is

$$\alpha_{H_2O} = \alpha_H + \alpha_c + \tfrac{1}{2}\alpha_{OH} \quad ,$$

where α_H is the amount of water loosely bound to the surface by hydrogen bonds, and α_{OH} is the amount of water released in dehydroxylation of the surface. At $T_{cal} > 500 - 600$ K, α_{H_2O} is $\sim \tfrac{1}{2}\alpha_{OH}$.

It was discovered in [4.93-95] that the isotherms of specific reversible adsorption of water on SiO_2 specimens subjected to exactly the same treatment showed a dependence upon the magnitude of the specific surface area. This is particularly clear in Fig.4.10: the adsorptive activity per unit of surface area of SiO_2 increases with a decrease in dispersity. The highest isotherm is that of water adsorption on an α-quartz monocrystal. Since the adsorptivity depends in turn upon the degree of hydration, α_{H_2O} (Fig.4.9), the data presented in Fig.4.10 give independent proof of the correlation between α_{H_2O} and s demonstrated earlier in Fig.3.27. Measurements of α_H (Table 3.4) also confirm this.

Quantitative relationships between adsorptivity with regard to water vapor and the degree of hydration of the surface were obtained for the oxides of Ge [4.100], Ti [4.98,101], Zn [4.99,102], Al [4.103], and for aluminosilicates [4.104].[4] These investigations are reviewed in detail in [4.106]. The characteristic feature of all of these oxides is a strong dependence of the reversible adsorption of water upon the concentration of hydroxyl groups on the surface. Similar but less pronounced relationships were observed for reversible adsorption of alcohols [4.107,108], benzene [4.109], and ammonia [4.110,111]. On these grounds certain researchers concluded that the hydroxyl groups on the surface are the main centers for the adsorption of molecules capable of forming hydrogen bonds with these centers. It will be shown below that this is not the case.

---

[4] In the 1950s and 1960s a large number of works were published which dealt with the adsorptive properties of powders of ground quartz, silicon, germanium, and other substances. They suggested values of adsorption which were abnormally high and did not agree with one another. This resulted, as was shown in [4.105,106], from neglect of the effect of particle aggregation in the course of crushing (Sect.3.1.1).

The concept of "absolute" adsorptive properties of a solid surface, which are treated rather like physicochemical constants has gained wide acceptance in works on adsorption. For instance, in [4.112,113] it is stated that the properties of the surface (per unit of area) of maximally hydrated specimens of SiO_2 (quartz, porous glass, aerosil, silica gel) are the same, regardless of their specific surface area and their origin, and can be described by "absolute" adsorption isotherms which are common to all specimens, as well as by common values for the heat of adsorption, the heat of wetting, and the concentration of hydroxyl groups on the surface. It is assumed that the different packing of silicon-oxygen tetrahedrons in amorphous particles of silica gel and in quartz crystals does not affect the adsorptive properties [4.114]. ZHDANOV [4.113] proposed an "absolute" water adsorption isotherm common to all specimens of SiO_2, regardless of their dispersity, which is supposed to describe the adsorption of water on any SiO_2 specimen. This isotherm obeys the BET equation (3.22) well, with $a_m = 6.4\ \mu\mathrm{mol} \cdot \mathrm{m}^{-2}$ and $C = 16$, which are considered to be physicochemical constants of general significance. On the basis of the "absolute" isotherm, the authors of [4.113,114] proposed a method of determining the specific surface area of SiO_2 specimens from a a single point on the isotherm.

All these concepts find little support in the experimental data discussed above. Investigations of the extent of hydration and the adsorptive properties of SiO_2 and GeO_2 surfaces per unit of area conclusively show that even if all specimens, including maximally hydrated ones, are identically treated, these properties depend on the size of the oxide particles. The constants C in the BET equation will undergo the corresponding changes. It is clear from Table 4.1 that the capacity of the monolayer varies greatly even with one and the same oxide, and depends upon α_{H_2O} and thus upon the dispersity of the specimen.

One reason for such good conformity among "absolute" adsorption isotherms is [4.105,106] that the values employed for specific surface s are calculated on the basis of the same isotherms for which the "absolute" quantities are later obtained. If one takes into account that adsorption of water takes place on discrete centers, then it is clear that s actually characterizes not the whole surface, but the number of active centers on the surface. If the absolute characteristics of adsorption are calculated using the value of s determined from the same isotherm, then

$$a_{abs}(\mu\mathrm{mol} \cdot \mathrm{m}^{-2}) \frac{a(\mu\mathrm{mol} \cdot \mathrm{g}^{-1})}{s(\mathrm{m}^2 \cdot \mathrm{g}^{-1})} = \frac{a}{a_M^{H_2O}\, \omega_{H_2O} N_A} = \theta \frac{1}{\omega_{H_2O} N_A} \quad , \tag{4.2}$$

where a_m is the capacity of the monolayer from the BET equation; ω_{H_2O} is the "landing site" of the H_2O molecule; N_A is the Avogadro number; and θ is the extent of occupation of the surface. Since ω_{H_2O} is taken in [4.112 - 114] to be constant and equal to 25 Å^2, a_{abs} depends only on the extent of occupation, θ. However, according to the BET equation (3.22), θ depends only weakly on C. As shown by HALA's and SCHAY [4.115], if the mean value of adsorption heat which enters the expression for C varies within 10%, the isotherms will remain affine and will coincide if reduced to the number of active centers. The conformity of "absolute" isotherms, according to [4.113], implies equal values of α_{H_2O} for different specimens which is equivalent to assuming that ω_{H_2O} has the same values for different specimens.

It follows that the sought-for conclusion, drawn from the fact of coincidence of "absolute" isotherms, inevitably follows directly from the initial assumptions. At the same time the experimental results, as shown in Sect. 3.2.4, do not confirm the supposition that α_{H_2O} is invariable. An example is given in Fig.4.11a, which shows "absolute" isotherms calculated from isotherms of H_2O for different oxides using values of s derived from the same isotherms. The isotherms almost coincide, and lie within 0.5 μmol • m^{-2} of one another. The isotherm for Qu-6, which has a very high constant C, lies somewhat higher. Calculating in this manner can lead to the conclusion that the hydrate covers of oxides of Si, Ti, and Zn are identical, which contradicts all the experimental facts. The adsorption isotherms of water, calculated with the values for the specific surface area obtained from nitrogen adsorption isotherms, differ significantly from one another (Fig.4.11b).

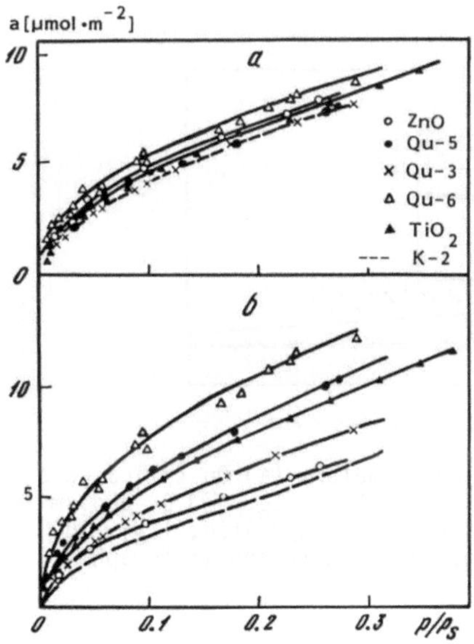

Fig. 4.11a,b. Specific adsorption isotherms of water vapor, calculated from the values of specific surfaces s_{H_2O} (a) and s_{N_2} (b) on various oxides

The surface area occupied by a molecule in the statistical Brunauer monolayer depends on the nature of the adsorbent and the degree of hydration. Its value for water molecules varies from 13 to 33 Å^2 (Table 4.1). In adsorption of more complex organic molecules, the value of ω differs significantly from the values obtained when the Van der Waals dimensions of the molecule are projected onto a plane, as was done, for example, in [4.116]. The orientation of molecules depends greatly on the extent of surface hydration [4.107,108, 117].

4.2.2 The Nature of Adsorption Centers and Mechanisms of Adsorption

Let us consider first the energetics of the adsorption process. The main energetic characteristic of the interaction between the surface and the gaseous phase is the differential heat of adsorption. Figure 4.12 presents the values measured in [4.118,119] for the differential heat of adsorption of water on silica gel specimens for which the extent of surface hydration was discussed in Sect.3.2.1 (Table 3.4). On the basis of the data presented in Figs.4.8,12, the values of differential adsorption entropy were calculated according to (2.21) (Fig.4.13). The state of liquid in equilibrium with its vapor was taken as the standard state. It is evident from Fig.4.12 that the values of adsorption heat q_d decrease with a decrease in the extent of hydration of the specimen surface, α_{H_2O}. The most pronounced dependence of q_d on α_{H_2O} is observed in the region of low occupation values. The connection discovered in [4.118] between q_d and the extent of surface hydration was confirmed in [4.121,122].

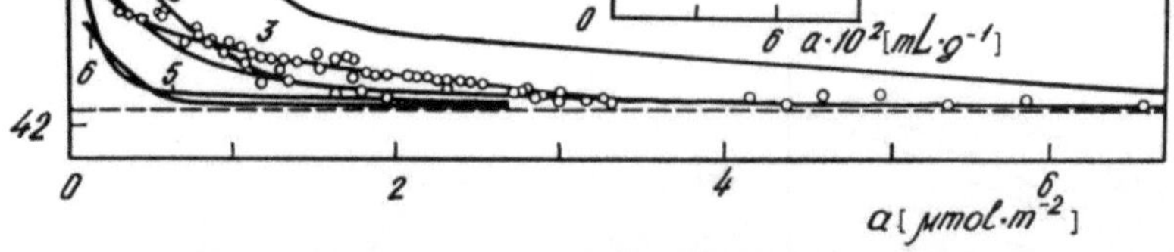

Fig. 4.12a,b. Differential heat of adsorption of water. (a) On silica gels (*1*) KSK-3, heat treated at 600 K; (*2*) K-2, T_{cal} = 600 K, α_{H_2O} = 3.01 μmol·m^{-2}; (*3*) K-2, T_{cal} = 600 K, α_{H_2O} = 2.8 μmol · m^{-2}; (*4*) K-2, rehydrated, T_{cal} = 1000 K; (*5*) K-2, T_{cal} = 1000 K; (*6*) oxidized graphite. (b) On hydrophobic "graphon" black [4.120]

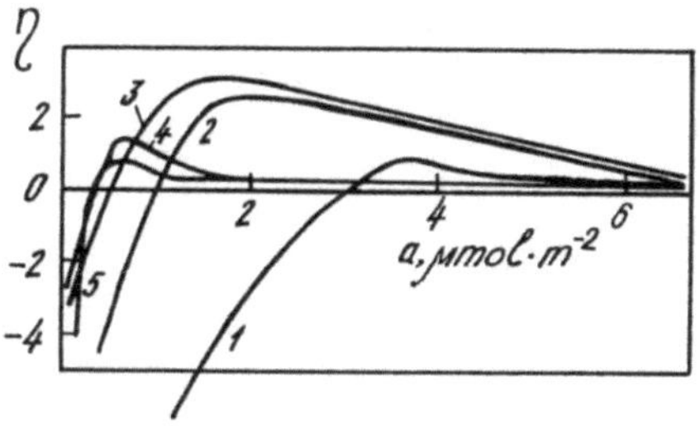

Fig. 4.13 Changes in entropy η (in arbitrary units) upon adsorption of water vapor on silica gels. (*1*) KSK-3, $T_{cal} = 600$ K; (*2*) K-2, $T_{cal} = 600$ K; (*3*) K-2, rehydrated, $T_{cal} = 1000$ K; (*4*) K-2, $T_{cal} = 1000$ K; (*5*) graphite

It can be seen from Fig.4.12 that the initial values of the adsorption heat are rather large (70 - 85 $kJ \cdot mol^{-1}$) indicating the existence of very active adsorption centers. As adsorption proceeds, the value of adsorption heat decreases and finally comes close to the heat of condensation L. The initial value of entropy $\Delta\eta$ in the low-occupation region ($a < 1$ $\mu mol \cdot m^{-2}$) corresponding to high values of adsorption heat is negative, which indicates a strong retardation of translational and rotational degrees of freedom of the adsorbed molecules. With increasing adsorption, the contribution of configurational entropy η_{conf} grows, and the entropy value becomes positive. Figure 4.13 shows that the configurational entropy is higher for more highly dehydrated specimens. According to investigations carried out by Zel'dovich and by Halsey and Taylor (Sect.2.1.4), if the potential barrier due to nonuniformity of the surface exceeds kT, then the configurational entropy η_{conf} for a nonuniform surface is lower than that for a uniform one, provided q_d follows exponential distribution over adsorption centers.

It must be noted, however, that this treatment of the changes in the heat and entropy of adsorption is rather arbitrary. Indeed, according to what was said in Sect.2.1.5, the experimentally observed changes in internal energy and entropy are equally related to the adsorbed substance and the solid, a fact which is often overlooked. For this reason, we do not consider the experimental values of q_d and $\Delta\eta$ to be exact characteristics of the adsorbed molecules, but view them rather as a kind of estimate.

Measurements of the differential heat of adsorption of water on other hydrated oxides also yield high initial values of adsorption heat: $q_0 = 70 - 85$ $kJ \cdot mol^{-1}$ (see Fig.4.14). Especially high values of adsorption heat were obtained with γ-Al_2O_3 dehydrated at $T_{cal} = 900$ K($q_0 > 170$ $kJ \cdot mol^{-1}$) [4.123-125]. Some researchers (e.g., [4.102,126]) attributed the high initial values of adsorption heat to the fact that the reaction proceeds as follows:

```
        O                 OH        OH
  \   / \   /             |         |
   Me     Me  + H2O → - Me - O - Me + q_hydr   .
  /   \ /   \             |         |
        O
```

Scheme 4.1

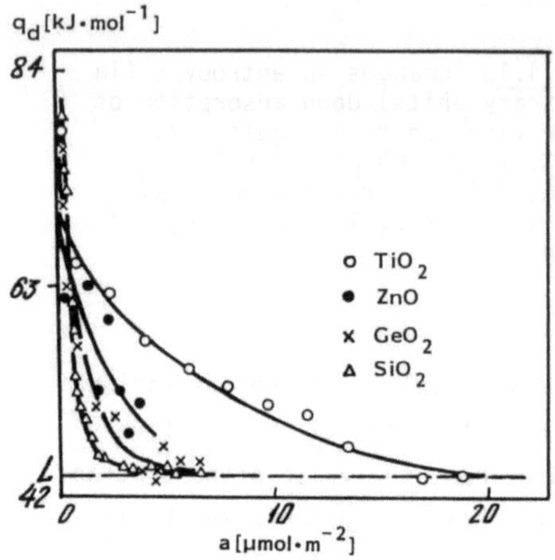

Fig. 4.14 Differential heat of adsorption of water on hydrated surfaces of various oxides

However, in the case of SiO_2 this does not accord with the dependence $q_d(\alpha_{H_2O})$. It follows from Fig.4.12 that as dehydration proceeds, the region of occupations corresponding to high values of q_d decreases, rather than increasing as ought to be expected according to the above scheme. Furthermore, simple thermochemical calculation yields a quite small value of hydration heat q_{hydr}. And indeed, direct calorimetric measurements [4.127] gave the value $q_{hydr} = 16.8\,kJ \cdot mol^{-1}$ for SiO_2. This value describes the overstrained character of the bonds in the oxygen bridge; this tension results from the repulsion between adjacent Si atoms. Studying the kinetics of hydration of SiO_2 reveals that this process occurs at a measurable rate only in saturated vapor or liquid water. At low vapor pressures the time constant of the reaction is many orders of magnitude greater than the time required to establish adsorption equilibrium in the calorimetric experiments.

An assumption widely accepted in the literature is that the formation of hydrogen bonds between the adsorbed molecules and the OH groups of the surface explains all the adsorptive properties of maximally hydrated oxide surfaces. This concept as applied to silica has been consistently developed by A.V. KISELEV since the 1950s [4.112,113,128,129]. The following models of adsorption are suggested in his works:

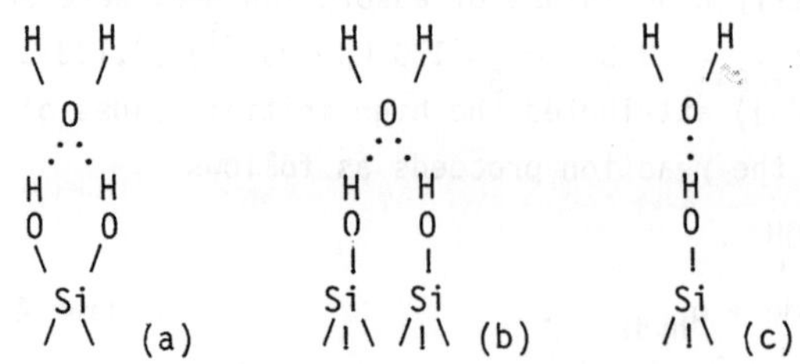

(4.3)

In the earlier works [4.112,113,128], Schemes (4.3a,b) were considered more probable, while in [4.129] preference is given to adsorption on isolated groups (4.3c). In the initial stage of occupation these schemes contradict the above-discussed calorimetric data. Indeed, were adsorption in the initial stage to proceed according to (4.3c), the value of q_d ($\sim 23-25\ kJ\cdot mol^{-1}$) would be lower than the condensation heat, a case which is typical of adsorption on hydrophobic surfaces (Fig.4.12b). This, however, contradicts all the experimental data on adsorption. In order to explain the high initial values of heat q_d ($\sim 70-85\ kJ\cdot mol^{-1}$), one must assume that complexes with 3 or 4 hydrogen bonds exist on the surface, which would be possible only if the surface were covered with narrow pores about 3 or 4 Å in diameter; there is no such experimental proof in the case of nonporous or macroporous silica specimens.

Let us now turn to the spectroscopic data. SIDOROV [4.130,131] was the first to question the independent role of OH groups as water adsorption centers in the low-occupation range. He proved that initial injections of water do not cause any noticeable changes in the intensity of the adsorption band of valence oscillations of isolated OH groups ($\nu'_{OH} = 3750\ cm^{-1}$). The adsorption here is accompanied by the appearance of the band $\nu_k = 3665\ cm^{-1}$, which is attributed to water molecules adsorbed on the more active centers termed centers the second kind by Sidorov. He also noticed that up to 70% of the OH groups remain on the surface of porous glass after heat treatment at temperatures up to 900 K. At the same time the adsorptive of the surface abruptly drops. As can be seen from Fig.4.8, the isotherms for specimens heat treated at 900-1000 K lose their convexity and become concave. Consequently, it is not the OH groups that are water adsorption centers in this range of occupations; rather the adsorbed water is held by quite strong bonds, and can be removed only by evacuation at 500 K (Sect.3.2.2).

FOLMAN and YATES [4.132] also demonstrated that the first injections of water have no influence on the intensity of the ν'_{OH} band. Furthermore, the adsorption of water was found to lead not to contraction, but to expansion of the porous glass specimen (Sect.2.1.5). This fact is taken by Folman and Yates to be proof that adsorption does not take place on the free hydroxyl groups.

Direct confirmation of the presence on the SiO_2 surface of adsorption centers which are more active than the OH groups was derived from an analysis of IR spectra in the range of deformation vibrations of water molecules, conducted with very pure silica gel specimens [4.133] and aerosil (Fig.3.19). The characteristic change in the spectrum during adsorption of small doses of water is the simultaneous appearance of bands $\nu_d = 1645\ cm^{-1}$ and $\nu_c = 3580\ cm^{-1}$

[4.133]. As Fig.3.19 shows, the intensity of the band of free hydroxyl groups ν'_{OH} remains much the same. With an increase in adsorption up to $\theta \sim 0.5$, the intensity of bands ν_d and ν_c grows, while that of the band ν'_{OH} changes but little. These facts support the view that nondissociative adsorption of water takes place on centers more active than the OH groups. A rise in pressure to $p/p_S = 0.6$ leads to the appearance of a wide band, $\nu_H = 3450\ cm^{-1}$, caused by the formation of H bonds between water molecules and the surface. At these occupations the heat of adsorption is $\sim 46-50\ kJ \cdot mol^{-1}$ (Fig.4.12), which is close to the energy of two H bonds. Similar behavior of the IR spectrum was observed in adsorption of water on an aluminosilicate (with 15% Al_2O_3) [4.133].

In order to gain additional information on the mechanism of adsorption of water, KVLIVIDZE et al. [4.134-136] investigated the NMR spectra. As can be seen from Fig.4.15, the NMR signal linewidths from adsorbed water molecules at equal occupations are not the same for different specimens. This is due to the presence of paramagnetic contaminants which reduce the relaxation time and broaden the line. The K-4 specimen turned out to be the purest, and this was later confirmed by mass spectroscopy. Consequently, we shall restrict our discussion to this specimen. The first injection of water (adsorption $a = 0.17$ $\mu mol \cdot m^{-2}$) does not change the linewidth of the NMR signal, which points to a rigid fixation of molecules. A similar result was obtained by MICHEL [4.137] with another very pure SiO_2 specimen. The adsorption of about 0.4 $\mu mol \cdot m^{-2}$ leads to a noticeable narrowing of the band, which indicates the appearance on the surface of more mobile molecules. Further adsorption causes monotonic narrowing of the linewidth. The values of adsorption heat decrease correspondingly (Fig.4.12), while the entropy of adsorption comes close to the entropy of the liquid phase (Fig.4.13).

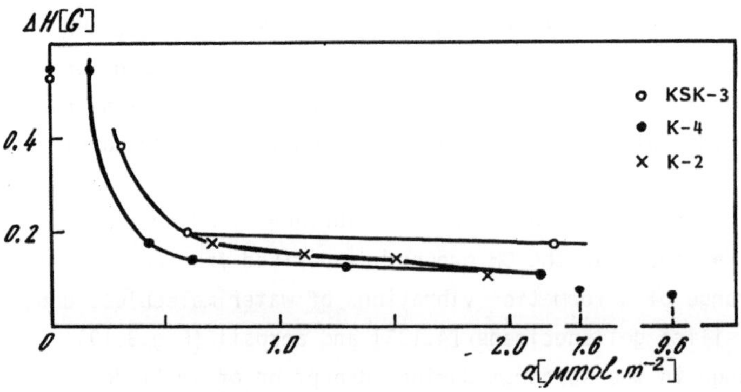

Fig. 4.15 Dependence of the linewidth of the NMR signal on the quantity of adsorbed water on silica gels

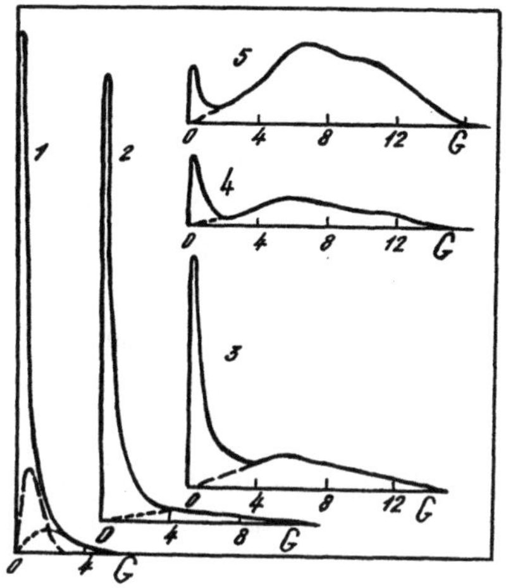

Fig. 4.16 The derivatives of NMR absorption signals from protons of water adsorbed on silica gel KSK-3 at 93 K

Detailed information on the mechanism of adsorption was obtained by KVLIVIDZE [4.135] in a study of the NMR of the adsorbed water phase in the low temperature range, where molecular movements are completely stopped. As noted in Sect.3.2.2, the NMR data point to the existence of at least three types of silanol patches on the surface, Model (3.25). From Fig.4.16 it is evident that the adsorption of water up to about 0.3 a_m (according to the BET equation, a monolayer of water is formed at $a_m = 6.2\ \mu mol \cdot m^{-2}$) does not lead to any noticeable change in the signal from single and paired OH groups, Models (3.25a and b). Even when the occupation reaches $\theta \sim 1.3\ a_m$, this signal is sufficiently intensive. Thus the NMR data also indicate that solitary OH groups do not play an independent role as primary centers of water adsorption. Together with the appearance of a wide component in the NMR signal from adsorbed water molecules, changes occur in the signal range that is farthest from resonance, which corresponds to patches of type III, Model (3.25c) with the greatest density of OH groups (Fig.4.16). Similar results were obtained with another silica gel specimen in [4.136].

The NMR data agree well with the above-mentioned low sensitivity of the band $\nu'_{OH} = 3750\ cm^{-1}$ in the IR spectrum in the initial region of water adsorption. A.V. KISELEV and LYGIN [4.129] proposed two hypotheses to explain this: (1) that the specimen is heated by the beam of IR radiation, resulting in the desorption of a number of water molecules bound to OH groups; (2) that adsorption takes place on active impurities (boron or aluminum) which diffuse from the bulk towards the surface when the specimen is heated. If this is so, then with an increase in T_{cal} the region of high values of adsorption heat ought to grow due to an increase in the concentration of surface impurities, rather than diminish as in Fig.4.12. As was mentioned in Sect.3.2.2, similar values for the concentration of nonhydroxyl active centers on SiO_2 surfaces were obtained with specimens of various origins (very pure silica gel and aerosil). ZHDANOV [4.126] considers it possible that ion radicals $\equiv Si^{+}$ or $\equiv Si-O^{-}$ play the role of active centers on the surface of partially dehydrated silica. However, no EPR signals from either $\equiv Si-O^{\cdot}$ and $\equiv Si^{\cdot}$ or from paramagnetic impurities were detected in the EPR spectra of the specimens of silica investigated by NMR and IR spectroscopic methods, and neither do such experiments confirm the assumption made by SIDOROV [4.130,131] that centers of kind II are valence-unsaturated atoms on the SiO_2 surface.

The idea that the hydrogen bond plays the predominant role in the adsorption of water was revived [4.138-140] by allowing that bonds may be established with the oxygen belonging to silanol groups or siloxane bridges, e.g.,

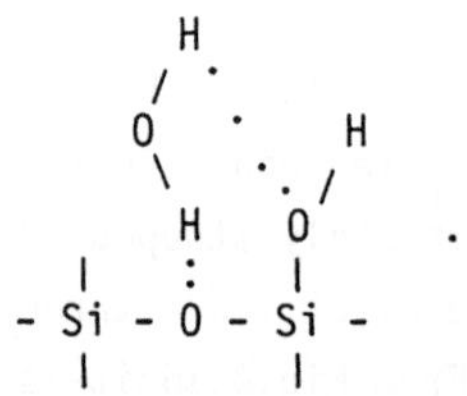

The hydrogen bond between the H_2O molecules and the oxygen of siloxane bonds will be quite weak, since the density of the electron p cloud on the oxygen atoms is very low owing to strong $d_{\pi}-p_{\pi}$ interactions in Si-O-Si bonds [4.141]. The existence of complexes of this kind does not agree with the high initial values of the adsorption heat of water.

Since 1962 we have been developing yet another approach to the mechanism of water adsorption on the surfaces of many oxides, including silica [4.119, 133, 134, 142]. We have assumed that the first doses of water are adsorbed directly on the hydrated Si atoms via a coordination mechanism, for instance according to the following models:

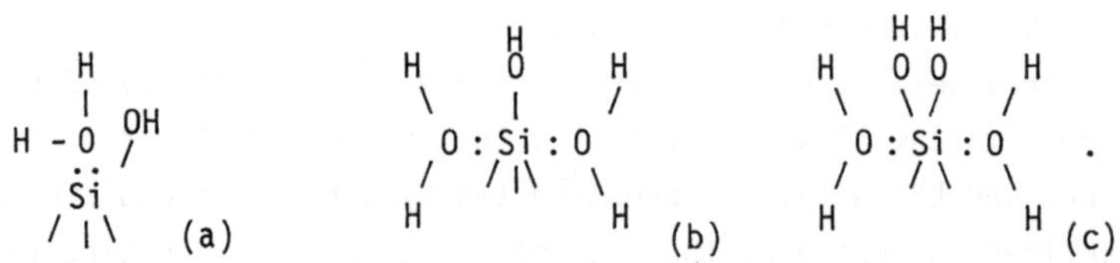

(4.4)

Modern quantum chemistry teaches that the tendency to form coordination bonds is not limited to typical transition elements [4.143]. Consider, for example, the fourth group of the periodic table. The carbon atom C is coordination saturated; its valence 4 coincides with the coordination number (c.n.) 4 in all carbon compounds with conventional bonds. The coordination number of silicon in most of its compounds is also 4; however, TREADWELL and WEILAND [4.144] and later WEYL [4.145] pointed out the possibility that the increase in the coordination number of silicon in $SiO_2 - H_2O$ systems results from the electrostatic attachment of water molecules, which ensure a better screening of the cation. Later it was demonstrated [4.143,146,147] that the unoccupied 3d orbitals of the silicon atom play an important role in the formation of siloxane bonds in silicates. The coordination number of silicon is known to reach 6 in some compounds, e.g., SiF_6, and under high temperatures and pressures a rutile like modification of SiO_2, so-called heavy quartz or stishovite [4.148], can be produced in which the coordination number of Si is also 6 (sp^3d^2 hybridization). The possibility that water can be coordination bound by SiO_4 tetrahedrons was proved in [4.149]. By analyzing the intensity of X-rays scattered from the surface of silica gel during its genesis, the authors of [4.149] showed the c.n.of Si in a gel to be equal to 5.8,which fully confirms Weyl's hypothesis. Drying the gel reduces the c.n. to 3.8, which is close to the normal value of 4 in SiO_2. All this, however, relates to the bulk phase.

It follows from what was said in Sect.3.2.3, that the coordination sphere of Si atoms on the surface is rather deformed due to the hydration of a number of surface tetrahedrons and the large variations in valence angles and bond lengths. The values of effective charges q_{eff} of the Si atoms in these tetrahedrons differ from those in the bulk. The coordination bond between a water molecule and the silicon atoms can be formed when an undivided pair of electrons is drawn from the oxygen in the H_2O molecule to the hybridized orbitals of Si, with 3d orbitals playing an important role. In the tetrahedral field the 3d orbitals split into a doubly degenerate system of e orbitals with decreased energy and a triply degenerate system of t_2 orbitals with increased energy. REVESZ [4.146,147] considered this from the standpoint of the most favorable spin configuration, and assumed that the π electrons of oxygen in the siloxane bond are distributed over $(e)^4(t_2)^0$ or $(e)^2(t_2)^0$ orbitals.

It is possible that the set of degenerate and partially occupied e and t_2 orbitals participates in the formation of coordination bonds with water. Note that the increase in the effective charge of silicon resulting from hydration and deformation of the surface tetrahedrons will lead to a decrease in the energy of e and t_2 orbitals.

In [4.150] it is considered more probable that the coordination bond is established not with the d orbitals, but with sp orbitals, whose energy is lowered due to hydration of the surface atoms of Si. Employing the Hukkel method within the framework of this hypothesis, the authors of [4.150] estimated the adsorption heat q_d and the change in the effective charge, Δq_{eff}, of silicon for the two possible schemes of coordination bond (4.4b) and (4.4c): $q_d = 155\ kJ \cdot mol^{-1}$ and $\Delta q_{eff} = 0.26$ for (4.4b); $q_d = 180\ kJ \cdot mol^{-1}$ and $\Delta q_{eff} = 0.31$ for (4.4c). Despite the qualitative nature of the model, the calculation gave a plausible explanation of the high initial values of the adsorption heat and the dependence of the adsorption heat on the extent of surface hydration. More recently, the coordination complexes between H_2O and Si were analyzed in a quantum-chemical CNDO approximation in [4.151], where the authors allowed the possiblity that H_2O molecules drift through hexagonal windows in the network of SiO_4 tetrahedrons and subsequent interaction occurs between the Si-OH-groups and the bulk phase of SiO_2. The coordination bond energies estimated for various configurations of H_2O molecules and Si-OH groups vary from 80 to 195 $kJ \cdot mol^{-1}$, and the bond lengths range from 2.1 to 2.4 Å. The charge transferred from water molecules to the surface fragment constitutes 0.15 to 0.2 of the unit charge. The population of atomic 3d orbitals changes little. The electrons on d orbitals are highly delocalized, and the approximate methods of MO LCAO can hardly be expected to give conclusive proof of the role played by d orbitals of Si in the formation of coordination bonds on the surface.

The difference between the adsorptive properties of surfaces of carbon on the one hand, whose atoms are incapable of forming coordination bonds, and silica on the other, were vividly demonstrated in [4.152] (Fig.4.17). The authors of [4.152] measured the differential heat of adsorption of water vapor on silica gel K-2 which was rehydrated after heat treatment at 1000 K and contained 5.2 $\mu mol \cdot m^{-2}$ of OH groups, and on oxidized graphite containing 0.8 $\mu mol \cdot m^{-2}$ of OH groups and 3.7 $\mu mol \cdot m^{-2}$ of oxygen on the surface. The concentrations of primary centers of adsorption on both surfaces, evaluated via the Dubinin-Serpinsky equation (2.44), turned out to be about the same, and thus the number of oxygen-containing groups capable of forming H bonds with water is presumably the same on both specimens. However, the initial sections of the adsorption heat curves for SiO_2 (Fig.4.17) lie much higher than those for graphite, which proves that adsorption of water on SiO_2 proceeds via the coordination mechanism.

According to NMR and IR spectroscopic data (Sect.3.2.2), the concentration n_c of coordination-bound water molecules on the surface of high-dispersion SiO_2 specimens is $(1-7) \cdot 10^{13}\ cm^{-2}$. The formation of a coordination bond will increase the coordination number of the silicon atom, and this in turn will cause changes in bond angles and lengths, resulting in expansion of the surface, which explains the FOLMAN and YATES experiments mentioned above [4.132](see also Sect.2.1.5), The expansion of porous glass which they ob-

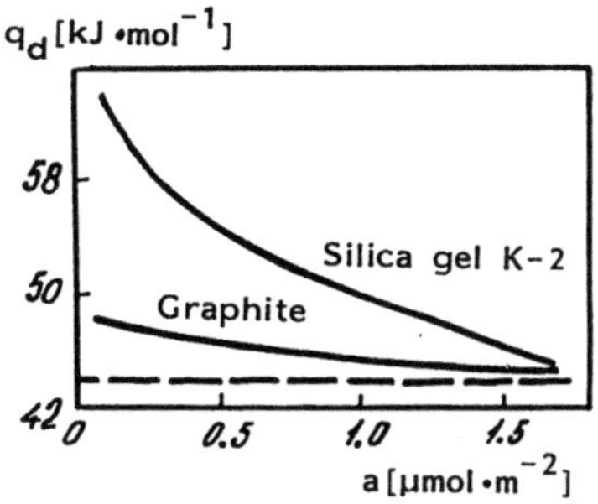

Fig. 4.17 The initial sections of the curves of the differential adsorption heat of water on silica gel K-2 and graphite

served in the initial stage of adsorption of H_2O and NH_2 occurs at occupations $n_c = 2.5 \cdot 10^{13}$ cm^{-2} (our estimate, made in [4.106]), which is close to the value of n_c based on spectroscopic data.

It is possible that an analysis of the IR spectra of skeleton oscillations of SiO_2 (in particular Si-OH in the range 960 - 980 cm^{-2}) [4.153] would yield the most direct information on the deformation of the silicon-oxygen tetrahedrons of the surface. This spectral region, however, has only been superficially investigated. Coordination-bound water, due to its low concentration, can be weakly manifested in this spectral region. HINO and SATO [4.154] discovered that the adsorption of water has a reversible influence upon the 960 cm^{-1} and 560 cm^{-1} bands. The shift of the 960 cm^{-1} band towards longer wavelengths is explained by the formation of hydrogen bonds. However, the shift in this direction is not in agreement with the results reported in [4.155], where the frequency of Si-OH vibrations is predicted to increase when an H-bond forms, because of the increasing contribution of $d_\pi - p_\pi$ interactions. It is possible that the shift in the frequency of Si-OH vibrations observed in [4.154] is related to a change in the coordination of the surface silicon atoms. TERTYKH and co-workers [4.156] observed a noticeable variation in the 980 cm^{-2} band which accompanied the decrease in intensity of the band of valence oscillations of coordination-bond H_2O molecules ($\upsilon_c = 3580$ cm^{-2}) in the course of dehydration of SiO_2.

The drawing of an undivided pair of electrons of the ligand molecule onto the Si atom, which occurs when a coordination bond is formed, results in a reduction in the positive charge of Si. This change in the effective charge of Si on the Si-SiO_2 boundary as a result of adsorption makes these atoms the centers for the capture of holes in the semiconductor. This fact is supported by numerous electrophysical experiments. The model of a coordination adsorption complex is the basis of the donor-acceptor mechanism by which the semiconductor surface is charged during adsorption. This will be discussed in detail in our forthcoming book [4.157].

It becomes apparent that spectroscopic, adsorptive, and electrophysical measurements confirm the model of coordination complexes

```
       OH       H
  \  /         /
 - Si . . . . O
  /            \
                H
```

(4.4a) which we proposed in 1962. Quantum-chemical calculations [4.150,151] performed at a later date prove that it is indeed possible that coordination bonds form between the molecules of water and the Si atoms in the silanol groups of silica.

Let us now consider the third element in the fourth group: germanium. The very fact that atoms of Ge can form oxides in tetradic (hexagonal modification) and hexadic (tetragonal modification) coordinations is proof of the ability of Ge atoms to form coordination bonds. Complex compounds of Ge are well known [4.158]. On the basis of investigations of the IR-spectra of water adsorbed on pure aerosil and on aerosil modified by treatment with $GeCl_4$, FINK and CAMARA [4.159] concluded that the water of initial injections forms coordination bonds with hydrated Si and Ge atoms according to a scheme similar to (4.4) above. Adsorbed pyridine molecules form bonds of similar strength with atoms of Ge [4.160]. Despite their proximity in the periodic table, the properties of tetrahedrally coordinated SiO_2 and GeO_2 differ [4.161]. SiO_4 tetrahedrons do not undergo such great deformation under mechanical stress; it is the angles that are changed. In the case of GeO_2, however, it is the parameters of the GeO_4 tetrahedrons that are changed, and this affects the energy spectrum of these oxides [4.157].

Titanium, an element from the side subgroup of group IV, is a typical transition element. $Ti(H_2O)_6^{3+}$ ions are classic examples of complex compounds. Investigations of the EPR spectra from partially reduced rutile specimens indicate that coordination bonds form between hyperstoichiometric Ti^{3+} ions and molecules of H_2O [4.162]. Note that Ti^{3+} ions are not the only centers of coordination bonds. The values of irreversible adsorption of water, α_{ir} ($\sim 10^{14}$ molecules $\cdot$ cm^{-2}), and the strength of the υ_c bands in the IR spectrum from the hydrate cover of rutile (see Table 3.4 and Fig.3.21) indicate that the number of primary centers of adsorption is much greater than the number of Ti^{3+} ions on the surface ($\sim 10^{12}$ centers $\cdot$ cm^{-2}; see also [4.162]).

Apparently, the dominant centers of water adsorption are hydrated Ti^{4+} ions. The IR spectra of rutile during adsorption of heavy water are presented in Fig.4.18 [4.162]. The first injection of water erases from the spectrum only

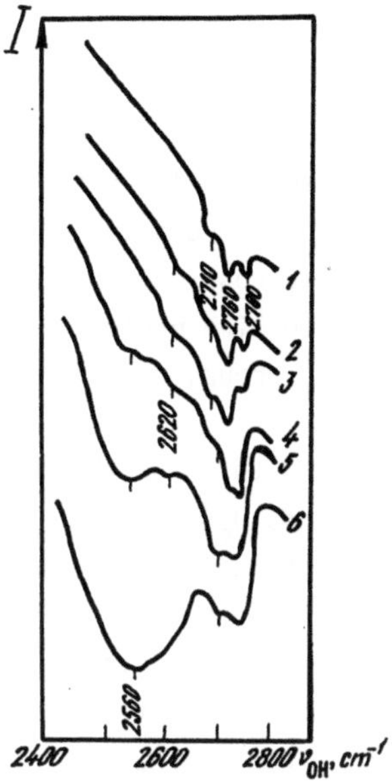

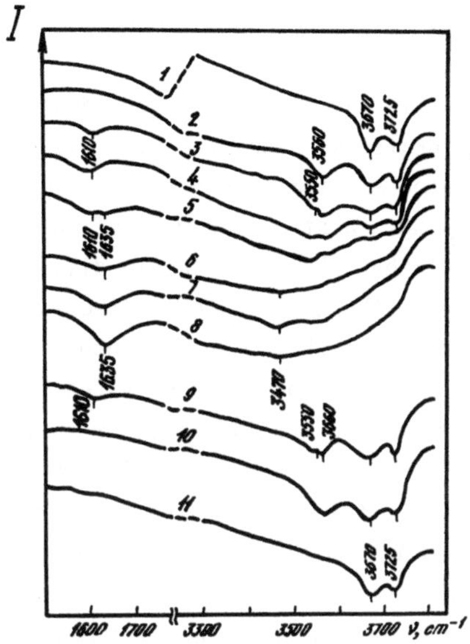

Fig. 4.18 IR transmission spectra (in arbitrary units) of a deuterized rutile surface. (*1*) During adsorption of D_2O; the specimen was vacuumized at 800 K. (*2-6*) After adsorption [4.162]. Adsorption a (in molec · cm^{-2}): (*2*) $1 \cdot 10^{14}$, (*3*) $3 \cdot 10^{14}$, (*4*) $5 \cdot 10^{14}$, (*5*) $7 \cdot 10^{14}$, (*6*) $10 \cdot 10^{14}$

Fig. 4.19 (*1*) IR transmission spectra from a particularly clean γ-Al_2O_3 specimen (in arbitrary units) vacuumized at 800 K, (*2-8*) After adsorption of H_2O at various values of p/p_S: (*2*) $5 \cdot 10^{-3}$, (*3*) $4 \cdot 10^{-2}$, (*4*) 0.1, (*5*) 0.2, (*6*) 0.22, (*7*) 0.5, (*8*) 1. (*9-11*) After desorption of H_2O at various T_{cal}: (*9*) 400 K, (*10*) 500 K, (*11*) 800 K [4.162]

the low-frequency band $\upsilon'''_{OH} = 2690$ cm^{-1} (in the region of OH oscillations 3540 cm^{-1}). In the region of deformation vibrations the latter is accompanied by a strong υ_d band, which indicates the nondissociative nature of the water adsorption. The possibility that coordination attachments form between water molecules and the surface of TiO_2 is supported by spectroscopic investigations [4.163,164]. GRIFFITH and ROCHESTER [4.165], on the other hand, failed to detect a band from firmly bound molecular water on rutile evacuated at 723 K. Direct proof of the presence of coordintion-bound H_2O on the surface of rutile was provided by an analysis of NMR spectra in [4.166].

Atoms of aluminum can enter oxides with coordination numbers 4, 5, and 6 in sp^3, sp^3d, and sp^3d^2 hybridization respectively. Detailed investigations of the IR spectra of γ-Al_2O_3 were carried out in [4.167,168]. From Fig.4.19 it is evident that the first injection of water onto a specimen dehydrated at

800 K restores the low-frequency band of hydroxyl groups ($\upsilon''_{OH} = 3560\ cm^{-1}$), while the band of deformation vibrations of water does not appear; this points to a dissociative mechanism of water adsorption. We pointed out in Sect.3.2.3 that the dehydrated surface of γ-Al_2O_3 is populated by very active EA centers – tricoordinated Al atoms. When these atoms interact with H_2O molecules, there is apparently such a strong drawing of undivided pairs of electrons onto p orbitals of Al that the H_2O molecule dissociates into H and OH. The proton can either attach to the oxygen atom in the lattice to form an OH group, or compensate the negative charge in the interstice (Sect.3.1.5).

The first injection of water elevates the c.n. of Al from 3 to 4. Another dose of water at $p/p_S = 0.04$ (Fig.4.19) gives rise to the bands $\upsilon_c = 3550\ cm^{-1}$ and $\upsilon_d = 1615\ cm^{-1}$ in the spectrum. The molecules of water presumably start being adsorbed on tetracoordinated Al atoms, whose coordination number is then increased from 4 to 5 [4.169]. As can be seen from Fig.4.19, this band is thermally quite stable, and is destroyed only by heat treatment at above 500 K. At great values of p/p_S the adsorption follows the H-bond mechanism. As in the case of SiO_2, the intensity of the high-frequency bands υ'_{OH} and υ''_{OH} remains practically the same during adsorption of H_2O up to $p/p_S \sim 0.2$.

A vast number of spectroscopic investigations have been conducted to analyze the adsorption mechanism of water in zeolites [4.129]. In contrast to pure oxides, the IR spectrum of water adsorbed in zeolites is rather complex, and consists of several narrow bands in the region of valence oscillations of OH groups of water molecules, and bands of deformation vibrations. It has been established that the position of the bands in the IR spectrum, the change in their shape after heat treatment, and the initial values of the adsorption heat of water strongly depend on the composition of the zeolite and the nature of the exchange cations. The thermal stability of adsorbed water molecules is much greater for zeolites with polyvalent ions, especially ions of transition elements.

In the literature, however, opinions diverge regarding the nature of the active centers responsible for adsorption of water and certain other molecules in zeolites. There is no doubt that in the initial stage of occupation a number of water molecules are localized directly at the exchange cations. This opinion is confirmed by the NMR data for Li, Na, and other cations [4.170-172]. The mobility of these molecules at room temperature, as evaluated with the aid of proton magnetic resonance, is quite high [4.173]. In zeolites with polyvalent ions and in decationized zeolites, interactions with Al^{3+} ions take on an important role. As has already been mentioned, the electron-acceptor properties of an atom depend largely on its coordination. In the zeolite lattice the coordination of an atom will depend indirectly on the nature, radius, and charge of the exchange cation. On the basis of IR and NMR spectroscopic data, the authors of [4.174] concluded that two types of primary centers of water adsorption may exist: cations in the small sodalite voids and Al^{3+} ions in the large voids.

From the above discussion it follows that the adsorption of water on hydrated oxide surfaces proceeds according to one of two mechanisms: in the initial stage molecular adsorption via the coordination-bond mechanism prevails, and then it is succeeded by molecular adsorption via the H-bond mechanism. On partially dehydrated surfaces these two mechanisms can be accompanied by activated dissociative adsorption, whose activation energy is quite high for certain oxides (e.g., SiO_2) and depends not only on the vapor pressure and temperature, but also on the extent of hydration (i.e., on α_{H_2O}). MORROW and CODY [4.175] suggested that in chemisorption of H_2O on a SiO_2 surface dehydrated at $T_{cal} > 400$ K, the role of active centers is played by the silicon-oxygen bridges, where one of the Si atoms is tricoordinated. The hydration of these surface fragments is accompanied by the formation of two types of hydroxyl groups with frequencies $\nu'_{OH(I)} = 3748\ cm^{-1}$ and $\nu'_{OH(II)} = 3741\ cm^{-1}$, the latter being connected with tricoordinated Si atoms. The peak intensities OH(I) and OH(II) are in a 10-to-1 relation at $T_{cal} = 1100$ K. It is our opinion that the concentration of these active centers has been overestimated. We failed to detect any EPR signals from trivalent Si at T_{cal} up to 1000 K [4.106].

Naturally, the ability to form coordination bonds is not limited only to water molecules. The number of molecules which can enter coordination bonds is quite high. SIDOROV [4.130,131] discovered that molecules of ammonia, alcohol, phenol, and chloroform can establish firm bonds with centers of the second kind. A. KISELEV and LYGIN [4.129] are inclined to think that adsorption on a touch of boron in porous glass is responsible for all these results. However, the initial values for the differential heat of adsorption on hydrated surfaces of very pure silica gel specimens are as high as 70 to 85 $kJ \cdot mol^{-1}$ for ammonia [4.111] and methanol [4.112], which, as in adsorption of water, cannot be explained by the formation of H bonds. The band of molecularly adsorbed methanol on a very pure aerosil persists in the IR spectrum up to $T_{cal} = 470$ K [4.129]. NMR investigations of ammonia adsorbed on SiO_2 revealed that the first doses of ammonia ($\sim 6 \cdot 10^{13}$ molecules $\cdot\ cm^{-2}$) fix themselves rigidly on the surface [4.176]. The number of primary adsorption centers is close to the concentration of coordination-unsaturated Si atoms estimated by other techniques (Sect.3.2.2).

Coordination bonds with ammonia and pyridine are manifested still more vividly in oxides of the next element, germanium. The IR spectra of pure germanium oxide [4.177] as well as of aerosil modified with $GeCl_4$ [4.159,160] display strong bands of coordination-bound ammonia. Interesting results were obtained from measurements of the differential heat of adsorption of NH_3 on germanium powders with particles covered by an oxide film of GeO_2 [4.178]. One can see from Fig.4.20, that evacuation of the specimen at $T_{cal} = 500$ K, which strips the coordination-bound water molecules from the surface, results in a sharp rise in the initial values of the adsorption heat of NH_3 ($q_d \sim 170$ kJ $\times$ mol^{-1}). At $T_{cal} = 700$ K the initial value of q_d is reduced to 40 kJ · mol^{-1}.

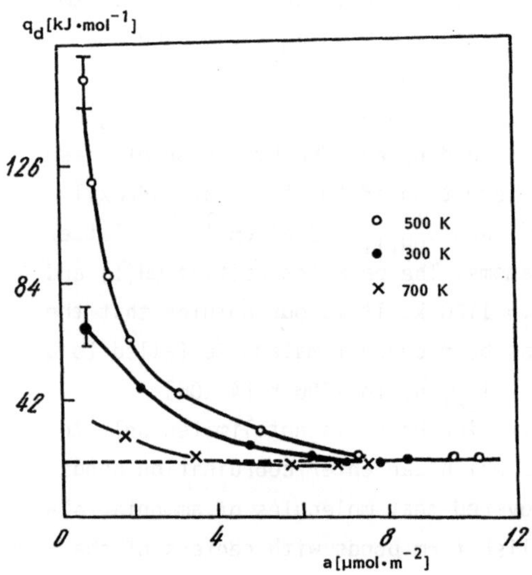

Fig. 4.20 Differential heat of adsorption of ammonia on the Ge real surfaces of specimens vacuumized at various temperatures

With heat treatment temperatures in the range of 300 to 500 K, the oxide film consists mainly of the elements of the hexagonal modification of GeO_2. Adsorption of NH_3 raises the coordination number of the Ge surface atoms from 4 to 5. Heat treatment at 700 K leads to considerable reconstruction of the film structure, and the tetragonal modification of germanium with c.n. = 6 begins to prevail. This means maximal coordination saturation for the majority of the surface atoms; the differential heat of adsorption then drops. These results conform well with electrophysical measurements on germanium surfaces [4.178].

The above-discussed results concerning adsorption of H_2O and NH_3 on the surface of oxides of Ge and Si do not confirm A. KISELEV and LYGIN's hypothesis

[4.129] that it is impossible for coordination bonds to be established with these elements of group IV because of the great difference in the energies of p and d orbitals. Actually this difference in energy can be lowered by hydration of the surface and distortion of the surface structural units.

With γ-Al_2O_3, aluminosilicates, and decationized zeolites, the formation of coordination bonds with various nitrogen-containing ligands (NH_3, pyridine, etc.) has been proved in numerous spectroscopic investigations (Sect.3.2.3). The initial adsorption heat of ammonia on γ-Al_2O_3 and aluminosilicates is 170 $kJ \cdot mol^{-1}$ [4.179,180], which is close to the values for Ge (Fig.4.20). Strong coordination bonds with oxygen-containing ligands (acetone and dimethyl ether) were reported in [4.181]. FILIMONOV [4.182] observed the formation of coordination complexes during adsorption of organic molecules with bonds such as $C-C\equiv N$, $>C-O$, and $\geqslant C-O$ oxides Mg, Zn, Ti, and Zr. Important conclusions were reached on the basis of the IR spectra of adsorbed molecules at increased temperatures. In the case of γ-Al_2O_3, the adsorption of oxygen-containing [4.181] and nitrogen-containing [4.183] ligands was demonstrated to be of an activated nature. LOW and coworkers [4.177] reported a noticeable increase in the intensity of the bands of coordination-bound water molecules ($\upsilon_c = 3550\ cm^{-1}$) when a GeO_2 specimen was heated.

Because an undivided pair of electrons of a ligand molecule is drawn onto the inner orbitals of the acceptor atom during the establishment of a coordination bond, some of the intramolecular bonds in the adsorbed molecule are weakened [4.184]. In the extreme case, a strong coordination bond can cause dissociation of the adsorbed molecule. Some authors observed the presence of dissociation products of ammonia and organic acids on partially dehydrated surfaces during adsorption of these substances. For instance, during adsorption of NH_3 on SiO_2 [4.185] and GeO_2 [4.159,177], NH_2 groups were detected by IR [4.159,177] and NMR [4.185] spectroscopy. The same groups were also discovered during adsorption of NH_3 on γ-Al_2O_3 [4.186]. In adsorption of organic acids on oxides of Ge, the bands of formiate and acetate ions were observed [4.187,188]. The same ions were also detected on Al_2O_3 and aluminosilicate, both in the IR and NMR spectra [4.189,190] and in inelastic electron tunneling (IET) spectra [4.191]. (The method of inelastic electron tunneling spectroscopy is discussed in detail in [4.157].) The latter method was employed in [4.192] to demonstrate that phenol and hydroquinone adsorbed on Al_2O_3 exist mainly in the form of $C_6H_5O^-$ and $C_6H_4(OH)O^-$ ions.

The overwhelming majority of researchers attribute the appearance of dissociation products either to chemisorption on oxygen bridges or to interaction between molecules and OH groups [4.159,177,187-192], e.g.,

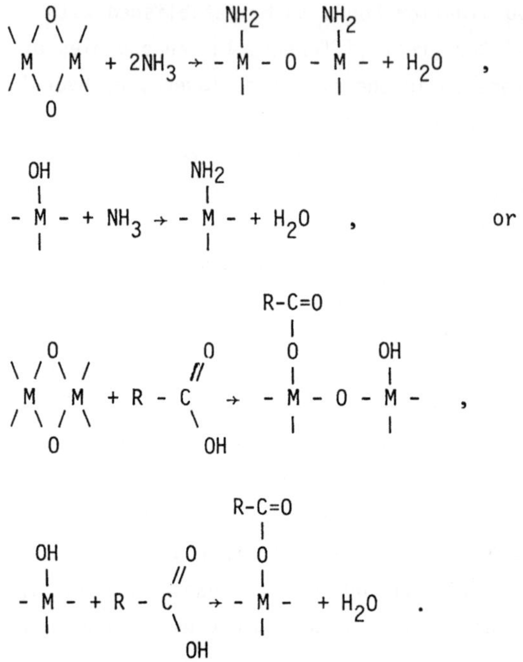

Scheme 4.2

However, as we shall see later one cannot exclude the possibility of dissociation via the coordination mechanism, which in some cases can yield the same dissociation products [4.193].

The possibilities of complex formation on the surface of an oxide are often analyzed on the basis of analogies drawn from the structure of bulk compounds of the corresponding elements, which is not always correct. The conditions under which surface complexes form in concentrations high enough to be registered spectroscopically depend on a number of factors [4.106,133], including (1) steric factors which allow the sonde molecule to enter the coordination sphere of the EA center, even though the sonde molecule may be rather large; (2) the symmetry of the wave functions of acceptor atom and ligand. In order for a sufficiently strong bond to be established, a correspondence must exist between the irreducible representations of interacting orbitals which would explain the utmost overlapping of wave functions. For this reason negative results on complex formation which are obtained with one kind of sonde molecules should not be generalized to all other ligand molecules. It is safer to take a concrete EA center and examine its properties with respect to a concrete sonde molecule.

For example, let us compare the adsorption of two molecules having a similar structure: H_2O and H_2S. As has already been demonstrated, H_2O molecules form strong coordination bonds with Ti atoms on the surface of partially reduced and hydrated rutile. Experiments on the adsorption of H_2S [4.194] under similar conditions revealed that in contrast to H_2O, H_2S molecules are attached to the surface very loosely, and are totally removed after evacuation at 350 - 400 K. Adsorption of H_2S at 300 K did not change the EPR spectrum of Ti^{3+} ions, nor did it have any influence on the conductivity of rutile.[5] The reason for this behavior of H_2S lies in the nature of the moleculear orbitals, which in H_2S are close to pure 1s - 3p bonds.

It is equally unwise to speak of the presence or absence of EA centers on the surface of a given oxide in general. As we have already noted, the electron-acceptor properties and the symmetry of the orbitals of the acceptor atom are determined first and foremost by coordination, which for a number of oxides depends on the heat treatment temperature, divergences from stoichiometry, etc. We used Ge and Al oxides as examples to demonstrate the considerable dependence of the coordination unsaturation of the surface atoms on T_{cal}. If the coordination number does not change, q_{eff} will depend on the deformation of the surface structure units and the hydration of the surface (e.g., SiO_2) which in turn will depend on the biography of the oxide, the conditions of treatment, the dispersity, etc. The electron-acceptor properties can be greatly enhanced by substituting a number of OH groups in the coordination sphere of acceptor atoms with other, more electronegative groups during the genesis of the oxide. In the case of SiO_2, the replacement of oxygen in the nearest coordination sphere of Si by fluorine increases the role of d orbitals in the formation of chemical bonds [4.195].

Note finally, that in many cases the transition from a weak donor-acceptor bond to a strong coordination bond and to the formation of a complex with transfer of charge (CTC) is an activated process. Experiments, some of which we have already discussed, show that the formation of a strong coordination bond or a CTC is facilitated by thermal excitation of the system or by irradiation with quanta of sufficient energy.

4.2.3 The Formation of the Adsorption Phase

So far we have been concerned mainly with the initial range of occupation, where the predominant mechanism of adsorption is the formation of coordination bonds. Let us now discuss the structure of the surface phase at various stages in the occupation of the surface by adsorbed molecules. The most detailed information here is provided by NMR.

[5] At increased adsorption temperatures ($\sim$500 K), dissociation of H_2S and the formation of Ti-SH groups were observed.

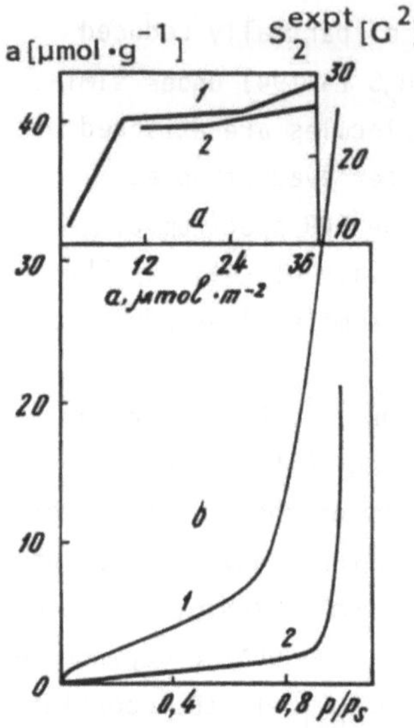

Fig. 4.21 (a) Dependence of the second moment of the NMR line, S_2^{expt}, upon water content. (b) Isotherms of the adsorption of water on hydrated silica gel at $T_{cal} = 500$ K (*1*) and on fluorinated SiO_2 (3% F by mass) at $T_{cal} = 750$ K (*2*)

Consider again the low-temperature NMR spectra of water adsorbed on macroporous silica gel [4.135,136] (Fig.4.16). A qualitative idea of the structure of the adsorbed phase can be obtained by comparing the second moment S_2^{expt} of the wide component of the NMR line of water with the theoretical value S_2^{theor} for idealized models of layer-wise occupation of the surface. Figure 4.21 shows the dependence of S_2^{expt} on the concentration of adsorbed water [4.136]. Near the monolayer according to BET ($a_m = 6.3\ \mu mol \cdot m^{-2}$) the value S_2^{expt} is $23 \pm 1\ G^2$, which exceeds the value S_2^{theor} for an ideal monolayer of water and is close to the value of S_2^{theor} for structures 2 or 3 water molecules thick (~22 - 23 G^2). Hence it follows that the adsorbed phase near the monolayer is composed of clusters of water molecules with an overall thickness 2 or 3 times the size of a molecule (Fig.4.22a). An earlier assumption regarding the formation of clusters of H_2O molecules was made on the basis

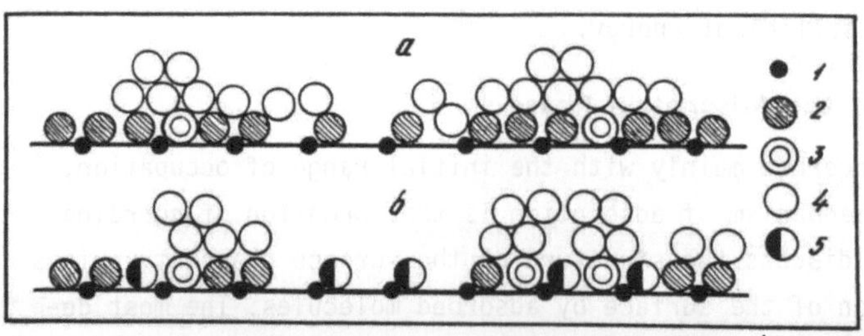

Fig. 4.22a,b. Model of the adsorbed water phase (occupancy close to a monolayer according to BET) on a hydrated (a) and a fluorinated (b) SiO_2 surface (*1*) Si atoms, (*2*) OH-groups, (*3*) Coordination bound H_2O molecules (*4*) H-bound H_2O molecules, (*5*) F-atoms

of data on the dielectric permitivity of adsorbed water [4.196] and on the overtone bands in the IR spectrum of the SiO_2 surface [4.197].

All these facts contradict the widely accepted view that the occupation of the surface occurs in a uniform layer-by-layer process in the course of adsorption. The idealized picture of the surface phase presented in Fig.4.22a resembles the statistical monolayer model, which forms the basis of the Brunauer-Emmett-Teller theory of polymolecular adsorption [4.198]. Owing to the clusterwise pattern of surface occupation, the landing sites of the molecules in a tightly packed monolayer (Sect.4.2.1) which are used in the calculations take on the meaning of effective values.

The centers of cluster formation are most probably the coordination-bound water molecules. As will be shown later, these molecules are quite highly protonized, and are more active as adsorption centers than the OH groups. Indeed, the values of adsorption heat near the monolayer much exceed the heat of condensation (equal to the energy of two H bonds) (see Fig.4.12). If one adopts as the number of coordination-unsaturated silicon atoms the estimated value of $n_c = 3 \cdot 10^{13}$ molecules $\cdot$ cm^{-2}, then the number of molecules in a single patch on the silica gel surface is about 10 or 12.

MICHEL [4.137], in studying the nuclear relaxation of water near a monolayer, discovered the existence of two phases whose times of spin-spin relaxation T_2 differ from one another. The correlation time for the more mobile phase is $\tau_s' = 2.7 \cdot 10^{-10}$ s; such is the state of 95% of the adsorbed molecules. For the other phase, $\tau_s'' = 2.3 \cdot 10^{-8}$ s; the remaining 5% of the molecules belong to it. The second phase was assumed in [4.106] to be composed of coordination-bound water molecules (Fig.4.22a), and the first phase of water molecules in a cluster, linked together by H bonds. High protonization of coordination-bound molecules facilitates intensive proton exchange inside the "swarm", which also agrees with the experimental data.

At occupations corresponding to the plateau on the curve S_2^{expt} (a) (Fig. 4.21) the clusters expand in tangential direction, while their thickness remains the same; $S_2^{expt} \approx$ const. Apparently the presence of solitary OH groups on the surface stimulates the propagation of clusters along the surface. Only at occupations $\theta > (4-5)a_m$ do the clusters begin to merge depending upon the concentration of hydroxyl groups on the surface. Consequently, as the surface undergoes dehydration, the merging of clusters will occur at higher and higher occupations.

The occupation of the surface in adsorption of ammonia has similar features. NMR investigations of ammonia adsorbed on two specimens of silica gel subjected to heat treatment at 500 and 1000 K also indicate the existence of two types of adsorption centers [4.176]. Calculations of the second moments of NMR signals from adsorbed ammonia at 90 K reveal that in the region of low occupations, the value of S_2^{expt} (1.6 G^2) is considerably lower than for solid ammonia at the same temperature. At 90 K the NH_3 molecule in solid ammonia rotates around the third-order axis, and $S_2^{expt} = 9.3\ G^2$. The smallness of S_2^{expt} for adsorbed ammonia is attributed to the precession of the rotation axis. As adsorption proceeds, the precession decreases due to interaction with the surrounding molecules, and at $\theta > 0.5 a_m$ the value of S_2 becomes close to the value typical of a molecule rotating around a stationary axis.

Studies of nuclear relaxation have provided valuable information on the mobility of adsorbed molecules, and there is now a large volume of experimental data on the influence which the surface has upon the relaxation time of adsorbed molecules (see the reviews [4.199,200]). The existence of several phases has been established for a number of adsorption systems; these phases are distinguished by their correlation times. From measurements of the temperature dependence of the relaxation times it was possible to estimate the activation energies for different types of motion of molecules within the surface phase. It should be noted, however, that the relaxation times are rather sensitive to the presence of paramagnetic impurities on the surface. Neglect of this circumstance is responsible for the numerous inconsistencies in the characteristics of the motion of adsorbed molecules reported in the literature. Measurements of the coefficients D of self-diffusion of water adsorbed on silica gels by means of the impulse field gradient method [4.201] and neutronography [4.202] revealed that in the monolayer at 300 K the value of D is just 5 or 10 times lower than for the bulk water, and the activation energy of motion is about 25 $kJ \cdot mol^{-1}$. The surface loses its influence on the mobility of molecules at a distance of 10 - 20 Å. A detailed discussion of the mobility and nature of molecular motion in polylayers is beyond the scope of this book, and the interested reader is therefore referred to the reviews [4.199,200].

For reasons of space we are also unable to address ourselves to questions relating to the orientation of adsorbed molecules. Much has been achieved through the use of diffraction methods (ELEED) and polarized irradiation in electron spectroscopy (UPSAR, ELSAR); see Sect.3.2.1. An exhaustive survey of the experimental data on ordered monomolecular layers can be found in [4.25]. For systems with comparatively weak adsorption bonds, where the adsorption can be completely violated by electron and UV beams, the use of polarized IR radiation is quite promising; its efficiency has been proved in theoretical calculations and experiments carried out by CHOLLET [4.203,204] on the orientation of adsorbed hydrocarbons.

Valuable information on the orientation of adsorbed molecules is also provided by radio spectroscopic investigations (EPR, NMR, ENDOR). Important data on the orientation of adsorbed organic molecules and the deformation of their electron shells has been obtained with the aid of high-resolution NMR on ^{13}C nuclei [4.205].

4.2.4 Proton-Donor Centers on the Surfaces of Oxides

On the surfaces of hydrated oxides one can expect the presence of three proton-containing groups which can play the role of Brönsted or proton-donor (PD) acid centers: OH groups, H_2O molecules, and hydroxonium H_3O^+ ions. The majority of researchers attribute the acidiferous properties of the surfaces to the existence of OH groups. Their proton-donor properties can be evaluated from the shift in the principal frequency of valence oscillations of OH groups in the IR spectrum ($\Delta\nu'_{OH}$) in the course of adsorption. The greater the value of $\Delta\nu'_{OH}$, the higher the acidity of the OH group and the higher the basicity of the adsorbed molecule. A detailed analysis of $\Delta\nu_{OH}$ shifts was performed in [4.182,206]. Table 4.2 presents some of the characteristic values of $\Delta\nu_{OH}$ shifts during adsorption of oxygen-containing and nitrogen-containing molecules. In the case of water and other oxygen-containing molecules the value of $\Delta\nu$ does not differ much from that corresponding to the interaction of these molecules in a liquid, which attests to the low acidic properties of the OH groups on the surface of SiO_2. The energy of hydrogen bonds between these molecules and isolated OH groups is $\sim 20-25\,kJ\cdot mol^{-1}$.

Presumably, binary $>Si(OH)_2$ groups will be more acidic [4.207]. Investigations of the adsorption of cations [4.207] and cross-polarization/magic-angle-spinning silicon-29 NMR study [4.208] confirm this assumption.[6] It follows from the NMR data discussed in Sect.3.2.2 that the surface can contain a rather high concentration of these groups, which are typical of the <100> face of β-cristobalite. However, the high initial values of the adsorption heat of water and ammonia on these groups can hardly be explained by adsorption on these groups, although they are more acid than the solitary groups. As has already been pointed out, the IR and NMR spectra do not display any variation in the intensity of the bands corresponding to these groups in the initial stage of occupation.

[6] It is interesting to note that the concentration of acid centers as reported in [4.208] increases with a decrease in dispersity, which confirms once again the correlation between specific adsorptive activity and particle size discussed in Sect.3.2.5.

Table 4.2. Frequency shift of valence oscillations of hydroxyl groups on the surface of silica due to the adsorption of certain molecules [4.182]

Adsorbate	$\Delta\upsilon_{OH}$[cm^{-1}]	Adsorbate	$\Delta\upsilon_{OH}$[cm^{-1}]
Water	330	Aniline	535
Phenol	345	Ammonia	800
Acetone	350	Pyridine	850
Methanol	350	Trimethylamine	990
Diethyl ether	460	Diethylamine	990
Dibutyl ether	485		

Great variations in $\Delta\upsilon'_{OH}$ are only observed during adsorption of strong organic bases (Table 4.2), where $\Delta\upsilon'_{OH}$ is $\sim 800 - 900\ cm^{-1}$. A proportional relationship between $\Delta\upsilon'_{OH}$ and the basicity pK_B of the adsorbed molecules was reported in [4.140,209,210] (Fig.4.23). NOLLER and coworkers [4.140] assumed that when the basicity pK_A of the adsorbed molecules D is greater than the pK_A of single OH groups (according to Brönsted), a proton is transferred, and DH^+ ions are formed. We shall see below that this assumption is not supported by experiment.

Molecules often employed as convenient sondes for proton-donor surface centers are ammonia, pyridine, and aniline, as well as certain acenes whose protonized forms display characteristic bands in the IR and electron spectra [4.211]. There is a large amount of experimental material on the use of these sonde molecules to mark PD centers on the surface of certain oxides. Sonde molecules such as ammonia, pyridine, and aniline give three groups of bands: during adsorption on EA centers (NH_3EA); during adsorption on PD centers with the formation of NH_4^+, PyH^+, and AnH^+ ions (NH_3PD); and upon the establishment of H bonds ($NH_3 - H$).

Investigations of the IR spectra of oxides of Si [4.133], Ti [4.162,163], and Al [4.167,212] and of aluminosilicates [4.133,212] have demonstrated that provided the oxides have been dehydrated at $T_{cal} = 700 - 800$ K, the adsorption of pyridine and ammonia gives rise to only one sharp band, NH_3-H (or Py-H). No complexes of NH_3 or Py with PD centers were detected, despite a sufficiently high concentration of OH groups on the surface at these T_{cal}, as indicated by strong υ_{OH} bands. This proves that the OH groups are not active enough as PD centers to form protonized molecules of the said bases, and thus the assumptions of NOLLER et. al. [4.140] seem doubtful. This is also in accord with the above-mentioned dependent role of OH groups on oxides in the initial stage of water adsorption.

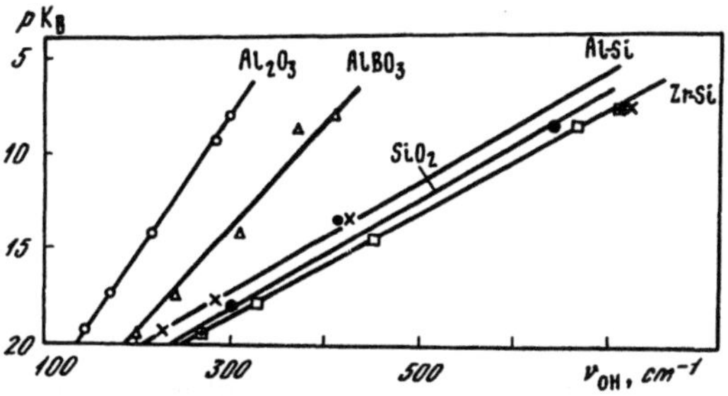

Fig. 4.23 Dependence of the shift in the band of OH oscillations, $\Delta\upsilon_{OH}$, upon the basicity pK_B of adsorbed bases for different catalysts [4.209]

The plots in Fig.4.23 for SiO_2, which is not catalytically active, and Al-Si and Zr-Si, which are active catalysts, have about the same slope, which indicates that groups with equal acidity exist on the surface of these catalysts. Naturally, the concentrations of such groups can differ. The lack of correlation between the proton-donor properties of OH groups and the catalytic activity is confirmed by experiments with catalysts poisoned by LI and Na ions [4.209]. The spectral manifestations of OH groups during the adsorption of bases were about the same for both active and poisoned catalysts. Some authors assume, on the basis of potentiometric measurements in SiO_2-electrolyte systems, that the ions of alkali metals replace the hydrogen in OH groups, and form Si-O-Me groups (e.g., [4.213,214], and references, therein). In principle, however, such measurements provide no information on the mechanism of adsorption. Spectroscopic investigations [4.209] revealed that treating oxides with LiOH and NaOH solutions does not influence the band of valence oscillations of OH groups. It is possible that the metal ions interact directly with coordination-unsaturated Si atoms on the surface, building fragments of silicide structures.

TRET'YAKOV and FILIMONOV [4.215], taking as the criterion for the proton-donor properties of OH groups on oxide surfaces the frequency shift $\Delta\upsilon_{OH}$ during adsorption of benzene molecules, which form π complexes with OH groups, demonstrated that there is no correlation between the acidity of these groups and the electronegativity of Me atoms of the given oxide. So it can be taken for a firmly established fact that the hydroxyl groups are not the predominant PD centers of simple oxides, as many researchers still consider them to be [4.129,140,213].

Various models of PD centers have been proposed in attempts to explain the high catalytic activity of aluminosilicates, which are industrial catalysts of great importance. In 1949 TOMAS [4.216] suggested that the role of PD centers is played by protons, which compensate the negative charge of Al atoms in tetrahedral coordination:

$$- \overset{|}{\underset{|}{Si}} - O - \overset{H^+}{\overset{|}{\underset{|}{Al}}} - O - \overset{|}{\underset{|}{Si}} - \quad . \qquad (4.5)$$

FRIPIAT et al. [4.217] believe the H_3O^+ ions to be possible PD centers. However, numerous investigations of IR [4.218] and NMR spectra (see the review [4.136]) have not revealed isolated protons or hydroxonium ions on the surface of silicates. Many authors (e.g., [4.219-221]) consider that the function of proton-donors in aluminosilicates is fulfilled by OH groups on the surface.

MIESSEROV et al. [4.222] attributed the increased protonization of OH groups on aluminosilicates to a shift of the electron density from proton to Al atom:

$$- O - \underset{|}{Al} - O - \overset{OH^+}{\overset{|}{Si}} \quad . \qquad (4.6)$$

BASILA [4.218] assumed the acidity to be due to the interaction between the molecule of the base adsorbed on the EA center (the Al atom in triple coordination) and the adjacent silanol group. Similar ideas regarding the enhancement of the proton-donor properties of Si-OH groups have been extended to zeolites [4.129]. This viewpoint, however, is not confirmed by studies of the IR and NMR spectra of adsorbed water: the first injections of water did not influence the intensity of the ν_{OH} band either with pure oxides and aluminosilicates (Sect.4.2.2) or with zeolites [4.182].

Numerous spectroscopic and catalytic investigations have convincingly proved that the adsorption of small quantities of water on partially dehydrated aluminosilicate catalysts [4.219,223] and decationized zeolites [4.224,225] promotes catalytic activity and proton acidity. The promotive action of water was also observed with certain cation zeolites, though to a somewhat lesser extent. Analysis of the IR spectra of adsorbed pyridine revealed that changes in T_{cal} for aluminosilicate catalysts cause a sharp rise in the intensity of the bands of PyEA complexes and a sudden drop in the intensity of those for PyH^+ complexes [4.226,227]. Figure 4.24 gives an example of such dependence for the case of decationized zeolite. The greatest variations in intensity are observed at $T_{cal} > 800$ K. The antibatic behavior of the concentrations of EA and PD centers gives reason to assume that the

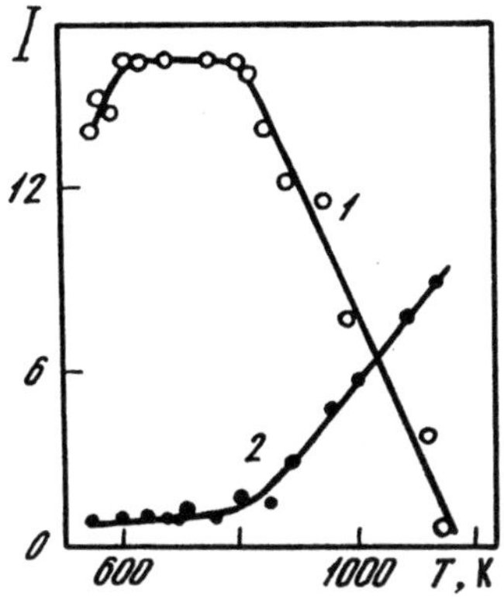

Fig. 4.24 IR transmission spectra of pyridine (in arbitrary units) characterizing (1) the PD centers (1545 cm^{-1}) and (2) the EA centers (1451 cm^{-1}) on a decationized zeolite [4.227]

PD centers are created at the expense of chemisorption of water on the EA centers on the aluminosilicate surface.

Based on an analysis of the IR and NMR spectra of adsorbed water and data on adsorption heat, we concluded that the role of PD centers on the surface of many oxides is played by coordination-bound water molecules [4.106,184, 228]. Because an undivided pair of electrons from the oxygen of the water molecule is pulled over onto the inner orbitals of the acceptor atom (Me), the O-H bond of the molecule is weakened, causing protonization. This is supported by experiment. The bands of NH_4^+ and PyH^+ in the IR spectrum are observed only when coordination-bound H_2O molecules are present on the surface ($\upsilon_c = 3550\ cm^{-1}$; $\upsilon_d = 1610 - 1640\ cm^{-1}$; see Sect.4.2.2). For instance, the following model applies to an Al atom in triple coordination:

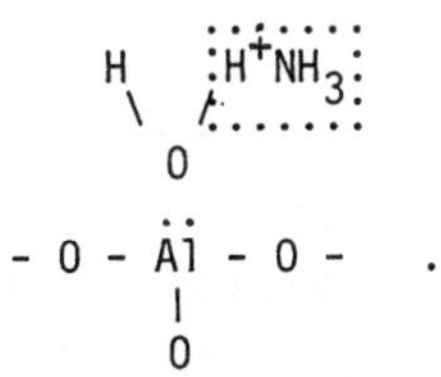

Table 4.3 presents the most typical results of investigations of EA and PD centers on oxide surfaces using sonde molecules. Certain discrepancies in the data on PD centers on γ-Al_2O_3 surfaces are easily discernible. The existence of PD centers on such surfaces was again confirmed recently by analyses of the luminescence spectra of adsorbed molecules of acridine [4.231] and the EPR spectra of different radicals [4.232].

From the viewpoint of the donor-acceptor mechanism of adsorption, the proton-donor activity of water molecules must correlate with the strength of their attachment to the surface, i.e., with the value of T_{max} (Table 4.3).

Table 4.3. IR spectroscopic data on the interaction of pyridine and ammonia with electron-acceptor and proton-donor centers on oxide surfaces

Adsorbent	EA centers		PD centers		Reference
	T_{cal} K	IR bands of coordination complexes	T_{max} [a] K	IR bands of sonde	
SiO_2(silica gel)	500	No	500	PyH^+	4.133
$mSiO_2$ nAl_2O_3 [b]	800	Py	700	PyH^+	4.133,212
γ-Al_2O_3	900	Py		No	4.133
	800	Py	600	PyH^+	4.168
	--	NH_3	--	NH_4^+	4.186,229
SiO_2 modified with Al	800	Py	--	PyH^+	4.159
TiO_2(rutile)	800	Py	700	PyH^+	4.162
TiO_2(anatase)	--	--	--	NH_4^+	4.230

[a] At T_{cal} T_{max} the ν_c and ν_d bands of water molecules and the bands of PyH^+ disappear from the spectrum

[b] 12% Al_2O_3

The most loosely bound and the least protonized are water molecules on pure oxides of Si and Al. The concentration and activity of the PD centers is determined by the activity of the EA centers, which in turn depends on the coordination number of the atom and the extent of hydration and deformation of the surface structure cells (Sect.3.2.3). This apparently is the main source of inconsistencies in reports on the presence or absence of PD centers on the surfaces of various oxides.

Additional information regarding the state of water on oxide surfaces can be derived from the mass spectra of thermal desorption. The most notable technique in this connection is the method of thermal desorption at a linear heating rate [4.233,234]. Although this technique has only been consistently developed in its application to metals, one can, by choosing the appropriate heating rate constants, obtain reliable information on the activation energy for desorption of particles from the surface of disperse dielectrics.

As was pointed out in Sect.3.2.3, the only product of desorption from the surface of a very pure SiO_2 specimen is water [4.235]. During heating of TiO_2 (rutile), on whose surface the coordination bonds of water molecules are stronger (Table 4.3), the mass spectra revealed, in addition to water, the release of a small quantity of hydrogen [4.236]. The mass spectra were obtained with a linear rate of heating the specimens, which were deuterized in order to improve the signal-to-noise ratio. As can be seen from Fig.4.25, the

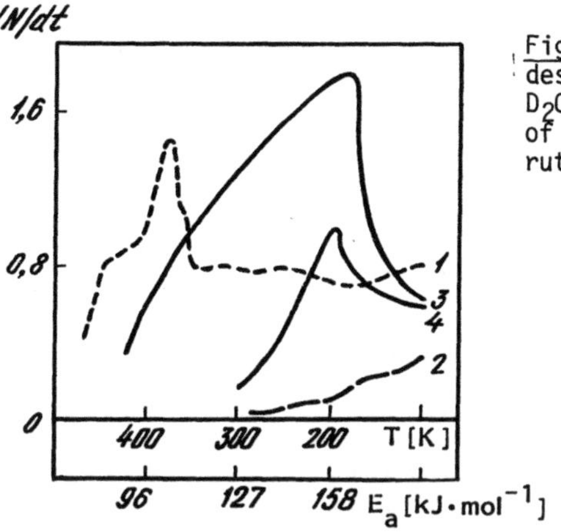

Fig. 4.25 Relative rate of thermal desorption dN/dt (in arbitrary units) of D_2O (*1,3*) and D_2 (*2,4*) from the surface of hydrated (*1,2*) and fluorinated (*3,4*) rutile

curve of the rate of thermal desorption of D_2O has a pronounced maximum corresponding to the desorption activation energy $E_d^{D_2O} = 101$ kJ · mol^{-1}. A similar value of $E_d^{H_2O}$ (105 kJ · mol^{-1}) was later obtained with the aid of a chromatographic technique [4.101]. At higher temperatures the mass spectra revealed the release of D_2, which is attributed by the authors of [4.236] to the destruction of coordination-bound water. This conclusion is supported by the fact that the D_2 band is repeatedly restored after rehydration of the specimens. The absence of a maximum prevents evaluation of the activation energy of desorption of D_2. However, calculations show the value of $E_d^{D_2}$ to be not less than 210 kJ · mol^{-1}. Preprogrammed thermal desorption of water from surfaces of rutile and anatase was also studied in [4.237].

Finally, let us discuss the nature of PD centers in zeolites. According to NMR data [4.173,174,238], in zeolites with alkaline-earth and rare-earth cations, as well as in decationized (dec.) zeolites, water molecules persist up to very high values of T_{cal} (800 K). Table 4.4, borrowed from the works of GVAKHARIYA [4.174,238], presents data on the mean proton-proton separations r_{H-H} in water molecules calculated from second moments of the NMR signal. In the zeolite with a single charged ion (NaY) the value of r_{H-H} is 1.58 Å, which differs little from that in liquid water (1.54 Å). Similar values of $\bar{r}_{HH}$ were obtained for the zeolite NaX [4.173] and for natrolite [4.239]. The value of r_{H-H} becomes considerably greater for zeolites with polyvalent cations and decationized zeolites, and there is a striking correlation between r_{H-H} and the thermal stability of water molecules (T_{cal}). With an increase in T_{cal} the molecules that remain in the zeolite are ones which are more and more deformed. There is also a definite correlation between r_{H-H} and the magnitude of the ionic potential e/r of the exchange cation. The greater e/r is, the more deformed the H_2O molecule is (Table 4.4).

Table 4.4. Average proton-proton distances in water molecules adsorbed in zeolites

Specimen	NaY	MgNaY		CaNaY		SrNaY	
T_{cal} K	500	500	700	500	700	500	700
r_{H-H} Å	1.58	1.59	1.65	1.59	1.63	1.58	1.62

Specimen	BaNaY		LaNaY		ScNaY		Y(dec.)	
T_{cal} K	500	700	500	700	500	700	500	800
r_{H-H} Å	1.58	1.62	1.62	1.66	1.63	2.2	1.57	1.66

It is interesting to compare these data with the results of investigations on the geometry of H_2O in bulk crystallohydrates using NMR and IR spectroscopy. SAFFAR [4.240] pointed out that depending on the type of anions and cations in the crystallohydrate lattice, the value of r_{H-H} for water molecules can vary from 1.52 to 1.61 Å. NMR investigations [4.241,242] revealed water molecules in crystallohydrates for which r_{H-H} ranged from 1.63 to 1.73 Å, which is close to the values of r_{H-H} for H_2O in zeolites (Table 4.4).

The observed increase of r_{H-H} in the adsorbed water molecules may be connected with an increase in the length of the O-H bond and an increase in the valence angle HOH. A study of the geometry of H bonds in solids has revealed that the HOH angle can reach 115°. If one takes 110°, as the average value of this angle, then the greatest value of r_{H-H} (1.67 Å) corresponds to an interatomic O-H separation of 1.02 Å, which is considerably greater than that in liquid water (0.95 Å).

The value of r_{H-H} for ScNaY zeolite stands alone in Table 4.4. In the NMR spectra of this zeolite dehydrated at $T_{cal} = 700$ K, a signal with the second moment $S_2 = 3.6\ G^2$ was observed, which can only be explained by assuming that the interacting protons responsible for the NMR signal stand 2.2 Å apart. This distance much exceeds the r_{H-H} value of water in crystallohydrates. GVAKHARIYA et al. [4.174,238] suggested that the observed NMR signal might come from water molecules in a predissociative state. And indeed, it was demonstrated in [4.244] with NMR that when crystallohydrates are thermally dehydrated, solid-phase hydrolysis of a number of water molecules occurs, producing pairs of ions $H^+ \ldots OH^-$. The r_{H-H} distance in such pairs was (e.g., for magnesium perchlorate) about 2 Å. Further proof of a predissociative state of water molecules in ScNaY zeolites was obtained by GVAKHARIYA, who detected the release of a considerable quantity of deuterium, as indicated by the mass spectra of thermal desorption of deuterized specimens. At lower T_{cal} (~500 K) as can be seen from Table 4.4, the value of r_{H-H} is smaller (~1.63 Å). A similar value of r_{H-H} for the same zeolite was reported in [4.244] (T_{cal} = 450 K).

Thus the NMR data indicate that firmly adsorbed water molecules in zeolites are highly deformed. As was pointed out in Sect.3.2.4, these molecules are adsorbed not on the cations themselves, but rather on the active EA centers: tricoordinated Al atoms. The cation only influences the coordination sphere of Al and thus controls its activity. This apparently explains the correlation between r_{H-H} and the nature of the exchange cation. The proposed model of the PD (Brönsted) center agrees well with the high catalytic activity of these zeolites. According to GVAKHARIY, catalytic activity in the reaction of cumene cracking declines in the following order: ScNaY > LaNaY ~ NaY(dec.) > MgNaY > SrNaY. At the same time the distance r_{H-H}, as indicated by NMR, decreases in the succession: 2.2 (ScNaY) > 1.67 (LaNaY) > 1.66 (NaY(dec.)) > 1.65 (MgNaY) > 1.63 (CaNaY) ~ 1.63 (SrNaY).

4.2.5 The Influence of Chemical Modification of Oxide Surfaces on the Protonization of Adsorbed Molecules

From the standpoint of the donor-acceptor mechanism of adsorption, the strength of the bond in the complex $D^{+\delta}M^{(q_{eff}-\delta)}$ and the extent of deformation of the molecule depend primarily on the effective charge q_{eff} of the acceptor atom. This can be controlled by modifying the immediate chemical environment of the M atom. For instance, according to [4.106,184,245], if OH groups in the first coordination sphere of the M atom are replaced by more electronegative halogen atoms, the result will be an increase in q_{eff} and in the activity of the M atom as EA center, which in turn leads to higher protonization of the water molecule adsorbed on this center (i.e., an increase in the activity of the PD center). For example:

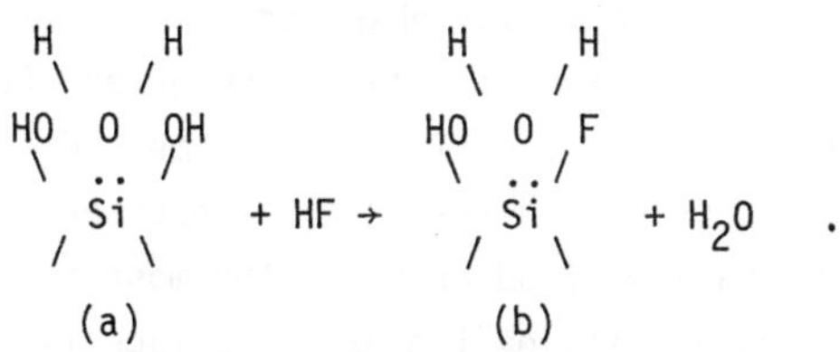

Scheme 4.4

In state (a) the H_2O molecule ought to be more highly protonized than in state (b). On the other hand, the decreased concentration of OH groups resulting from fluorination of the surface must cause a considerable reduction in the activity of the surface with regard to adsorption according to the mechanism of formation of H bonds with OH groups on the surface.

In order to study the interaction between molecules of various structures and certain functional surface groups, silica gels and aerosils have been synthesized in which OH groups were substituted to varying degrees by methoxy, methyl, ethyl and other saturated radicals, as well as by radicals having acid and basic properties. The majority of investigations have been aimed at clarifying the influence of chemical modification on physical Van der Waals adsorption of saturated compounds and on the adsorption of molecules capable of forming H bonds or π complexes. Fluorination and methylation were shown to make the surface hydrophobic, and to make the convex isotherms of water adsorption concave. A detailed review of all these investigations can be found in [4.246,247]. The mechanism by which chemical modification influences the activity of EA and PD centers of the surface has received less attention.

CHAPMAN and HAIR [4.248] were the first to discover that treating porous glass with ammonium fluoride leads to the substitution of a large fraction of OH groups by F, and to increase Brönsted acidity of the surface; the latter manifested itself through an increase in the catalytic activity of SiO_2 in the reaction of cumene cracking. Chapman and Hair assumed that the electronegative F atom pulls the electron cloud over from neighboring atoms, which results in protonization of the OH groups:

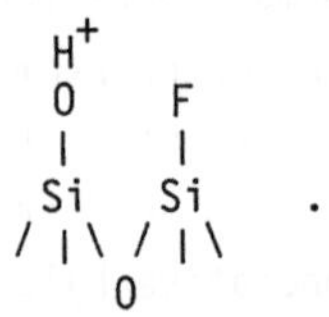

This is not obvious since it would demand the redistribution of the electron density in two adjacent tetrahedrons linked mainly by σ bonds. CANT and LITTLE [4.249] pointed out that this scheme, if realized, would cause a shift in the valence oscillations of OH groups, which is not actually observed.

A sharp increase in proton acidity and catalytic activity was observed after fluorination of γ-Al_2O_3 [4.250]. The IR spectra of fluorinated specimens displayed intensive bands of PyH^+ ions not observed with the initial specimen. From the viewpoint of the coordination mechanism of adsorption, the most plausible reason for the increase in the proton acidity of the oxide surface is an increase in the q_{eff} of Al and Si atoms due to the penetration of their coordination sphere by fluorine atoms, and a subsequent increase in the protonization of water molecules acting as PD centers, see Model 4.4. These hypotheses were examined in [4.136,251-253] with simultaneous NMR spectroscopy of fluorine and protons on fluorinated specimens of silica gel $SiO_2(F)$ and γ-$Al_2O_3(F)$.

Let us consider first the data on SiO_2. Fluorination with a surface concentration of 3% F by mass resulted in an abrupt drop in the intensity of the signal from single OH groups, see Model (3.25), and a nearly threefold decrease in the intensity of the signal from OH(II) and OH(III) groups. The latter is in in conformity with the behavior of the IR spectra [4.251,254], where fluorination of SiO_2 caused a drop in the intensity of the OH band (ν'_{OH} = 3750 cm^{-1}). Analysis of the shape of the NMR line of ^{19}F nuclei (Fig.4.26) reveals that the atoms of F replace the OH groups on the surface of SiO_2, and that their topography resembles the topography of protons (Sect.3.2.2). The surface disposition of fluorine atoms is indicated by the fact that with an increase in the fluorine concentration up to 5.5% by mass, the second moment of the NMR signal from ^{19}F nuclei ($^{F}S_2 = 0.5\ G^2$) is much lower than the value of $^{F}S_2$ for the bulk compound SiF_4 ($^{F}S_2 = 7.2\ G^2$ [4.255]). The first stage of fluorination is apparently the formation of coordination bonds between HF and silicon of the silanol groups ($\nu_c^F = 905\ cm^{-1}$ [4.156]). When T_{cal} exceeds 500 K, this band is erased from the spectrum and replaced by the band $\nu_F = 950\ cm^{-1}$, which is typical of Si-F groups. The formation of these groups is confirmed by ESCA data [4.256].

An important result was obtained in investigations of the NMR spectra of protons on a fluorinated silica specimen SiO_2(F). Figure 4.27 shows that the spectrum has an intensive wide component from water molecules. This signal persists in the spectrum after evacuation of specimens at T_{cal} up to 750 K,

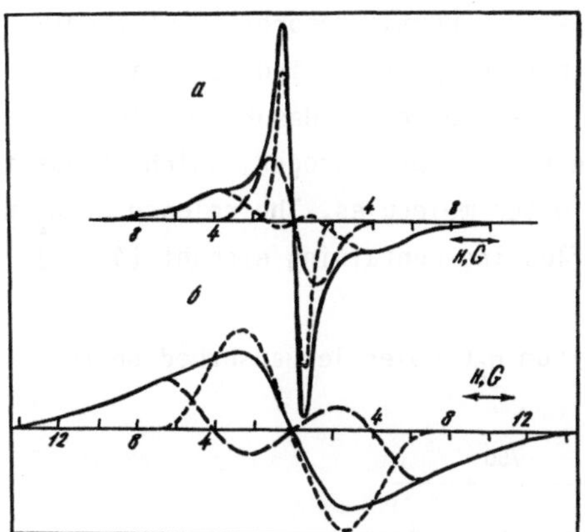

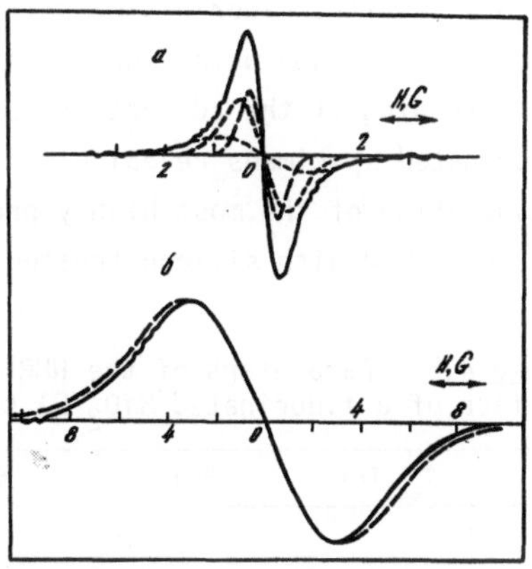

Fig. 4.26 The derivatives of NMR ^{19}F absorption signals from fluorinated specimens of (a) SiO_2 (3% F by mass, T_{cal} = 500 K) and (b) γ-Al_2O_3 (2% F by mass; T_{cal} = 600 K). The dashed lines represent the components of the signal

Fig. 4.27 The derivatives of NMR proton absorption signals from fluorinated specimens of (a) SiO_2 (T_{cal} = 750 K) and (b) γ-Al_2O_3 (6% F by mass; T_{cal} = 800 K). The dashed lines correspond to signals from H_2O and OH groups

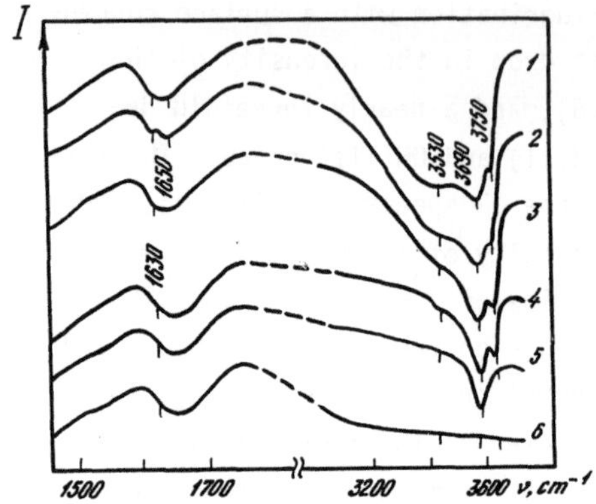

Fig. 4.28 IR transmission spectra (in arbitrary units) of fluorinated silica gel (3% F by mass) vacuumized at various temperatures [4.251]

which indicates how strongly the water molecules are bound to the surface. The proton-proton distance r_{H-H} in the water molecules, evaluated on the basis of experimental values of S_2, was $\sim 1.7 - 1.8$ Å, which confirms that they are highly protonized. The IR spectra of this specimen (Fig.4.28) displayed bands of coordination-bound water with a very high thermal stability (up to $T_{cal} = 700$ K) [4.251]. Recall that on hydrated SiO_2 they persist up to $T_{cal} =$ 500 K (Fig.3.19). The adsorption of pyridine gave rise to intensive bands of PyH^+. The data on concentrations n_c and proton-proton separations r_{H-H} for coordination-bound water molecules on the surface of fluorinated $SiO_2(F)$ (3% by mass) are presented in Table 4.5.

Recent investigations of the mass spectra of products of thermal desorption from surfaces of fluorinated and hydrated specimens of SiO_2 [4.245] confirm the NMR results. Fluorination shifts the peak of thermal desorption of firmly coordination-bound water towards higher temperatures, which indicates a strengthening of the adsorption bonds. The mass spectra of desorption from fluorinated specimens reveal a considerable release of hydrogen, which is due to dissociation of the most highly protonized water molecules. The release of H_2 was also observed with silicon treated with a fluorine-containing etchant [4.257].

Table 4.5. Parameters of the NMR signals from H_2O molecules adsorbed on the surface of a fluorinated $SiO_2(F)$ specimen

T_{cal} [K]	500	600	750
Parameters			
$n_c \cdot 10^{-14}$ molec·cm^{-2}	0.90	0.87	0.84
$H_2O\,S_2$ [G^2]	13.0±0.5	11.7±0.5	10.5±0.5
r_{H-H} [Å]	1.74	1.75	1.80

Another interesting feature of fluorinated SiO_2 specimens is a considerable weakening of the bonds between the remaining OH groups and the surface. Analysis of the IR and NMR spectra reveals that the least thermally stable are single hydroxyls OH(I), which are totally removed from the spectrum at T_{cal} = 600 - 800 K (Fig.4.28). On hydrated specimens (Sect.3.2.2), in contrast, these groups have the highest thermal stability. It is very probable that the atoms of fluorine replace some oxygen atoms in the siloxane bridges, which facilitates the rotation of adjacent tetrahedrons (Fig.3.22) and lowers the dehydration temperature as compared with untreated specimens.

The NMR signal of ^{19}F from fluorinated γ-Al_2O_3 is wider than that from SiO_2(F) because of the greater contribution made to the second moment of the fluorine line by interactions of Al and F nuclei. It was found that with concentrations below 2% by mass the fluorine atoms occupy the surface, while at higher concentrations they penetrate the surface layer to form fluorides. In the proton spectrum of γ-Al_2O_3 (Fig.4.27) it is possible to single out a wide component of the NMR signal of water which remains up to T_{cal} = 750 K. The IR spectrum displays bands of coordination-bound water (υ_c and υ_d), as well as intensive PyH^+ bands up to T_{cal} = 800 K [4.258]. The proton-proton separation r_{H-H} derived from the value of $S_2^{H_2O}$ ranges from 1.67 to 1.70 Å [4.253]. As with SiO_2(F) and zeolites, the most thermally stable water molecules are those which are most deformed.

Similar conclusions can be drawn from data on fluorinated specimens of TiO_2 (rutile) [4.162,212]. Analysis of the IR spectra revealed that fluorination elevates the temperature at which the υ_c and υ_d bands are destroyed from 650 to 750 K. While untreated rutile specimens displayed bands of PyH^+ ions, fluorinated ones also showed ions of a weaker aniline base (AnH^+). An increase in proton acidity upon chlorination and fluorination under certain conditions was also observed in [4.259,260]. The increased strength of the coordination bonds between water and a fluorinated TiO_2 surface, as well as increased protonization, are also confirmed by mass spectrometry (Fig.4.25). The activation energy of the desorption of water, $E_d^{H_2O}$, increased upon fluorination from 100 to 163 $kJ \cdot mol^{-1}$, while the activation energy of the release of deuterium, $E_d^{D_2}$, in contrast decreased from 210 to 155 $kJ \cdot mol^{-1}$.

All these facts support the mechanism suggested at the beginning of this section for the influence of chemical modification on the protonization of coordination-bound water molecules. On a qualitative level all this is in agreement with the known bulk properties of fluorine-containing compounds. The replacement of oxygen by fluorine in the coordination sphere closest to atom M always results in an increase of the q_{eff} of the central atom M. For instance, the value of q_{eff} is 0.78 for AlO_6 but 1.32 for AlF_6, and the ionicity of the Al-O bond is 63% while that of Al-F is 72% [4.261]. This was convincingly demonstrated by a comparison of q_{eff} for oxides and fluorides based on data obtained with X-ray photoelectron spectroscopy [4.262]. The effective charge q_{eff} of SiF_4 is also higher than in SiO_2, which is confirmed by the former's much greater ability to form complexes. SiF_4 complexes with Py, NH_3, and other amines are well known [4.263].

The strong influence which chemical modification has upon the strength of the coordination bonds of molecules and upon the extent of molecular deformation could be one reason for the seemingly contradictory interpretations of the IR spectra of hydrate covers of oxides. Since the concentration of coordination-unsaturated centers on the surface is not too high ($\sim 10^{13}$ centers $\cdot$ cm^{-2} in the case of SiO_2), the presence of even a small amount of halogen molecules on the surface will affect the activity of these centers.

Finally, let us consider the influence of fluorination on the mechanism of water adsorption, which proceeds via the formation and subsequent merging of clusters to form a polymolecular film (Sect.4.2.3). In accordance with all the published results [4.247], fluorination of the SiO_2 surface was found to lower the adsorptive activity of the surface with respect to water vapor (Fig.4.21), which must be attributed to a decrease in the concentration of OH groups on the surface. Indeed, while the value of α_{OH} for an initial specimen hydrated and evacuated at $T_{cal} = 500$ K was $\sim 12\ \mu mol \cdot m^{-2}$, the value for a specimen containing 3% F by mass and evacuated at the same temperature was 4.6 $\mu mol \cdot m^{-2}$, and 1.2 $\mu mol \cdot m^{-2}$ for a specimen evacuated at 750 K. At these temperatures, however, a considerable quantity of coordination-bound water remains on the surface of $SiO_2(F)$, with α_c not less than 1.4 - 1.5 $\mu mol \cdot m^{-2}$.

The adsorption of H_2O was measured on a $SiO_2(F)$ specimen dehydrated at 750 K, when many coordination-unsaturated centers are blocked by highly protonized water molecules which can act as adsorption centers (Fig.4.21). For partially hydrophobic surfaces the Dubinin-Serpinsky equation (2.46) yields a rather good estimate of the number of adsorption centers. For the specimen in question such computation gave the value 2.5 $\mu mol \cdot m^{-2}$, which is close to the sum of $\alpha_{OH} + \alpha_c = 2.6\ \mu mol \cdot m^{-2}$. This can be taken as a direct indication that coordination-bound water molecules are active centers of subsequent water adsorption.

A comparison of the second moment of protons of adsorbed water on the surface of hydrated and fluorinated SiO_2 specimens (Fig.4.21a) reveals that the merging of water clusters and the formation of a polymolecular film occurs at close values of occupation in the two cases: $a > 5a_m$ ($\sim 30\ \mu mol \cdot m^{-2}$). According to the adsorption isotherms, however, (Fig.4.21b), with a fluorinated specimen this process is shifted towards greater relative pressures of water vapor.

The formation and tangential propagation of clusters ought to depend on the concentration of OH groups on the surface (Fig.4.22). Although upon fluorination the number of primary adsorption centers increased by a factor of 3 or even 5, the number of OH groups decreased nearly tenfold. This is apparently the reason why merging of clusters on fluorinated surfaces occurs at much greater values of p/p_S than on hydrated surfaces.

5. Excited States in Adsorption and Catalysis

One of the major problems in the study of the mechanism of adsorption and catalysis relates to the transfer paths of energy released during the interaction between the molecules and the solid surface. Indeed, the heat of chemisorption of certain molecules on a solid surface can amount to a hundred kilojoules per mole. This is more than enough for the excitation of vibrational degrees of freedom (with $E \sim 30-60\ kJ \cdot mol^{-1}$), and in some cases also for electron excitation of the molecules. This energy can be channeled both to the formation of a stronger bond of the adsorbed molecule, and to chemical transformations.

Many researchers have considered the possibility of using the chemisorption energy and the energy of the surface reaction to accelerate the reaction itself. Problems of energy exchange between the adsorbed molecules and the solid in the course of adsorption and catalysis were analyzed in pioneering works by LANGMUIR [5.1] and RIDEAL and Taylor [5.2]. In 1934 ROGINSKY and ZEL'DOVICH [5.3] suggested the existence of a so-called presorptive state (Sect.2.2.2), which precedes the establishment of a strong chemisorptive bond. This active state may lead both to steady chemisorption and to the catalytic reaction. Since then, the mechanism by which the energy of catalytic reactions is used has occasionally been examined. In 1953 SEMENOV and VOEVODSKY [5.4] proposed the so-called radical chain theory of catalysis. According to this theory, the excess free energy of catalytic reactions and of homogeneous chain reactions is carried by free radicals, which bring fresh quantities of the initial substances into the reaction. The aggravation theory was based on the assumption that the increase in catalytic activity with the increasing molecular weight of the catalyst is due to its increased ability to retain excitation energy obtained in one way or another [5.5]. The concept of resonance in adsorption and catalysis is based on the closeness of the oscillatory terms of the adsorbed molecule and the solid (or the proximity of the oscillatory and electron levels) [5.6].

All of these concepts, as well as other ones not mentioned here, have in common the assumption that a molecule on the surface, having acquired excess energy in the course of chemisorption or a surface chemical reaction, can persist in the excited state for a while before transferring the energy to the solid. This energy is utilized in further transformations on the surface.

In most cases the theory of a transition state failed to predict the reaction rate on the basis of the known structure of the reacting molecules. Gas kinetics already deals with the task of calculating reaction cross sections (reaction rate constants) as functions of the energy of various degrees of freedom of the reacting molecules. In the very near future, the theory of chemisorption and heterogeneous catalysis will be confronted with a similar problem which belongs to so-called "level" kinetics. It will include measuring and calculating the rate of energy exchange between the adsorbed molecule and the surface, the speed of energy transfer along the surface, and the reactivity of vibrationally and electron-excited molecules on the surface. Progress in theoretical approaches and the development of novel experimental techniques have now made it possible to begin a systematic study of the role of excited states in chemisorption and catalysis [5.7,8].

5.1 Vibrationally Excited Molecules on Solid Surfaces

5.1.1 The Lifetime of Vibrationally Excited Molecules on the Surface

The creation of vibrationally excited molecules is known to be the main path of activation in homogeneous gaseous-phase reactions, and this process has therefore received much attention in experimental studies [5.9]. In order for the reaction to take place, sufficient vibrational energy must be accumulated on the bond which is to be ruptured. Similar excitation ought to exist in heterogeneous processes. If one assumes that the processes of adsorption and desorption proceed via a stage in which vibrationally excited molecules are created, then this stage must be taken into account both in deriving the isotherms and the kinetics of adsorption, and in describing the kinetics and the mechanism of catalytic reactions (Chap.2).

Let us analyze the consequences which arise if the creation of vibrationally excited molecules and the corresponding relaxation processes are taken into account in the derivation of the conventional Langmuir adsorption isotherm.

First we write down a very simple stepwise scheme of reversible adsorption of particle A on an active center z with the formation of excited state (Az)*:

$$A + z \underset{k_2}{\overset{k_1}{\rightleftharpoons}} (Az)^* \quad , \qquad (Az)^* \underset{k_4}{\overset{k_3}{\rightleftharpoons}} Az \quad .$$

The occupancy of active centers is

$$\Theta = \frac{\frac{k_1}{k_2}\frac{k_3}{k_4} p_A}{1 + \left(\frac{k_1}{k_2}\frac{k_3}{k_4} + \frac{k_1}{k_2}\right)p_A} \quad , \tag{5.1}$$

where k_1, k_2, k_3, k_4 are the rate constants of elementary steps, and p_A is the pressure of component A. Putting $k_1/k_2 = a$; $k_3/k_4 = b$, we get

$$\Theta = \frac{abp_A}{1 + a(b + 1)p_A} \quad . \tag{5.2}$$

If $b \gg 1$ (i.e., the equilibrium constant of the relaxation stage is large), then (5.2) becomes

$$\Theta = \frac{abp_A}{1 + abp_A} \quad , \tag{5.3}$$

and is identical with the Langmuir isotherm.

The case $b \ll 1$ (i.e., processes with a very small equilibrium constant of the relaxation stage) has rather interesting implications. For instance, at very high pressures

$$\Theta = b / (b + 1) \quad , \tag{5.4}$$

i.e., the maximal coverage attainable is less than 1. To check this relationship, one needs a complete set of adsorption isotherms at various temperatures. Since factor b apparently decreases with temperature, the value of Θ also ought to diminish with a rise in temperature. Naturally, these judgments apply to electron-excited molecules as well.

Since (4.2) does not differ formally from the Langmuir isotherm, the kinetic equations of catalytic processes, according to the law of mass action, will not differ from those obtained with neglect of vibrationally excited particles.

Consider an example of adsorption kinetics in which the creation of excited particles is taken into account (for the sake of simplicity, desorption is neglected).

Excitation neglected

$$A + z \xrightarrow{k_1} Az$$

$$p = p_A^0 \exp(-k_1 t) \quad . \tag{5.5a}$$

Excitation taken into account

$$A + z \underset{k_{-1}}{\overset{k_1}{\rightleftharpoons}} (Az)^*$$

$$(Az)^* \xrightarrow{k_2} Az$$

$$p_A = p_A^0 \exp\left(-\frac{k_1 k_2}{k_{-1} + k_2} t\right) \quad . \tag{5.5b}$$

It is apparent that if the relaxation is rapid ($k_2 \gg k_{-1}$), then (5.5a) coincides with (5.5b). The general form of these equations of monomolecular kinetics is the same, but the values of constants, activation energies, and pre-exponential factors differ.

The creation of vibrationally excited molecules must be taken into account when analyzing desorption by the flash method. Consider, for example, monomolecular desorption from the surface by analogy with monomolecular dissociation in the gaseous phase:

$$Az \underset{k_{-1}}{\overset{k_1}{\rightleftharpoons}} (Az)^* \xrightarrow{k_2} A + z \quad .$$

The formula for the reaction rate has the form

$$v = \frac{k_1 k_2}{k_{-1} + k_2} Az \quad . \tag{5.6}$$

Consequently, by varying the rate at which the specimen is heated, we can control the limiting stage of desorption, and thus determine all the necessary constants of this process.

Theoretical works dealing with the mechanism of transfer of vibrational energy from the adsorbed molecule to the solid are scarce [4.10-19], and there is no established opinion on the mechanism of "vibration relaxation".

At low surface occupancies, vibrational excitation of the adsorbed molecule is relaxed via direct energy transfer to the solid; at high occupancies the energy can be channeled within the adsorbed layer.

The most simple and natural transfer of energy from the adsorbed molecule to the lattice is possible when the quantum of vibrational energy is equal to the energy of one of the phonons of the crystal lattice. The lifetime of such excitation can be estimated as follows: In real crystals the free path of a phonon λ_{ph} is $\sim 10 - 10^2$ lattice constants. Dividing λ_{ph} by the speed of sound ($\sim 10^5\ cm \cdot s^{-1}$ in crystals) we find the lifetime to be $\sim 10^{-10}$ s. The upper limit of the phonon spectrum in MgO, for example, corresponds to about

600 cm^{-1}. Thus, this mechanism is probable for excited states with energy ($E \leq 0.1$ eV) and has a lower, though not negligible probability for $600 \leq E \leq 1200$ [cm^{-1}], assuming a two-phonon process.

A similar estimate can be made if energy is assumed to be transferred over the local oscillation of a kind of "impurity" center. According to MARADUDIN [5.15], who considered a molecular crystal, such an oscillation has a lifetime of $\sim 10^{-8}$ s, if its frequency satisfies the condition $\omega_{loc} \leq 2\omega_D$, where ω_D is the Debye frequency. This calculation was carried out under the assumption that the molecular weight of the "impurity" molecule is comparable to that of the atoms or molecules of the lattice. If, however, the energy of excitation of the molecule is much greater than $h\omega_D$, the transfer of excitation must involve a large number of phonons. The probability of a many-phonon process is

$$p_n \sim e^{-S} \cdot \frac{S}{n!} \quad , \tag{5.7}$$

where S is the dimensionless Stokes loss (in the optical sense). The number of phonons taking part in the energy transfer is estimated by the formula

$$n = \Omega / \omega_c \sim (10 - 20) \quad , \tag{5.8}$$

where Ω is the intramolecular frequency of oscillations ($\sim 10^3$ cm^{-1}) and ω_c is the uppermost frequency of intermolecular oscillations in a molecular crystal ($\sim 10^2$ cm^{-1}).

KOZHUSHNER and SHUB [5.17] calculated the probability of creation of vibrationally excited molecules, assuming quasi-classical motion of the molecule in the adsorption potential pit U(R), represented by the Morse potential (Fig.5.1):

$$U(R) = D(e^{-2\alpha R} - 2e^{-\alpha R}) \quad , \tag{5.9}$$

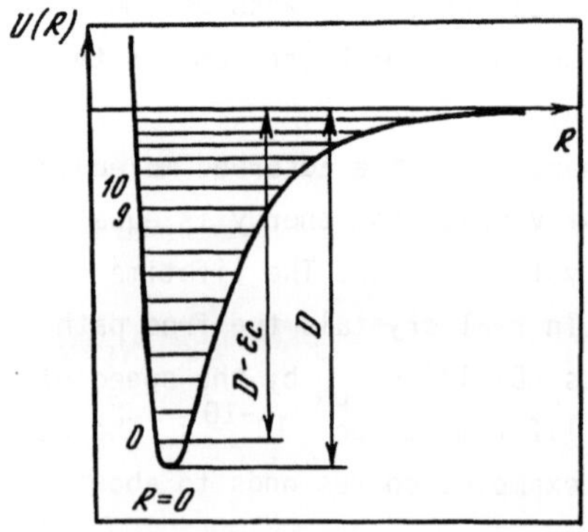

Fig.5.1. The potential energy of a molecule, described by the Morse function (5.9). U is the energy; D the depth of the potential pit, which is approximately equal to the dissociation energy; D - ε_0 is the true energy of dissociation at 0 K; ε_0 is the level of zero oscillations; and R the interatomic distance in a molecule

where R is the coordinate; D is the energy of dissociation (depth of the potential pit); α is a constant characteristic of the given molecule; and $R=0$ is the equilibrium position of the molecule.

During adsorption the energy of the translational movement of the molecule as a whole is transferred to the internal oscillations of the molecule. For this process to be possible, the depth D of the potential pit must be greater than $\hbar\omega_0$, where ω_0 is the frequency of intramolecular oscillations: $D > \hbar\omega_0$. If, however, $D < \hbar\omega_0$, then excitation of intramolecular oscillations is hardly possible, even in an already adsorbed molecule. In this case the oscillation participated in by the adsorbent atom closest to the adsorbed molecule must build up the energy $\hbar\omega_0 \gg kT$, which is very improbable. Moreover, such a fluctuation of energy, if $D < \hbar\omega_0$, will cause desorption of the nonexcited molecule, and the probability of desorption, p_d, will be much greater than the probability of excitation, p_e:

$$\frac{p_d}{p_e} \sim e^{\frac{\hbar\omega_0 - D}{RT}} \gg 1 \quad . \tag{5.10}$$

Consequently, the concept of formation of vibrationally excited molecules is adequate only if the energy of adsorption is sufficiently high.

Most probably the molecule, upon hitting the surface and losing an amount of kinetic energy equal to about kT, passes to a sufficiently high level and then relaxes, jumping from one level to another in the potential pit; this may give rise to vibrations. In the calculation of the probabilities of vibrational excitation of the molecule, p_e, and of relaxation transition p, the main contribution to the transition matrix element comes from the region of greatest variation in potential, that is, the region of repulsive potential. The latter is represented by the Morse potential.

The total probability of transition from a certain adsorption level is proportional to the time the molecule occupies this level. This time is determined by the thermal relaxation rate, i.e., the rate of transition between the two adjacent adsorption levels E and $E-\hbar\omega_0$, accompanied by the transfer of energy to oscillations of adsorbent atoms (lattice phonons). At the same time, oscillations of the molecule with energy $\hbar\omega$ start. The exact equality $n\hbar\omega = \hbar\omega_0$ (n is an integer) is not always fulfilled. However, since the adsorption levels are broadened due to interaction with phonons, and additional degrees of freedom exist in the molecule, the state of resonance can be considered to be always attained. The probability of transition depends little on the magnitude of energy transferred to those additional degrees of freedom.

The final expression for the probability that a vibrationally excited molecule will be created in the course of adsorption has the form

$$p_e = \frac{m_a}{\mu} \frac{D}{\hbar\omega_0} x^2 e^{-x} \frac{1}{1 + \bar{n}(T)} \quad , \qquad (5.11)$$

where m_a is the mass of the lattice atom; μ is the equivalent mass for the intramolecular oscillation; $\bar{n}(T)$ is the Bose distribution function of phonons with frequency ω over lattice temperature T;

$$x = \frac{2\pi\sqrt{2m\hbar\omega_0}}{7\hbar\alpha} \left(\frac{\hbar\omega_0}{D}\right)^{1/2} \quad ; \qquad (5.12)$$

m is the mass of the molecule; and α is the same constant as in (5.9). The dependence of p_e on the mass of the molecule enters through x, which also depends on ω_0 according to (5.12).

Estimated via (5.12), the probability that a vibrationally excited CO molecule will be created during adsorption upon Pt was found to be at least 10^{-5}.

Under the same assumptions, the expression for the lifetime was also obtained:

$$\frac{1}{\tau} = \pi\sqrt{D} \frac{1}{\alpha} \frac{m^{3/2}\omega_0}{\hbar\omega} \exp\left(- \frac{\pi\sqrt{2m\hbar\omega_0}}{\hbar\alpha}\right) \quad , \qquad (5.13)$$

where the designations are the same as in (5.11). The lifetime of the same CO molecule on Pt, estimated via (5.13), is $\tau_{min} \approx 10^{-5}$ s.

Knowing the probability of formation of vibrationally excited molecules and their lifetime on the surface, one can calculate the equilibrium concentration of excited molecules (adsorption equilibrium). With reasonable values for the adsorption parameters (adsorption heat $\sim 80\ \mathrm{kJ \cdot mol^{-1}}$, concentration of active centers $10^{15}\ \mathrm{cm^{-2}}$, T = 300 K, nearly complete coverage) this concentration is $\sim 10^{12} - 10^{13}\ \mathrm{cm^{-2}}$.

We see that even at adsorption equilibrium the concentration of excited adsorbed molecules can be rather high, and these molecules can play a certain role in heterogeneous catalysis and other surface chemical reactions.

Simple kinetic estimates reveal that the calculated lifetimes are sufficient for the molecules to take part in the catalytic reaction, even if only direct hits of molecules from the gaseous phase are taken into account. For instance, at a pressure of 1 atm, the number of collisions between molecules from the gaseous phase and a single surface atom is about $3 \cdot 10^8$ per second. Consequently, if the lifetime is greater than 10^{-8} s, the kinetics of excitation and relaxation must be taken into account when calculating the overall reaction kinetics.

At high occupations, energy transfer between the adsorbed molecules becomes more probable. Under certain conditions an adsorbed molecule can transfer its energy to a similar nonexcited molecule by means of resonance. According to the Förster-Dexter theory, the probability that resonant energy will be transferred from donor D to acceptor A is

$$p_{D\to A} \simeq \frac{1}{\tau_0}\left(\frac{R_0}{R}\right)^m \quad , \tag{5.14}$$

where τ_0 is the lifetime of the excited state; R is the distance between interacting molecules; R_0 is the minimal distance at which this interaction will take place; and m takes on values of 6, 8, or 10 for dipole-dipole, dipole-quadrupole and quadrupole-quadrupole interactions respectively. In real systems, resonant energy transfer is reasonably probable at distances of 40 - 70 Å for dipole-dipole and 8 - 15 Å for dipole-quadrupole interactions [5.20,21].

MORGAN and SOMORJAI [5.22] used the concept of resonant energy transfer to explain a steep increase in the rate of oxidation of CO on Pt observed when the occupancy exceeds a certain level. This increase may occur after the formation of islands of adsorbed molecules on the surface ("patches"). When the CO molecules are adsorbed on adjacent centers within a cluster, resonant energy transfer from one adsorbed molecule to another becomes possible. In this way, the energy accumulated within a cluster can become sufficient to cause excitation of molecules, e.g.:

```
          O  ħω  O  ħω  O* ħω  O  ħω  O
          C  →   C  →   C  ←   C  ←   C
          |      |      |      |      |
   - Pt - Pt  -  Pt  -  Pt  -  Pt  -  Pt - Pt -    .
```

The conditions of resonance are well satisfied for both harmonic and anharmonic (in the case of low excitation, when the frequencies of the two oscillators are close to each other) oscillator. The time of transfer of excitation from one molecule in the nth oscillatory state to another, nonexcited one, is [5.14]:

$$\tau_n = \frac{n^2}{n!\, x_e^{n-1}} \quad , \tag{5.15}$$

where x_e describes anharmonicity.

For a molecular crystal these excitations were shown by OVCHINNIKOV [5.14] to be able to migrate for a long time without giving away their energy. In this case the lifetime of the excited state will be limited by the possible deexcitation processes, which include: (1) emission of radiation with the molecule passing to the (n - 1)th level which takes $\sim 10^5$-10^6 s for crystals of solid O_2, N_2, and H_2, and $\sim$1 - 10 s for CO, NO, etc.; (2) transition of the molecule to the (n - 1)th level with transfer of one quantum to the neighboring molecule and absorption of one or more phonons from the lattice. This process depends strongly upon n and T. For example, at $n = 7$ and $T = 30$ K such a transition takes about 10 s. The values of time obtained are quite high; note, however, that they refer to molecular crystals at very low temperatures.

In theoretical studies by KOZHUSHNER et al. [5.17,23] it was demonstrated that there are four main physical mechanism of energy transfer from molecules adsorbed on the surface.

1) Energy transfer to phonons as discussed above. This mechanism is most typical of adsorption on dielectrics. If there is an opportunity of exchanging an oscillatory quantum for one or two lattice phonons, then the characteristic lifetime of the excitation state will be in the range of $10^{-7} - 10^{-10}$ s, and show little dependence upon temperature. With the many-phonon mechanism the relaxation time can be of the order of seconds, and decrease considerably with a rise in temperature.

2) Energy transfer in adsorption on metals. This is generally determined by the probability that the energy will be taken by free electrons. If the multipole, exchange, and polarization interaction between the intramolecular oscillations and the conduction electrons are taken into account, the lifetime of the excitation state will depend little on the type of metal or the magnitude of the oscillatory quantum $h\omega_0$ of the molecule, in contrast to the previously discussed mechanism. The lifetime virtually does not depend on temperature, and is $\sim 10^{-8} - 10^{-10}$ s, depending on the nature of the interaction and the equilibrium distance between the molecule and the surface. The mechanism of relaxation on semiconductors with a forbidden bandwidth lower than $h\omega_0$ is similar to relaxation on metals.

3) Energy transfer in semiconductors having a forbidden bandwidth greater than $h\omega_0$. The energy can either be transferred to free electrons or expended on the ionization of impurities. In [5.26] we have observed the energy transfer from excited dye molecules to the deep electron centers in dielectric-semiconductor structures. The relaxation time will depend strongly upon temperature, and the energy can also be channeled to the excitation of polarons and plasmons.

4) Energy transfer to the translational motion of the same molecules. This mechanism is possible when weak bonding between the molecule and the surface is combined with a low probability of many-phonon energy dissipation. The characteristic lifetimes can be as great as $10^{-6} - 10^{-4}$ s. The probability of transition per oscillation in the adsorption potential pit (Fig.5.1) in this case is comparable to the probability of Landau-Teller oscillation-transla-

tional exchange in gases. This mechanism opens up the possibility of low-temperature heterogeneous separation of isotopes.

5.1.2 Heterogeneous Vibrational Relaxation

As early as the beginning of this century, Knudsen, using so-called thermal accommodation coefficients, experimentally determined the energy exchange in collisions of atoms and molecules with the surface. If the mean energy of an incident particle is W_0, that of an outgoing particle is W_1, and the energy of the molecule on the surface is W_2, then the accommodation coefficient is

$$\gamma = \lim_{W_1 \to W_0} \frac{W_2 - W_0}{W_1 - W_0} \quad , \quad \text{or} \quad \gamma = \lim_{T_1 \to T_2} \frac{T_2 - T_0}{T_1 - T_0} \quad . \tag{5.16}$$

The latter expression contains temperature instead of energy. In general, the accommodation coefficient of a molecule on the surface is composed of three separate accommodation coefficients for the exchange of translational (γ_{trans}), rotational (γ_{rot}), and vibrational (γ_{vib}) energies. The techniques of measuring the accommodation coefficients are described in [5.24,25]. Most of them measure the coefficients without special attention to energy exchange with different degrees of freedom. Not until recently did reports on partial accommodation coefficients begin to appear [5.25-35], in connection with studies of the behavior of vibrationally excited molecules on the surface.

Pioneering work on the heterogeneous relaxation of vibrationally excited molecules was done by MORGAN and SCHIFF [5.27], who measured the accommodation coefficient for heterogeneous vibrational relaxation of nitrogen on the surface of Pyrex glass ($\gamma_{vib} = 4.5 \cdot 10^{-4}$).

In recent years a number of novel techniques have been developed to study the behavior of vibrationally excited molecules on surfaces.

GERSHENZON et al. [5.28,29], and KOVALEVSKY and SHUB [5.30] investigated heterogeneous vibrational relaxation of excited molecules of nitrogen and deuterium on different surfaces. Vibrationally excited N_2^* molecules were produced in the region of UHF discharge. They were further directed onto a precatalyst layer of TiO_2, which effectively removes atoms and electron-excited molecules from nitrogen. Some vibrationally excited N_2^* molecules are also removed though to a lesser extent. The excited molecules proceeded into the reactor, whose walls were covered by the relaxation catalyst under investigation. The concentration of vibrationally excited molecules in the outgoing flow was determined with the aid of titration: about 1% CO_2 was injected into the flow, and the intensity of IR radiation emitted by the CO_2 molecules was measured.

By virtue of quick resonant exchange between CO_2 and N_2 according to the reaction

$$N_2^* + CO_2 \rightleftharpoons N_2 + CO_2^* \quad ,$$

this intensity is a measure of the concentration of N_2^* molecules in the flow. A similar technique was employed to study heterogeneous vibrational relaxation of D_2^* molecules, which can also be effectively titrated by adding CO_2 to the flow.

The values of accommodation coefficients γ_{vib} obtained for N_2 and D_2 at 300-400 K range from 10^{-4} to 10^{-2}, and the corresponding effective first-order rate constants $k_{eff} = \gamma_{vib}\bar{v} / 2R$ range from 1 to 10^2 s^{-1}. (Here $\bar{v}$ denotes the mean thermal velocity of the molecule, and R is the radius of the container.) These constants were shown in [5.29] to be close to the constants of diffusional destruction, so that in determining γ_{vib}, one must take the diffusion and the flow velocity profile into account. In most cases the diffusion coefficient of vibrationally excited molecules turned out to be lower than that of nonexcited molecules, due to resonant exchange in the gaseous phase. Two types of temperature dependence of γ_{vib} were obtained. In order to explain the first kind of dependence (decrease in γ_{vib} with a rise in temperature, and no dependence on pressure at low temperature in vibrational relaxation of N_2^* on silver and Teflon), use was made of the mechanisms of physical adsorption of excited molecules and the adsorption-phonon mechanism of relaxation [5.28-30]:

$$N_2^* + z \underset{k_{-1}}{\overset{k_1}{\rightleftharpoons}} N_2^*z \quad , \qquad N_2^*z \underset{k_2}{\longrightarrow} N_2z \quad ,$$

where N_2^* is a vibrationally excited nitrogen molecule; z is the center of physical adsorption; and k_1, k_{-1}, and k_2 are the rate constants of elementary steps. The vibrational relaxation coefficient is defined as the ratio of the relaxation rate of excited molecules to the number of collisions between these molecules and the surface. Employing the method of quasi-stationary concentrations (assuming stationarity of N_2^*z), one easily obtains:

$$\gamma_{vib} = \frac{-d[N_2^*] / dt}{[N_2^*]\bar{v} / 4} = \frac{4k_1k_2(1 - \Theta)}{(k_2 + k_{-1})\bar{v}} \quad , \qquad (5.17)$$

where Θ is the occupancy. If the rate of deactivation is much greater than the rate of desorption, i.e., $k_2 \gg k_{-1}$, then

$$\gamma_{vib} = 4k_1(1 - \Theta) / \bar{v} \quad , \qquad (5.18)$$

and the relaxation coefficient is equal to the sticking coefficient k_s (2.62). Usually γ_{vib} is much lower than k_s. Consequently, $k_2 \ll k_{-1}$, and the limiting step is usually the transfer of an oscillatory quantum to the solid. Then

$$\gamma_{vib} = \frac{4k_2k_1}{k_{-1}\bar{v}}(1 - \Theta) \tag{5.19}$$

and, since $K = k_1 / k_{-1} = \Theta / (1 - \Theta)$,

$$\gamma_{vib} \simeq 4k_2\Theta / \bar{v} \quad . \tag{5.20}$$

The temperature dependence of the accommodation coefficient will then be similar to the function $\Theta(T)$, and the occupancy Θ decreases with a rise in temperature according to the exponential law $\sim\exp(q_d / RT)$, where q_d is the adsorption heat.

The characteristic lifetime of oscillatory excitation on the surface, τ_{vib}, can be estimated by using the value obtained for γ_{vib} and the occupancy Θ determined from the values of adsorption heat:

$$\tau_{vib} = \frac{1}{k_2} = \frac{4\Theta}{\gamma_{vib}\bar{v}} \quad . \tag{5.21}$$

For relaxation of nitrogen on silver, $\tau_{vib} \simeq 10^{-6}$ s [5.30]; for relaxation of nitrogen on molybdenum glass, $\tau_{vib} \sim 10^{-8}$ s [5.27]. This value is close to the calculated time of two-phonon decay of local oscillation [5.15]. BLACK et al. [5.31] obtained values of γ_{vib} which were 2 or 3 orders of magnitude higher; however, the values which they used for the heat of physical adsorption of N_2 were rather too high.

In relaxation of nitrogen on a quartz surface, γ_{vib} displayed the second kind of dependence: it decreased with pressure and increased with temperature [5.28]. This was explained with the aid of the mechanism of chemisorptive oscillatory relaxation [5.29]:

$$N_2^* + z_x \underset{k_{-1}}{\overset{k_1}{\rightleftharpoons}} N_2^*z_x \quad , \quad N_2^* + z_x \underset{k_{-2}}{\overset{k_2}{\rightleftharpoons}} N_2z_x \quad , \quad N_2^* + z_x \xrightarrow{k_3} N_2z_x \quad ,$$

where z_x is the center of activated chemisorption. The vibrational energy is lost due to the formation of chemical bonding with the surface. In this case $k_3 \gg k_{-1}$, i.e., the lifetime of the adsorbed molecule is great compared with the time needed to give off a quantum ($1 / k_{-1} > 1 / k_3$). Using the method of quasi-stationary concentrations, one easily gets

$$\gamma_{vib} = \frac{4n_xk_{S_1}}{\bar{v}} = \frac{k_{-2}}{\left\{k_{S_1}\frac{[N_2^*]}{[N_2]} + k_{S_2}\right\}[N_2] + \frac{4k_{-2}n_x}{\bar{v}}} \quad , \tag{5.22}$$

where n_x is the concentration of chemisorption centers [cm^{-2}], and k_{S_1} and k_{S_2} are the sticking probabilities for excited (N_2^*) and ground-state (N_2) nitrogen molecules. At $1 - \Theta_x \ll 1$, where $\Theta_x \sim 1$ is the occupancy of chemisorption centers, one gets

$$\gamma_{vib} = \frac{k_{S_1}}{k_{S_1}\frac{[N_2^*]}{[N_2]} + k_{S_2}} \frac{4n_x k_{-2}}{\bar{v}} \frac{1}{[N_2]} \quad . \tag{5.23}$$

The experimental value of the activation energy, estimated from the temperature dependence of γ_{vib}, is $\sim 25\ kJ \cdot mol^{-1}$. By finding the ratio of this value to the value of desorption activation energy (k_{-2}), one can estimate the number of chemisorption centers on quartz. If the preexponential factor for k_{-2} is close to normal ($10^{11} - 10^{12}\ s^{-1}$), then $n_x \sim 10^9\ cm^{-2}$. This low value of n_x is probably related to defects on the quartz surface (Chap.3); their concentration is hard to measure by conventional physical techniques.

For dipole molecules of HCl, CO, CO_2, and N_2O on various kinds of glass [5.29,32-34], higher values of γ were obtained: $\sim 10^{-1} - 10^{-2}$.

Among clean surfaces, those of metals are most active in heterogeneous vibrational relaxation; the energy exchange here follows the second mechanism discussed in Sect.5.1.1. Semiconductors are less active ($\gamma_{vib} = 1 \cdot 10^{-3}$ for vibrational relaxation of nitrogen on a silicon surface at room temperature), and dielectrics still less so: $\gamma_{vib} = 10^{-4} - 10^{-5}$ for N_2 and D_2 on SiO_2 or silica glass. The values of γ_{vib} can become greater in the presence of an adsorbed phase.

Heterogeneous vibrational relaxation of D_2 on the surface of a molecular crystal (solid CO_2) was studied in [5.17]. In the temperature range 77 - 125 K the vibrational relaxation coefficient was shown to be

$$\gamma_{vib} = 1.45 \cdot 10^{-4} \exp(200/RT) \quad . \tag{5.24}$$

With a rise in temperature the probability of energy transfer decreases. In [5.19] it was demonstrated that the main mechanism of relaxation here is the adsorption of D_2 on solid CO_2 and the transfer of vibrational energy from deuterium to the asymmetric valence oscillation of the CO_2 molecule, with simultaneous excitation of rotational and translational degrees of freedom of the deuterium molecule. In these weakly-bound molecules the energy is expended on the excitation of the molecule being desorbed, rather than being transferred to the lattice phonons. Such phenomena must be typical of cases of weak adsorption (the fourth mechanism of energy transfer in Sect.5.1.1).

5.1.3 The Róle of Vibrationally Excited Molecules in Catalysis

The kinetics of fast catalytic reactions is strongly dependent upon the rate of creation and heterogeneous relaxation of vibrationally excited molecules. Consider, for example, a quick process of recombination of atoms on a solid surface, according to KOVALEVSKY and SHUB [5.30]:

$$\text{(a)}\quad z + A \to zA \quad , \qquad \text{(c)}\quad [zA_2]^* \to zA_2 \quad ,$$

$$\text{(b)}\quad zA + A \rightleftharpoons [zA_2]^* \quad , \qquad \text{(d)}\quad zA_2 \to z + A_2 \quad .$$

At the stage (b) it is probable that vibrationally excited molecules will form, with subsequent relaxation of vibrational energy. If this assumption is correct, the overall process will depend on the rate of reaction, and a correlation ought to be observed between the vibration accommodation coefficients γ_{vib} and the corresponding coefficients of the recombination of atoms, β. The data obtained are presented in Table 5.1. It can be seen that γ_{vib} and β are symbatic, and even the numerical values are close in some cases. This regularity is naturally approximative, since the values of γ_{vib} are determined for the relaxation of molecules from the first vibrational level to the ground level, while recombination of atoms can create high oscillatory levels, and relaxation will involve transition from, say, the 10th to the 9th level (Fig.5.1).

If the bonding with the surface is weak, the vibrationally excited molecules can be desorbed from the surface without relaxation. For instance, a considerable increase in the concentration of vibrationally excited molecules in the gaseous phase was observed when deuterium atoms were passed

Table 5.1. Coefficients of the recombination of atoms, and vibrational relaxation coefficients of molecules on various surfaces

Gas	Surface	β	γ_{vib}
N_2	Pyrex	$3 \cdot 10^{-5}$	$1.5 \cdot 10^{-4}$
	Quartz	$8 \cdot 10^{-1}$	$7 \cdot 10^{-4}$
	Teflon	$2.9 \cdot 10^{-5}$	$6 \cdot 10^{-4}$
	Copper	$2.3 \cdot 10^{-3}$	$1 \cdot 10^{-3}$
	Silver	$1.5 \cdot 10^{-3}$	$1.1 \cdot 10^{-2}$
H_2	Pyrex	$1.5 \cdot 10^{-5}$	$1 \cdot 10^{-4}$
	Teflon	$3 \cdot 10^{-4}$	$1 \cdot 10^{-4}$
D_2	Quartz	$7 \cdot 10^{-4}$	$9.5 \cdot 10^{-5}$
	NaCl	$4.1 \cdot 10^{-4}$	$6.3 \cdot 10^{-4}$

through a platinum helix, and thus the formation of vibrationally excited molecules during a surface catalytic reaction was proved [5.35]. This result is important for an understanding of the phenomena encountered in heterogeneous-homogeneous catalysis, as well as chain reaction in general. It is often assumed a priori in the theory of chain reactions that the wall (i.e., solid surface) destroys atoms and radicals, and breaks the chain reaction. The data reported imply that the disappearance of atoms and radicals can be accompanied by the creation of vibrationally excited molecules and thus initiate a new chain of transformations in the gaseous phase.

KISLYUK et al. [5.36] studied the recombination of H atoms on Pt and found that hydrogen molecules take away about 70% of the total energy at 293 K. The probability that hydrogen atoms recombine on the surface of tungsten was calculated in [5.37]; it was shown that about 56% of the energy is concentrated on inner degrees of freedom in the form of vibrational and rotational excitation. The possibility of energy transfer to the lattice phonons was not, however, taken into account.

Apparently, in the above-described experiments by GERSHENZON et al. [5.27-29] and KOVALEVSKY and SHUB [5.30], a large fraction of excited N_2 or D_2 molecules titrated with the aid of CO_2 are formed from N or D atoms persisting in the discharge plasma.

A convenient technique of investigating the elementary stages of chemisorption and catalysis (in particular the energy exchange between the adsorbed molecules and the solid) is the method of molecular beams, which has been consistently developed [5.38-45]. A beam of reactant molecules injected through a small aperture into the vacuum chamber hits the target catalyst. The vacuum chamber is continuously evacuated ($p < 10^{-8}$ torr), and collisions of molecules within the beam or with the walls are eliminated. Diffusional hindrances are also avoided.

The advantage of the molecular beam method is that the following can be determined: (1) the minimum lifetime which a molecule on the surface must have in order for the reaction to take place; (2) the probability of reaction when a single collision takes place with the surface; (3) the dependence of the reaction upon the energy of the molecules in the beam, the temperature, and the energy distribution (including translational, vibrational, and rotational components) between the surface, the reactants, and the reaction products. Sources of beams of excited molecules have been especially developed for these purposes. With this method, one can also study the dependence of the reaction rate upon the quantity of adsorbate deposited on the target.

A convenient technique for studying the mechanism of multistage reactions and the rate constants of elementary steps is the method of modulated beams.

The beam of reactant molecules passes through an interrupter (a rotating toothed disc), and the phase shifts (delay times) of the release of various reaction products after collision of the molecular beam with the surface are measured. The theory and practice of modulated molecular beams was developed by MADIX and others [5.41-45].

If the molecules hitting the surface do not undergo a chemical change, two extreme cases can be observed: (1) mirror reflection of the molecules (the angle of incidence equals the angle of reflection); (2) adsorption of molecules, followed by the establishment of thermal equilibrium with the surface and evaporation. In this case the intensity of the reflected beam is proportional to cosine θ, where θ is the angle with the normal to the surface:

$$I = I_0 \cos\theta \quad . \tag{5.25}$$

The law (4.25) characterizes complete thermal accomodation of molecules.

It is a very common opinion in adsorption and catalysis (e.g., [5.46]) that the desorbed molecules obey the equilibrium Maxwell-Boltzmann distribution law. Studies of the angular distribution of desorbed molecules have proved the faultiness of this notion.

PALMER, SMITH and others [5.40,48] demonstrated that the distribution of molecules of CO_2 desorbed from the <111> Pt face declines from the law of cosine (5.25). This proves that the CO_2 molecules depart from the surface with excessive translational energy. ERTL [5.49] confirmed the results reported by PALMER and SMITH [5.40] and investigated the angular distribution of CO_2 molecules resulted from oxidation of CO on the <111> Pd face. CO_2 molecules desorbed from surfaces of platinum or palladium at 1150 K were found, when moving normal to the surface, to possess an excessive translational energy corresponding to a temperature of 3000 K (or 30 $kJ \cdot mol^{-1}$). ERTL [5.49] suggested that the CO_2 molecules, having crossed the potential barrier when created, leave the surface without falling into the potential pit.

Another explanation of the absence of thermal accommodation in this case [5.50] involves the concept of a preadsorption state (precursor; see Sect. 2.2.2):

$$CO_{2(\text{chem.})} \rightleftharpoons CO_{2(\text{prec.})} \rightleftharpoons CO_{2(\text{gas})} \quad .$$

In the precursor state the molecules can migrate freely along the surface, acquiring or giving off excess energy. From the precursor state the molecule can (1) pass into a tightly bound chemisorbed state (apparently of little activity); (2) enter a chemical (catalytic) reaction; (3) be desorbed, carrying away excessive energy. Weakly bound molecules are more capable of leaving the surface in an excited state than tightly bound ones.

The molecular beam method is currently the best technique for studying the dependence of the reaction rate upon various types of excitation of gas molecules. However, works dealing directly with this question are still scarce.

PRADO-SILVA et al. [5.51] studied the dependence of the reaction rate on the vibrational and translational components of the gas temperature and on the temperature of the catalyst in isomerization of cyclopropane into propylene on mica. A molecular beam of cyclopropane was directed onto a mica target heated to varying temperatures. By injecting inert gas molecules into the beam between the exit aperture and the target, the translational temperature could be lowered while the vibrational component remained high. The rate of isomerization was found to be independent of the translational component of the gas temperature. The dependence of the reaction rate on the translational temperature component corresponded to an activation energy of 270 $kJ \cdot mol^{-1}$, and the dependence on the surface temperature to an activation energy of 83 $kJ \cdot mol^{-1}$. RYSKIN [5.52] questioned the conclusions reached in [5.51], because the value 270 $kJ \cdot mol^{-1}$ corresponds to gas-phase isomerization of cyclopropane.

EHRLICH and STUART [5.53] used the method of molecular beams to study methane decomposition on the iridium tip of an electron projector. The evolution of the picture on the luminescent screen allowed them to monitor the rate of growth of the carbon film on the iridium tip, and thus the rate of the catalytic reaction. A comparison of the decomposition rates of methane on the one hand, and deuterated methane and ethane on the other hand, revealed an anomalously high isotope effect, which was attributed to the participation of vibrationally excited molecules in the reaction. In this case, the decomposition reaction apparently activates the deformation vibration of the methane molecule, as assumed by the authors. More recently, however, other explanations have been proposed [5.54] which do not involve the concept of vibrationally excited molecules.

The intensive development of laser technology provided a stimulus for work on the selective excitation of separate oscillatory modes of adsorbed molecules using laser radiation. This method was employed, for instance, by BASOV et al. [5.55], who demonstrated that the rate of condensation of vibrationally excited CO_2 molecules on the surface of condensed carbon dioxide depends upon the vibrational component of the gas temperature. It was proposed that this effect should be put to use in the separation of isotope-substituted $^{12}CO_2$ and $^{13}CO_2$ molecules. There have been reports by laser physicists stating that the rates of reactions involving vibrationally excited and nonexcited molecules are not the same (e.g., [5.56]). In some cases, however, it is impossible to separate selective (quantum) effects and purely thermal effects of laser radiation.

So far there is no conclusive proof that the reaction rate depends on whether the molecules are vibrationally excited or not, but there is experimental evidence that vibrationally excited molecules are created in adsorptive and catalytic processes.

It is very probable that vibrationally excited molecules participate in the so-called associative mechanism of heterogeneous catalytic reactions [5.57]. Note, however, that stepwise mechanisms also do not exclude the possibility that excited particles take part in the elementary steps; only the numerical values of the rate constants will then be different.

5.2 Electron-Excited Molecules in Adsorption and Catalysis

5.2.1 The Creation in Electron-Excited Molecules

Energy transfer from adsorbed particles to the solid can result in electron excitations in the solid itself: transition of electrons to higher levels (in particular into the conduction band), creation of surface excitons, etc. These phenomena are well known; they have been analyzed in a large number of theoretical works on adsorption and catalysis. Various experimental techniques, such as conductivity measurements and optical spectroscopy, have been used to investigate them.

A question which has received less attention is whether electron-excited molecules are created in adsorption and catalysis. The creation of excited molecules is most likely in highly exothermal reactions, e.g., during recombination of atoms $H+H(q=430\ kJ\cdot mol^{-1})$, $O+O(q=490\ kJ\cdot mol^{-1})$, and $N+N$ $(q=940\ kJ\cdot mol^{-1})$. And indeed, the creation of electron-excited nitrogen molecules was observed by HARTECK and co-workers [5.58-61] during recombination of N atoms on metals. First, electron-excited nitrogen molecules at high oscillatory levels in a metastable state $N_2^*(A^3\Sigma_u^+)$ with an energy of 6.17 eV $(584\ kJ\cdot mol^{-1})$ are created on the surface of Ni, Co, or Ag. Then they leave the surface, diffuse into the gaseous phase and, colliding with other molecules, pass into the state $N_2(B^3\Pi_g)$.

These states are vibrationally excited to levels $v'=8$ and 6, since the potential energy curves of terms $N_2(B^3\Pi_g)$ and $N_2(A^3\Sigma_u^+)$ intersect near $v'=8$ and 6. The $N_2(B^3\Pi_g)$ molecules produced can then be detected by intensive glow.

The formation of electron-excited molecules during heterogeneous recombination of N atoms was investigated by TABACHNIK and SHUB [5.62]. They found this process to take place only on clean metallic surfaces; an oxide film kills the catalyst. Excited N_2^* molecules are produced only on metals which adsorb nitrogen in molecular form. Ta, W, and the like, which adsorb nitrogen in dissociate form, do not produce any escaping excited N_2^* molecules.

The oxygen molecule, which is triplet in the ground state, has singlet low-energy excited states $^1\Delta_g$(94.5 kJ · mol^{-1} or ∼1 eV) and Σ_g^+(147.5 kJ · mol^{-1} or ∼1.65 eV). The formation of electron-excited oxygen molecules $^1\Delta_g$ on HgO [5.63] and Σ_g^+ on Ni [5.64] was detected during recombination of atoms O + O.

Some idea of the possible role of electron-excited molecules in surface reactions can be obtained from experiments on the relaxation of electron-excited molecules emerging from the bulk onto the surface. The probability of deexcitation of a singlet molecule $O_2(^1\Delta_g)$ on Pyrex glass at room temperature is $\sim 2 \cdot 10^{-5}$ [5.65]. Under same conditions, more highly excited singlet oxygen $O_2(^1\Sigma_g^+)$ relaxes with a probability of $2 \cdot 10^{-3} - 4 \cdot 10^{-2}$, or 2 - 3 orders of magnitude faster [5.66]. Systematic data on the relaxation of singlet oxygen $^1\Delta_g$ on various surfaces were obtained by RYSKIN and SHUB [5.67]. The probability of relaxation of $O_2(^1\Delta_g)$ on various surfaces at room temperature varied from 10^{-2} for metals to 10^{-5} or 10^{-6} for dielectrics. The highest probability of relaxation of $O_2(^1\Delta_g)$ was observed with silver ($5 \cdot 10^{-2}$). Most metals in the experiments reported in [5.67] were undoubtedly covered with oxide film. Silver is distinguished from other metals by the large quantities of molecular oxygen on its surface. One can therefore assume that the high destruction rate of singlet oxygen on silver is due to the resonant transfer of excitation:

$$O_2^* + O_2 \rightarrow O_2 + O_2^* \quad .$$

One might then ask whether the catalytic activity of many oxidation catalysts can be explained by the presence of electron-excited O_2 molecules on their surfaces. Although there are almost no data on surfaces which are of interest as catalysts, it can be assumed that $O_2(^1\Delta_g)$ will take part in surface reactions. SPIRIDONOV [5.68] suggested that an anion radical O_2^- created on the surface will react with hydrocarbons via intermediate formation of $O_2(^1\Delta_g)$; the formation of $O_2(^1\Sigma_g^+)$ is considered improbable.

The phenomenon of adsorboluminescence discovered at the Institute of Chemical Physics of the USSR Academy of Sciences by RUFOV et al. in 1966 [5.69] is also among the processes of electron excitation on the surface which result from energy transfer via the solid.

There is now a large volume of data on the luminescence caused by adsorption of O_2, CO, and NO on various oxides, as well as luminescence during catalytic oxidation of H_2, CO, and methanol, and decomposition of N_2O [5.70,71]. Luminescence has also been observed during adsorption of O_2 on metals [5.72,73], but this glow is probably due to the presence of an oxide film on the metal surface. The experiments conducted at the Institute of Chemical Physics failed to detect adsorboluminescence on metals cleansed in a ultra-high vacuum.

The mechanism of adsorboluminescence is not yet fully understood, nor is it known why the luminescence is most easily observed in adsorption of oxygen. One possible explanation is that electron-excited O_2 molecules have a comparatively low energy which quite readily finds a resonant level in one of the solid's defects, into which it can pass and be irradiated. Nonradiative transitions are, however, more common. The quantum yield of adsorboluminescence, defined as $\eta = N_{ph}/N_a$, where N_{ph} is the number of photons emitted and N_a is the number of adsorbed molecules, is only about 10^{-6}.

Luminescence accompanying adsorption of O_2 on MgO has been most thoroughly investigated [5.70,71]. The glow was found to appear at low occupancies ($\sim 10^{10}$ molecules $\cdot$ cm^{-2}). The introduction of transition-metal oxides (NiO and MnO_2) into MgO reduced the quantum yield of luminescence, while admixture of MoO_3 increased η a little.

The activation energy of adsorption was determined from the temperature dependence of the intensity of the glow (under the assumption that the intensity of the glow is approximately proportional to the rate of adsorption of O_2 on MgO). The value obtained was about 19 kJ $\cdot$ mol^{-1}.

Figure 5.2 presents the so-called isotherms of adsorboluminescence of O_2 on MgO [5.70]. By integrating the kinetic curves of adsorboluminescence $N = \int I\,dt$, where I is the number of photons per second, one can estimate the number N of adsorbed photoactive molecules. Plotting N versus pressure (Fig. 5.2) and subjecting these curves to the Freundlich equation, RUFOV [5.70] derived the number of photoactive centers on the surface of MgO. The value

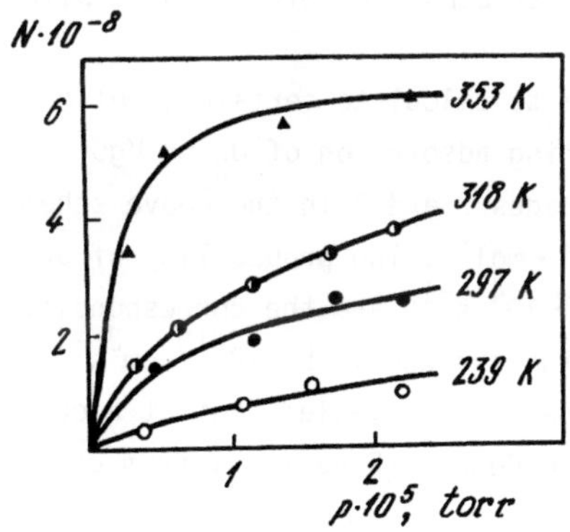

Fig.5.2. The isotherms of adsorboluminescence: the dependence of the number of emitted photons on the pressure of O_2 during adsorption of O_2 on MgO

obtained turned out to be extremely small ($\sim 3 \cdot 10^5$ cm^{-2}), and virtually unmeasurable by any other method. With such a small number of active centers, one would expect the results to be poorly reproducible and the value obtained to be strongly influenced by different impurities. In fact, however, this result was found to be well reproducible, and doping MgO with various impurities in general had little effect on the luminescence.

The results obtained can be described by the following kinetic scheme, which accounts for the electron-acceptor properties of oxygen and the diffusion of a number of molecules along the adsorbent surface [5.74]:

$$1)\ O_2 + S_1 \xrightarrow{k} O_2S_1 \quad ,$$

$$2)\ O_2 + S_0 \xrightarrow{k_A} (O_2 \ldots S_0)_{diff} \quad ,$$

$$3)\ (O_2 \ldots S_0)_{diff} + S_2 \xrightarrow{k_t} O_2S_2 \quad ,$$

$$4)\ (O_2 \ldots S_0)_{diff} + F \xrightarrow{W} (O_2^- \ldots S_0)_{diff} + \ldots F^+ \quad ,$$

$$5)\ (O_2^- \ldots S_0)_{diff} + S \xrightarrow{k_4} O_2^-S + h\nu \quad ,$$

where k is the adsorption rate constant; k_A is the adsorption rate constant for mobile oxygen (precursor; see Sects.2.2.2,5.1.3); and k_t is the trapping constant of mobile oxygen on surface defects.

In this scheme oxygen is assumed to be firmly adsorbed on S_1 centers; adsorption on S_0 leads to mobile oxygen molecules $(O_2 \ldots S_0)_{diff}$, which are either captured on S_2 centers, or accept an electron from an F center in MgO and become migrating O_2^- molecules. The radiative stage is not the limiting one; it is apparently connecting within the capture of a mobile O_2^- molecule by some surface defect. The role of electron donors is played by F centers distributed in the bulk of the crystal. From there the electron tunnels to the mobile O_2 molecule.

The kinetic data available made it possible to calculate certain kinetic parameters which characterize luminescence during adsorption of O_2 on MgO [5.74]. The values of activation energy for stages 2 and 3 in the above scheme were found to be $E_A = 27$ kJ $\cdot$ mol^{-1} and $E = 15$ kJ $\cdot$ mol^{-1}. The probability of an elementary diffusional jump is greater than $5 \cdot 10^4$ s^{-1}, and the corresponding coefficient of the diffusion of oxygen along the surface is $D > 10^{-10}$ cm^{2}s^{-1}. Comparison with the experimental data on luminescence revealed that at a quantum yield of 10^5 cm^{-2}, electrons are taken from donors (F centers) from depths of up to 20 Å from the adsorbent surface. Hence the cross-section of the capture of an electron by an O_2 molecule on the surface can be estimated as $\sigma = 10^{-13} - 10^{-14}$ cm^2.

The mechanism of excitation in the model proposed in [5.74] relies on the existence of weakly bound molecules on the surface (precursors), as opposed to the model of direct excitation of impurity centers discussed in a number of other works [5.75,76]. For instance, the spectrum of glow observed in [5.76] after adsorption of O_2 on the system (CaO + Bi) was characteristic of luminescent impurities (e.g., of Bi^{3+}).

Systematic investigations of adsorboluminescence spectra [5.71] on pure MgO and on MgO with admixed oxides of V, Mn, Mo, Cr. Fe, and Al revealed that in the range studied (4000 - 7000 Å) these spectra are not influenced by impurities and are apparently determined by structural defects in MgO. The spectrum of adsorboluminescence on pure MgO displayed a wide maximum (2.2 eV) at 5600 Å, due evidently to structural defects in MgO.

The photoluminescence spectra of MgO excited in the absorption band of an F center display either two bands (3700 and 4800 Å) or one (3700 Å), depending on the conditions under which the specimen is treated. These bands are usually [5.77] attributed to an F center (4800 Å) and an F^+ center (3700 Å). This is also confirmed by the fact that the 4800-Å band is extinguished by the injection of O_2, while the band of the F^+ center is intensified. This can be explained by the capture of an electron from the F center by an O_2 molecule:

$$O_2 + F \rightarrow O_2^- + F^+ \quad .$$

Studies of thermoluminescence reveal that in both photoexcitation of MgO and excitation of MgO by adsorption, the thermoluminescence spectra are the same, and are similar to the luminescence spectra of O_2 on MgO. The temperature-dependent maxima in photoexcitation differ from those in excitation by adsorption of O_2, which indicates that the traps involved in luminescence are different. (The nature of the traps depends on how the specimen is treated.) However, the electron transition leading to emission is the same in both cases [5.75]. The connection between adsorboluminescence and defects is also indirectly confirmed by the correlation between the intensity of the absorption band of F centers in MgO diffuse reflection spectra and the intensity of adsorboluminescence of O_2. It must be noted, however, that the measured concentration of F centers is 6 orders of magnitude greater than the concentration of luminescence centers.

BREUSSE et al. [5.78] observed luminescence during adsorption of O_2 on TiO_2, as well as in catalytic oxidation reactions. Their results were interpreted from the viewpoint of idealized concepts of the electron theory of adsorption and catalysis. There was no agreement between the proposed equations and experiment.

It is possible that cooperative processes cause a summation of separate excitations in the solid [5.70,79], and that light is emitted due to atomic regrouping of the solid surface after adsorption (Sect.5.3).

Pulsed injection of gas onto a solid also gives rise to triboluminescence, a phenomenon connected with the formation of unstable centers with excess energy on the surface. Triboluminescence is usually attributed to the formation of centers with high electric potential due to mechanical stresses (friction of gas over the surface, scratching and crushing of crystals, etc.), with subsequent electric discharge. Analysis of triboluminescence spectra shows, however, that electric discharge phenomena alone cannot be responsible for all the distinctive features of this effect. It is quite possible that we are also dealing with electron-excited states here [5.62,80].

A highly intensive type of luminescence is so-called radical recombination luminescence [5.81]. Recombination of atoms O + O, N + N, etc., on crystal phosphors such as ZnS-Mn, $CaSO_4$ - Eu, and the like, causes an intense glow in the region of the absorption band of impurities in the solid, which indicates electron excitation of these centers. In some cases the bands connected with the energy level of the adsorbed gas were also observed in the emission spectrum. Here the mechanism of excitation is not clear either. It is quite possible, that the emission results from heating of the surface caused by the heat released during recombination of atoms. The results of work on luminescence during heterogeneous recombination of H + H and O + O atoms on oxides are most often interpreted from the VOLKENSTEIN viewpoint [5.82] on electron phenomena in adsorption and catalysis, without taking into account the distinctive features of real semiconductor surfaces of adsorbents and catalysts.

Recombination luminescence was observed by KOVALEVSKY [5.83] during the interaction of hydrogen and deuterium atoms on NaCl films. The spectrum of luminescence corresponded to the D line of sodium. The intensity of the glow depended on the thickness of the film; the thinner the film the greater it was. Heating films thicker than 0.2 - 0.3 mm even to $400^{\circ}C$ did not cause luminescence. The recombination coefficient for deuterium atoms, measured in the absence of glow on sufficiently thick films, was $\beta = 4 \cdot 10^{-4}$. The recombination coefficient increased greatly on thin films, and glow appeared. This increase in the recombination coefficient accompanying the glow may be due to the appearance on the NaCl surface of Na atoms in an electron-excited state. These results confirm the conclusion that the bulk of the solid participates in the dissipation of energy released during recombination of atoms.

In order to explain the mechanism of luminescence of MgO during adsorption of O_2, this phenomenon was compared to radical recombination luminescence [5.71]. However, the regularities exhibited by luminescence in atomic oxygen differed from those for molecular oxygen. The number of centers is greater in recombination luminescence than in adsorboluminescence; the glow spectrum is different; and deexcitation is faster in molecular oxygen than in atomic oxygen.

An alternative explanation of adsorboluminescence is that excited molecules, e.g., the singlet oxygen $^1\Delta_g$, are formed during adsorption and subsequently deexcited. However, experiments with the injection of singlet oxygen $^1\Delta_g$ did not help to reveal any laws of adsorboluminescence. Nevertheless, these experiments did lead to the discovery of considerable difference in the adsorptivity of ground-state (triplet) and electron-excited (singlet) oxygen. As opposed to triplet oxygen, singlet oxygen $^1\Delta_g$ was adsorbed on MgO with an activation energy of practically zero [5.84]. This result confirms the assumption that electron-excited and nonexcited molecules have different adsorptive properties.

Another phenomenon, exoelectronic emission, is also explained on the basis of recombination of radicals on the surface as the energy source for electron excitation [5.85]. The quantum yield of exoelectronic emission is 2 - 3 orders of magnitude lower than that of adsorboluminescence. Exoelectronic emission and adsorboluminescence have certain features in common. It is possible that both phenomena are produced by similar processes of excitation. Direct proof of exoelectronic emission being caused by excited molecules was obtained in [5.86]: the emission of electrons from ZnO was induced by a flow of excited CO_2 molecules.

The formation of electron-excited molecules on the surface can also result in the creation of ions. The authors of [5.87] detected ionization of the products of exothermal decomposition of trimeric acetone peroxide on the surface of a hot emitter (oxidized tungsten) from the ion current (positive and negative ions) in a mass spectrometer. Feeding the products of thermal decomposition into the mass spectrometer from a separate reactor did not give rise to ion currents. This was explained by the authors as being due to the formation of electron-excited reaction products on the surface, with subsequent ionization. Recombination of hydrogen and oxygen atoms on a luminophor surface led to the emission of positively and negatively charged particles [5.81].

5.2.2 Electronic Excitation and the Reactivity of Adsorbed Molecules

A detailed discussion of the reactivity of adsorbed molecules would go beyond the scope of this book. We shall therefore direct our attention only to certain aspects of this problem.

One of the most widely accepted theories of the reactivity of molecules is based on the WOODWARD-HOFFMANN orbital symmetry rules [5.88]. According to this theory, the lack of correspondence between the symmetry of the orbitals of the initial reactants and that of the reaction products creates an energy barrier. Electron excitation can cancel this exclusion and lower the activation energy of the reaction, for instance during interaction between a molecule and a solid. The application of the Woodward-Hoffmann rules to catalysis is discussed in [5.89,90].

Excited electrons can be less tightly bound to the nucleus, since they are farther from it than in the ground state. These electrons are more smeared in space and consequently more reactive. The presence of excited electrons lowers the potential barrier of the reaction. It can also happen that an electron pair breaks when one of the electrons passes into the excited state — then the unpaired electrons can take part in radical-type reactions. Such a mechanism was proposed in [5.91] as an explanation of the increased activity of doubly bound oxygen of $\geq M=O$ oxides which results from photoexcitation in oxidation catalytic reactions.

UGO [5.92] pointed out a striking similarity between the properties of coordinated or chemisorbed molecules on the one hand, and those of photochemically excited molecules on the other hand. He referred to so-called soft catalysts, i.e., catalysts with atoms or ions of transition elements functioning as active centers and interacting with π ligands (olefins, CO, oxygen, etc.). In these "soft" complexes of transition metals and π ligands, the geometry and the charge distribution of the ligand molecule will be similar to those of an isolated molecule of the same substance in the first electron-excited singlet state. For instance, a CS_2 molecule in the ground electron state is linear, and has a C - S bond length of 1.55 Å. In the lowest excited triplet state ($^3A_2$) the molecule is curved, with a C - S bond length of 1.64 Å and a bond angle of 135°. The same value of bond angle S - C - S and bond length C - S are observed in $(PPh_3)McS_2$ complexes where M = Pt, Pd.

A nonexcited acetylene molecule has a linear structure: $H-C\equiv C-H$ (the $C\equiv C$ bond length is 1.20 Å). The first excited state $C_2H_2(^1A_4)$ has *trans*-structure:

```
          H
         /
    C ≡ C
   /
  H
```

(the C ≡ C bond length is 1.38 Å, the H - C ≡ C angle is 120°). The excited *cis*-state

```
    C ≡ C
   /     \
  H       H
```

has a bond length of 1.38 Å and a bond angle of 142°. In complexes of C_2H_2 with metals in zerovalence or the lowest oxidation state [Ni(0), W(0), Ir(I), Pt(0)] the bond length is ~1.28 - 1.36 Å, and the bond angle is ~140° - 149°. If the metal is in a higher oxidation state, the changes are smaller: in complexes of C_2H_2 with Pt(II) the bond length is 1.24 Å and the bond angle is 163°. This is because in high oxidation states metals display lower coordination properties. In "rigid" complexes with non-transition metals the variations in the ligand's geometry are also small.

Deviations from coplanarity in complexes of metals with olefins are also caused by a partial passing of electrons onto the first excited (π^*-antibonding) orbital of the olefin. Spectral data on the excited states of ethylene indicate that excitation causes an increase in the separation between carbon atoms C = C (1.69 Å compared with 1.34 Å), and that the molecule is no longer planar, but has the configuration of a triangular pyramid at each carbon atom.

In oxygen complexes of Ir(I) the O - O bond length varies from 1.30 to 1.65 Å, while in complexes of Rh(I) the bond length is 1.46 Å. The bond length in molecular O_2^- and O_2^{2-} ions (1.49 Å) is considered by Ugo to be insufficient to explain the properties of coordinated oxygen molecules. The distances actually observed in complexes are best matched by the O - O bond length in triplet excited oxygen molecules $^3\Sigma_u^-$(1.42 Å) and $^3\Sigma_u^+$(1.60 Å). The bonding of oxygen in such complexes is then similar to $\pi^4\pi^{*2} \rightarrow \pi^3\pi^{*3}$ in a spin-paired oxygen molecule.

The reactivity of molecules in an adsorbed or coordinated state is, according to Ugo, also similar to the reactivity of electron-excited (e.g., photochemically) molecules, whence the catalytic properties of the relevant metals also follow. For instance, reversible poisoning of metallic catalysts in the lowest oxidation state is connected with reversible dissociation of sulfides in the electron-excited coordinated state and with a subsequent irreversible reaction of electron-excited sulfide compounds. In homogeneous complexes the reaction can follow the mechanism

$$Pt(PPh_3)_2 + H_2S \rightleftharpoons Pt(PPh_3)_2(H_2S) \rightleftharpoons (PPh_3)_2Pt\begin{matrix}/H\\ \backslash SH\end{matrix} \xrightarrow{+O_2} \left[(PPh_3)_2Pt\begin{matrix}/S\\ \backslash\end{matrix}\right]_2 + H_2O \quad ,$$

$$Rh(PPh_3)_2Cl + CS_2 \rightleftharpoons Rh(PPh_3)_2(CS_2)Cl \xrightarrow{\text{acceptor of sulfur}} Rh(PPh_3)_3(CS)Cl \quad ;$$

and on the surface of a metal or oxide the reaction can follow the scheme

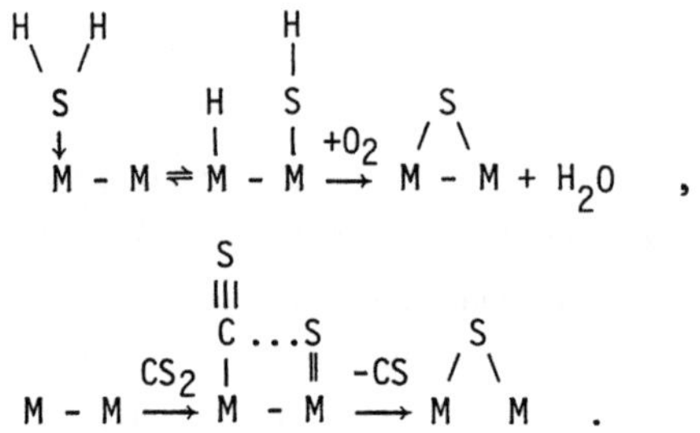

These reactions are similar to photochemical dissociation of H_2S or CS_2 excited by a quantum $h\nu$ of wavelength λ:

$$H_2S \xrightarrow[h\nu]{\lambda = 2000 - 2550\ \text{Å}} HS + H \quad ,$$

$$CS_2 \xrightarrow[h\nu]{\lambda = 2100 - 2400\ \text{Å}} CS + S \quad .$$

A similar explanation can be offered for the activation of water in homogeneous complexes:

$$Pt(PEt_3)_3 + H_2O \rightleftharpoons [PtH(PEt_3)_3]OH \quad ,$$

on the surface of metals and oxides:

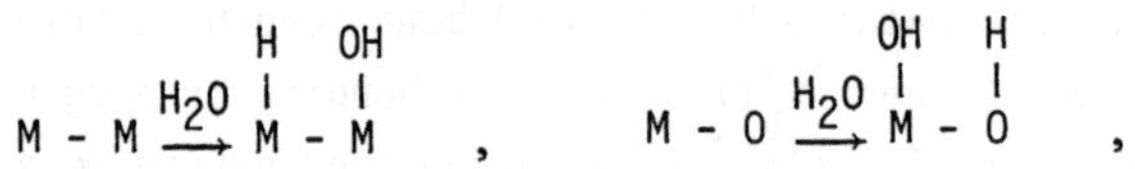

and photochemically:

$$H_2O \xrightarrow[h\nu]{\lambda = 1356 - 2420\ \text{Å}} H + OH \quad .$$

Hydrogen activates more readily than water; according to Ugo, this is because hydrogen shows a greater tendency towards the creation of excited states, although the photochemical reaction proceeds with quanta of shorter wavelengths:

$$H_2O \xrightarrow[h\nu]{\lambda = 1000 - 1500\ \text{Å}} 2H \quad .$$

In homogeneous complexes the reaction follows the scheme

$$\mathrm{Pt(PEt_3)_3 + H_2 \rightleftharpoons (PEt_3)_3Pt}\begin{matrix}/\mathrm{H}\\ \backslash \mathrm{H}\end{matrix}\ ,$$

and on the surface

$$\mathrm{M - M + H_2 \rightarrow \begin{matrix}H & & H\\ | & & |\\ M & - & M\end{matrix}\ ,\quad M - O + H_2 \rightarrow \begin{matrix}H & & H\\ | & & |\\ M & - & O\end{matrix}\ .}$$

In spite of its somewhat formal nature, the analogy between catalysis and photochemical excitation observed by Ugo is very promising. It emphasizes coordinated-related alterations in the structure and electron state of a molecule which may cause it to become activated.

5.3 Atomic Reconstruction of the Solid in Adsorption and Catalysis

This chapter opened with a discussion of one of the possible ways in which chemisorption energy or the energy released in the exothermal stage of catalysis is utilized, namely in the creation of excited molecules with increased reactivity. An alternative way in which this energy might be used is in atomic reconstruction of the solid surface, which results in a more favorable configuration of the active centers of chemisorption and catalysis.

Reconstruction of the surface layer of a catalyst in the course of the reaction is a phenomenon which has long been known. Recall, for instance, the classic work by ROGINSKY, TRET'YAKOV, and SHEKHTER [5.93] on electron microscopy of the surfaces of platinum and palladium as catalysts in oxidation reactions.

Use of the ELEED technique recently led to the discovery of two-dimensional phase transformations on the surfaces of heterogeneous catalysts (Sect.3.1.3). In certain cases the process of adsorption, even before a monolayer has completely formed, passes through a number of successive phase transformations which are also reflected in the shape of the adsorption isotherm. For instance, Fig.5.3 shows the structures of the surface layer in adsorption of oxygen on the <110>Ni face at 373 K [5.94]. Despite the narrow range of occupancies (0.555 - 0.625), four different two-dimensional structures can be discerned, with sudden transitions from one to another. In adsorption at low temperatures the number of structures is apparently lower, and the formation of the new two-dimensional phase occurs through the growth of two-dimensional nuclei (clusters) [5.94]. However, here also a two-dimensional structure is formed, and not just a "statistic" adsorptive layer.

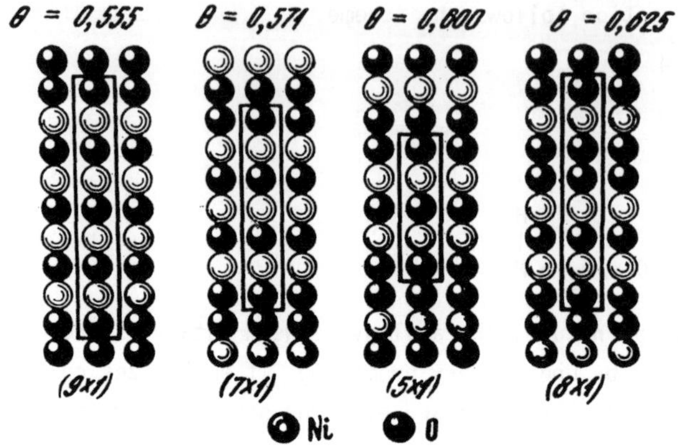

Fig.5.3. Ordered structures formed in adsorption of oxygen on <110>Ni face at 370 K at monolayer occupancies of 1/2 - 2/3 [5.94]

The mechanism of two-dimensional phase formation on the surface of metals was analyzed in detail by USTINOV [5.95,96]. According to his data, which were obtained by the electron projector method, ELEED, and thermal desorption, at medium temperatures the process of chemisorption goes through two successive stages: the sticking of gas particles to the surface, and migration of those particles towards the ordered adsorbate patch. The preadsorption state (precursor) is a reversibly adsorbed gas which maintains a dynamic equilibrium with the patches and disappears when the space surrounding the adsorbent is evacuated [5.97]. The presorption state exists at higher pressures and lower temperatures than gas over a clean surface. The supersaturation here causes condensation, resulting in the formation of patches of two-dimensional phases. The greater the rate at which gas molecules stick to the surface (in comparison with the diffusion rate on the surface), the greater is the oversaturation of the two-dimensional precursor gas. The fact that the sticking coefficient is independent of Θ (Sect.2.2.2) then implies that the limiting step of chemisorption is adsorption into the precursor state. Should the second step be the limiting one (surface migration of molecules to the patch), the sticking coefficient ought to decrease linearly with Θ. Statistical chemisorption without phase formation will be observed to take place at low temperatures, when oversaturation is high and the mobility of adsorbed molecules is low.

Crystallization patches of the two-dimensional adsorbate phase are formed, as a rule on dislocations, flaws, and foreign atoms, as well as on the edges and vertices of microcrystals.

USTINOV [5.95,96] also analyzed regularities in the transformations of two-dimensional adsorbed phases at high occupations. For instance, on almost all cubic $\langle 100 \rangle$ faces of metals a phase transition occurs at $\Theta = 0.5$. In investigations on thermal desorption (and also in measurements of the work function, chemisorption heat, etc.) this is indicated by the existence of two phase maxima denoted β_1 and β_2. The phase transition $\beta_1 \rightarrow \beta_2$ at $\Theta = 0.5$ displays the characteristic features of a phase transition of the second kind. A (2×2)-type structure, in which the adsorbed molecules form a checkered pattern over the metal atoms M, passes into a (1×1)-type structure, with adsorbed molecules over every atom on the surface:

```
    A   A   A                   A A A A A
    M M M M M                   M M M M M
     A   A                     A A A A A
   M M M M M     + nA →        M M M M M     .
  A   A   A                   A A A A A
  M M M M M                   M M M M M
   A   A                     A A A A A
 M M M M M                   M M M M M
```

In the final structure $\beta_1 + \beta_2$ with Θ close to unity, the adsorbed molecules are indistinguishable.

Ampel data are now available on the formation of two-dimensional surface phases on metals during adsorption and on the abrupt transformations of these phases. However, a detailed discussion of adsorption on metals would go beyond the scope of this book. The data on oxides are not so vast, because of difficulties encountered in the application of the ELEED technique. Some recent work on ZnO monocrystals using ELEED [5.98] has confirmed in general the conclusions drawn from data on the adsorption of simple gases on metals.

Experiments on the diffraction of low-energy electrons contradict the statistical concepts of Langmuir adsorption. Adsorption (and consequently catalysis) ought to be envisaged as a chain of abrupt transformations, with the surface being reconstructed at the expense of energy released in the process of adsorption or catalysis. Each instance of such reconstruction leads on the one hand to changes in the configuration of active centers on the surface, which in turn results in a change in the catalytic activity, and on the other hand to a loosening of surface atoms, which can exhibit increased activity during reconstruction and take part in the catalytic process. Consequently, surface phase transitions are the most probable reactions of creation or destruction of active centers on the surface of catalysts. Taking the kinetics of these processes into account will lead to qualitative changes in the form of kinetic equations.

ERTL [5.49,99,100] studied the kinetics of oxidation of CO on different faces of Pd monocrystals, monitoring the structure of the adsorbed layer by ELEED. His results gave rise to entirely new concepts about the nature of elementary acts of adsorption and catalysis. It was found that at various occupancies different adsorption phases are formed which consist of O atoms, CO molecules, or mixed phases of O + CO. Oxidation of CO can occur on the boundary between patches of O and CO phases, or within a patch of a mixed O + CO phase. Once an occupancy of about 0.25 has been attained, the phase of O atoms abruptly becomes more active in terms of interaction with CO. This transition was accelerated by small additions of CO (as well as H_2), and under certain conditions the kinetics followed the equation

$$w = k\Theta_O p_{CO} \quad , \tag{5.26}$$

where Θ_O is the occupancy of the surface by oxygen. Equation (5.26) is usually interpreted as an indication that the catalytic reaction follows the so-called Rideal (sticking) mechanism (Sect.2.2.2). Actually, the reaction occurs on the surface.

Similar kinetics were observed by ROZENTULLER et al. [5.101] in investigations of the oxidation of CO on the surface of $BaTiO_3$ near the Curie point. The adsorption of donors (CO, H_2) lowers the Curie point, while the adsorption of acceptors (O_2) raises it several degrees. On the other hand, the high-temperature paraelectric phase of $BaTiO_3$ is more active in the oxidation of CO and the ferroelectric phase. An equation of the same type as (5.26) is valid for oxidation of CO on $BaTiO_3$ near the Curie point, because adsorption of CO shifts the Curie point towards a value more favorable for the catalytic reaction.

ROZOVSKY and coworkers [5.102] studied the kinetics of the reduction of oxides of Fe, Mo, Cu, Zn, and V, and demonstrated that two-dimensional phases are formed here as well. Figure 5.4 shows the dependence of the rate of reduction. These results were obtained with the pulse technique in a study of the interaction of CO with Fe_2O_3. It is evident that the rate of reduction is a nonmonotonic function of the extent of reduction. First the reaction proceeds on defects and their neighbors, and the reaction rate decreases as the number of available defects diminishes. Then the nuclei of the solid product phase appear, and an interface emerges between solid phases, causing an increase in the reaction rate. Finally, the reaction slows down as nuclei merge together and the solid reactant runs out. Note that all these processes

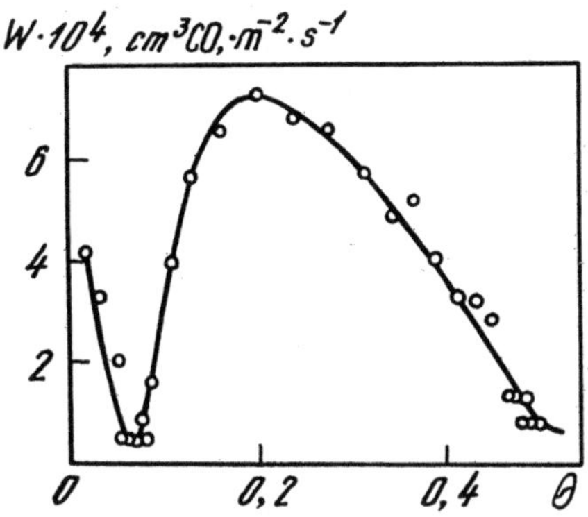

Fig.5.4. The kinetics of reduction of the surface of iron oxide by carbon monoxide at 453 K [5.102]

can occur in the surface layer of the oxide. The maximum on the curve in Fig. 5.4 and on similar curves describing the reduction of other oxides corresponds to reduction amounting to less than a monolayer. The kinetic laws differ for different oxides. In some cases (e.g., reduction of cuprous oxide) the reaction autoaccelerates up to the formation of a monolayer.

Critical phenomena in oxidation of propylene on supported MoO_3/MgO catalysts were reported in [5.103]. As the occupation of the surface proceeds, the heat of adsorption of oxygen at 473 K drops from 280 to 80 $kJ \cdot mol^{-1}$. If, however, $C_3H_6 + O_2$ is injected on the surface, then at first the value of adsorption heat is exactly equal to the heat of adsorption of pure oxygen. When it reaches 147 $kJ \cdot mol^{-1}$, an intensive release of heat begins, which indicates that oxidation of C_3H_6 is occuring, and the products of oxidation appear in the gaseous phase. These results indicate that the catalytic reaction starts after the formation of a certain structure due to adsorption of O_2.

Another type of two-dimensional phase transition was observed in [5.104]. On the surface of solid $Co_xMg_{1-x}O$ solutions, isolated Co^{2+} ions and clusters of Co^{2+} ions exist and interact via oxygen bridges Co - O - Co. The rate of oxidation of CO is greater on clusters than on isolated Co^{2+} ions. It was shown that the cluster size depends on temperature. An increase in temperature gives rise to a two-dimensional phase transition: big clusters are split into small ones. Adsorption of CO can impede this two-dimensional phase transition and stabilize the cluster in a nonequilibrium state on the surface.

BERMAN and one of the authors of this book [5.105,106] examined regularities in the kinetics of adsorption and other chemical reactions on surfaces taking two-dimensional phase transitions into account. The kinetics was described on the basis of a model of the formation and growth of nuclei (clusters) developed primarily for three-dimensional phase transitions [5.107].

The kinetics of the formation of clusters of substance A can be considered to be a set of reversible successive reactions of the following type:

$$A + A \rightleftharpoons A_2 \quad ,$$

$$A_2 + A \rightleftharpoons A_3 \quad ,$$

$$\cdots\cdots\cdots \quad ,$$

$$A_{i-1} + A \rightleftharpoons A_i \quad ,$$

$$\cdots\cdots\cdots \quad ,$$

$$A_k + A \rightleftharpoons A_{k+1} \quad ,$$

where the subscript k denotes the "critical cluster" whose formation triggers the growth of the next phase.

In the simplest case, if interactions between clusters are neglected, the rate of creation of nuclei in the stationary state is [5.108,109]:

$$v_{cl} = kC^m \quad , \tag{5.27}$$

where k and m are constants, and C is the concentration of the adsorbed substance. The validity of (5.27) was confirmed experimentally in [5.110].

Note that the value of the factor m depends strongly upon the system in question. For condensation of water, $m = 37$; in other cases it is much lower. Data on the formation of critical clusters during oxidation of metals are scarce. It has been shown [5.111], however, that the oxidation of lead is limited by the step

$$2Pb + O_2 \rightarrow Pb_2O_2 \quad .$$

Hence $m = 2$ in this case. By analogy with (5.27) for the conventional formation of nuclei, the rate of creation of nuclei of the two-dimensional phase on a solid surface can be described as

$$\frac{dN}{dt} = k\Theta^n \quad , \tag{5.28}$$

where dN/dt is the number of nuclei formed per unit time and unit surface; Θ is the occupancy of the surface by the adsorbed substance; n is the number of particles necessary for the formation of a nucleus; and k is a constant[1]. We will assume also that the rate of growth of a two-dimensional cluster along each axis is proportional to the concentration of adsorbed particles:

[1] The conditions necessary for the formation of surface associations of molecules were analyzed by Dubinin, Serpinsky, and Voskresensky (Chap.2). In (5.26) we have observed the energy transfer from the exited dye molecules to the deep electron centers in a semiconductor-dielectric system.

$$v_x = k_x\Theta \quad , \quad v_y = k_y\Theta \tag{5.29}$$

Then by time t, the area of a nucleus born at time τ will be

$$S(t,\tau) = k_x k_y \left(\int_\tau^t \Theta dt\right)^2 \quad . \tag{5.30}$$

The total area of all the nuclei by time t will be

$$S(t) = kk_x k_y \int_0^t \Theta^n(\tau) \left(\int_\tau^t \Theta dt\right)^2 d\tau \quad . \tag{5.31}$$

From (5.31) we get the reaction rate

$$\frac{dS(t)}{dt} = 2kk_x k_y \Theta \int \Theta^n(\tau) \left(\int_\tau^t \Theta dt\right) d\tau \quad . \tag{5.32}$$

If a change in the reaction rate due to possible overlapping of nuclei is taken into account, and if in place of assumption (5.29) it is assumed that the occupation of the surface by adsorbed particles is constant in the course of the process, the kinetic equation for the formation of the new phase on the solid surface will be:

$$\frac{d\Theta_{n.ph.}}{dt} = k(1 - \Theta_{n.ph.})\Theta^{n+2} \frac{t^2}{2} \quad , \tag{5.33}$$

where $\Theta_{n.ph.}$ is the occupancy of the surface by the new phase. Equation (5.33) describes the kinetics of the formation of a two-dimensional phase on a solid surface. If it is assumed that active centers are created or destroyed in the course of the surface phase transition, then (5.33) will describe the non-stationary stage in the development (activation) of the surface, and the curves depicting a change with time in the adsorptive or catalytic activity will have periods of induction and sections of autoacceleration. In the simplest cases of kinetics in which phase transitions on the surface are taken into account, the reaction rate in the stationary state can be described as

$$v = v_1 v_2 \quad , \tag{5.34}$$

where v_1 is the rate at which the phase composition of the catalyst changes [e.g., according to (5.33)], and v_2 is the reaction rate on a reconstructed surface. The characteristic feature of such kinetics is the existence of abnormally high reaction orders ($n > 3$). Equations of type (5.33,34) well describe the experimental curves, which have a period of induction and sections of autoacceleration and strong inhibition of the reaction. An equation of type (5.34) was employed in [5.112] to account for the influence of the reaction medium on the catalyst. KRYLOV and BERMAN [5.50,105,106] also analyzed more complicated cases of creation and destruction of active phases on the surface, and adsorption and catalysis on these phases. Thus the creation of surface phases during adsorption and catalysis can result in the occurrence of critical phenomena such as an abrupt increase or drop in the reaction rate after a certain occupation has been reached, and anomalous kinetic laws. The models proposed here do not claim to give a detailed description of systems, exhibiting two-dimensional phase transitions, and of course they are not the only ones capable of explaining the formation of multiple states in heterogeneous catalysis. A detailed analysis of the peculiarities of nonstationary kinetics in chemical systems can be found in [5.113-118].

Reconstruction of the surface at the expense of energy released in the exothermal act of catalysis or adsorption can lead to the creation of structures that can exist only under conditions of adsorption and catalysis. They can also be termed "excited structures", but the excitation here spreads over a large part of the surface area.

In conclusion, let us point out that the questions dealt with here are part of a rapidly developing branch in the theory of heterogeneous catalysis which has received the name "catalytic act dynamics" [5.118,119]. Instead of considering the quasi-stationary state of the catalyst surface with regard to the composition of the reacting medium (as was often done previously), it analyzes the processes of interaction between the components of the reacting medium and the surface and between the substances in the adsorption layer, as well as changes in the subsurface layer of the catalyst and the energy exchange between the reacting substances and the catalyst. An important aspect of this branch of study is that these processes are considered on an atomic and molecular scale. Work in this field promises great progress in our understanding of the processes of adsorption and catalysis.

References

Chapter 1

1.1 O.V. Krylov: *Catalysis by Nonmetals* (Academic, New York 1970)
1.2 J.R. Smith (ed.): *Theory of Chemisorption*, Topics Current Phys., Vol.19 (Springer, Berlin, Heidelberg, New York 1980)
1.3 V.F. Kiselev: *Surface Phenomena in Semiconductors and Dielectrics* (Nauka, Moscow 1970) [in Russian]
1.4 V.F. Kiselev, O.V. Krylov: *Electronic Processes in Adsorption and Catalysis*, Springer Ser. Chem. Phys. (Springer, Berlin, Heidelberg, New York, Tokyo) (to be published)
1.5 V.F. Kiselev, O.V. Krylov: *Adsorption and Catalysis on Transition Metals and Their Oxides*, Springer Ser. Chem. Phys. (Springer, Berlin, Heidelberg, New York, Tokyo) (to be published)

Chapter 2

2.1 J.R. Smith (ed.): *Theory of Chemisorption*, Topics Current Phys., Vol.**19** (Springer, Berlin, Heidelberg 1980)
2.2 J.E. Lennard-Jones: Trans. Faraday Soc. **28**, 333 (1932)
2.3 A. Clark: *The Theory of Adsorption and Catalysis* (Academic, New York 1970)
A. Clark: *The Chemisorptive Bond. Basic Concepts* (Academic, New York 1974)
2.4 P.M. Gundry, F.C. Tompkins: Trans. Faraday Soc. **52**, 1609 (1956); **53**, 218 (1957)
2.5 H.B.G. Casimir, D. Polder: Phys. Rev. **73**, 360 (1948)
2.6 B.W. Ilyin: *Die Natur der Adsorptionskräfte* (Deutscher Verlag der Wissenschaften, Berlin 1954)
2.7 J.D. Hirschfelder, C.F. Kurtiss, R.B. Bird: *Molecular Theory of Gases and Liquids* (Wiley, New York 1964)
2.8 H.C. Lonquet-Higgins: Discuss Faraday Soc. **40**, 7 (1965)
2.9 R. Heller: J. Chem. Phys. **9**, 154 (1941)
2.10 J.J. Chessick, A.C. Zettlemoyer: Can. J. Chem. **33**, 251 (1955)
L. Galatry, C. Girard: Mol. Phys. **49**, 1085 (1983)
2.11 H. Hoinkes: Rev. Mod. Phys. **52**, 933 (1980)
2.12 D.M. Young, A.D. Crowell: *Physical Adsorption of Gases* (Butterworths, London 1962)
2.13 A.V. Kiselev, D.P. Poshkus, A.J. Grumadas: J. Chem. Soc. Faraday Trans. **75**, 1281 (1979)
2.14 J.P. Hobson: Crit. Rev. Solid State Sci. **4**, 221 (1974)
2.15 A.V. Kiselev: Zh. Fiz. Khim. **35**, 233 (1961)
A.V. Kiselev: in *Adsorbtsia v mikroporach* (Adsorption in Micropors) ed. by M.M. Dubinin, V.V. Serpinski (Nauka, Moscow 1983) pp. 5-10

2.16 O. Sinanogly: Adv. Chem. Phys. **1**, 315 (1964)
2.17 N.D. Sokolov: in *Osnovnye problemy teorii fizicheskoi adsorbtsii* (Basic Problems in the Theory of Physical Adsorption) ed. by M.M. Dubinin (Nauka, Moscow 1970) pp. 82
2.18 V.V. Serpinsky: in *Osnovnye problemy teorii fizicheskoi adsorbtsii* (Basic Problems in the Theory of Physical Adsorption) ed. by M.M. Dubinin (Nauka, Moscow 1970) pp. 84
2.19 T. Takaishi: in *Progress in Surface Science*, **6**, 43 (Pergamon, Oxford 1975)
2.20 V.F. Kiselev: *Poverkhnostnye yavleniya v poluprovodnikakh i di̊elektrikakh* (Surface Phenomena in Semiconductors and Dielectrics) (Nauka, Moscow 1970)
2.21 J.R. Sams: Trans. Faraday Soc. **60**, 149 (1964)
2.22 E.A. Guggenheim: *The International Encyclopaedia of Physical Chemistry and Chemical Physics* (Pergamon, Oxford 1960) Vol. 1, p.79
2.23 F.F. Volkenstein: *The Electronic Theory of Catalysis on Semiconductors* (Pergamon, New York 1963)
2.24 H. Hellman: *Einführung in die Quantumtheorie* (Leipzig, 1937)
2.25 L.P. Gor'kov, L.P. Pitaevsky: Dokl. Akad. Nauk SSR **151**, 822 (1963)
2.26 C. Herring, N. Flicker: Phys. Rev. **134 A**, 362 (1964)
2.27 W.J. Carr, M. Ashkin: J. Chem. Phys. **42**, 2796 (1965)
2.28 N.R. Kestner: Discuss. Faraday Soc. **40**, 174 (1965)
2.29 L. Salem: Discuss. Faraday Soc. **40**, 150 (1965)
2.30 O. Sinanogly, K.S. Pitzer: J. Chem. Phys. **32**, 1279 (1960)
2.31 G.V. Tsitsishvili: Soobshch. Akad. Nauk Gruz. SSR **26**, 153 (1961) [in Russian]
2.32 T. Kichara: Adv. Chem. Phys. **5**, 2 (1963)
2.33 H. Margenau, W.G. Pollard: Phys. Rev. **60**, 128 (1941)
2.34 J. Bardeen: Phys. Rev. **58**, 727 (1940)
2.35 E.S.R. Prosen, R.G. Sacks: Phys. Rev. **61**, 65 (1942)
2.36 K.F. Woiciechowski: In *Progress in Surface Science* 2, 14 (Pergamon, Oxford 1972)
2.37 C. Mavroyannis: Mol. Phys. **6**, 593 (1963)
2.38 A.D. McLachlan: Mol. Phys. **7**, 381 (1964)
2.39 E.M. Lifshitz: Dokl. Akad. Nauk SSSR **97**, 643 (1954); **100**, 879 (1955); Zh. Eksp. Teor. Fiz. **29**, 94 (1955) [in Russian]
2.40 F.C. Brooks: Phys. Rev. **85**, 92 (1952)
2.41 C.A. Coulson: *Valence*, 6th ed. (Oxford University Press, Oxford 1961)
2.42 J.C. Slater: *Electronic Structure of Molecules* (McGraw-Hill, New York 1963)
2.43 J.N. Murrell, S.F.A. Kettle, J.M. Tedder: *Valence Theory* (Wiley, London 1965)
2.44 H.B. Gray: *Electrons and Chemical Bonding* (W.A. Benjamin, New York 1965)
2.45 I.B. Bersuker: *Elektronnoe stroenie i svoistva koordinatsionnykh soedinenii* (Electron Structure and the Properties of Coordination Compounds) (Khimiya, Leningrad 1976)
2.46 G.M. Zhidomirov, A.A. Bagaturjanz, I.A. Abronin: *Prikladnaja kwantovaja khimija Khimija* (Applied Quantum Chemistry) (Khimia, Moscow 1979)
2.47 R.S. Mulliken, W.B. Person: *Molecular Complexes* (Interscience, New York 1969)
2.48 D. Shopov, A. Andreev: *Khimicheskaya svyaz' pri adsorbtsii i katalize*, T.1, Metally (The Chemical Bond in Adsorption and Catalysis, Vol.1, Metals) (Bulgarian Academy of Sciences, Sofia 1975)
2.49 K.H. Johnson, R.P. Messmer: Int. J. Quantum Chem., Symp. 10, 147 (1976)
2.50 D.D. Eley: Discuss, Faraday Soc. **8**, 34 (1950)
2.51 M. Green, M.J. Lee: in *Solid State Surface Science*, Vol.1, ed. by M. Green (Marcel Dekker, New York 1969) p.133
2.52 S.M. Tsenter: Zh. Eksp. Teor. Fiz. **8**, 682 (1938) [in Russian]

2.53 F.F. Volkenstein: Zh. Fiz. Khim. **21**, 1317 (1947) [in Russian]
2.54 T.B. Grimley: Adv. Catal. **12**, 1 (1960)
T.B. Grimley: in *The Chemical Physics of Solid Surfaces* and *Heterogenous Catalysis*, Vol. 2, ed. by D.A. King, D.P. Woodruff (Elsevier, Amsterdam 1983) Chap.2
2.55 J. Koutecky: Trans. Faraday Soc. **54**, 1038 (1958)
2.56 J. Horiuti, T. Toya: in *Solid State Surface Science*, Vol.1, ed. by M. Green (Marcel Dekker, New York 1969) p.1
2.57 V.L. Bonch-Bruevich: Zh. Fiz. Khim. **27**, 662 (1953) [in Russian]
2.58 D.A. Dowden: J. Chem. Soc. (London) 242 (1950)
2.59 P.B. Weisz: J. Chem. Phys. **21**, 1531 (1953)
2.60 H.J. Engell, K. Hauffe: Z. Elektrochem. **57**, 762, 773 (1953)
2.61 P. Aigrain, C. Dugas: Z. Elektrochem. **56**, 363 (1952)
2.62 M. Boudart: J. Am. Chem. Soc. **74**, 1531 (1952)
2.63 J.E. Germain: J. Chem. Phys. **51**, 263, 691 (1954)
2.64 H.J. Krusemeyer, D.G. Thomas: J. Phys. Chem. Solids, **4**, 78 (1958)
2.65 K. Hauffe, R. Stechemesser: in *Symposium on Electronic Phenomena in Chemisorption and Catalysis on Semiconductor*, ed. by Th. Wolkenstein, K. Hauffe (De Gruyter, Berlin 1969) p.1
2.66 F.S. Stone: in Coloq. Quim. Fis. Proc. Superfic. Solid, Madrid, 1965, p.1
2.67 S.R. Morrison: *The Chemical Physics of Surface* (Plenum, London 1977)
2.68 V.L. Bonch-Bruevich, V.B. Glasko: Dokl. Akad. Nauk SSSR **124**, 1015 (1959) [in Russian]
2.69 J. Koutecky: Phys. Rev. **108**, 13 (1957)
2.70 J. Koutecky, M. Tomashek: Phys. Rev. **120**, 1212 (1960)
2.71 J. Koutecky: in *Excited States in Quantum Chemistry*, Proc. NATO Adv. Stud. Inst., ed. by J. Spice (D. Reidel, Dordrecht 1978) p.533; Prog. Surf. Membr. Sci. **11**, 1 (1976)
2.72 S.G. Davison, J.D. Levine: *Surface States* (Academic, New York 1970)
2.73 R. Gomer: in *Surface Science: Recent Progress and Perspectives*, ed. by T.S. Jayadevaiah, R. Vanselov (CRC, Cleveland 1974)
2.74 S.K. Lyo, R. Gomer: in *Interaction on Metal Surfaces*, ed. by R. Gomer, Topics Appl. Phys., Vol.4 (Springer, Berlin, Heidelberg 1975)
2.75 T.N. Rhodin, G. Ertl (eds.): *The Nature of the Surface Chemical Bond* (North-Holland, Amsterdam 1979)
2.76 T. Higuchi, T. Ree, H. Eyring: J. Am. Soc. Chem. **79**, 1330 (1957)
2.77 K.H. Johnson, R.P. Messmer: J. Vac. Sci. Technol. **11**, 236 (1974)
2.78 I.P. Batra, O. Robaux: Surf. Sci. **49**, 653 (1975)
2.79 R.C. Baetzold: J. Chem. Phys. **68**, 555 (1978)
2.80 N.D. Sokolov: in *Sbornik plenarnykh dokladov na Vsesoyuznoi konferentsii po mekhanizmu geterogennykh kataliticheskikh reaktsii* (Collection of the Plenary Reports of the All-Union Conference on the Mechanism of Heterogeneous Catalytic Reactions) (Institut khimicheskoi fiziki AN SSSR, Chernogolovka 1977) p.66
D.W. Bullet: in *Surface Properties and Catalysis*. Proc. NATO Adv. Stud. Inst. Lille (Reidel, Dordrecht 1983) p.47
2.81 A. Laforque, J. Rousseau, B. Imelik: Adv. Chem. Phys. **8**, 141 (1965)
2.82 K.H. Johnson: *The New World of Quantum Chemistry*, ed. by B. Pullman, R. Parr (Reidel, Dordrecht 1977) p.317
2.83 J.C. Slater, K.H. Johnson: Phys. Today **27**, 34 (1975)
2.84 A.N. Terenin: Zh. Fiz. Khim. **14**, 1362 (1940) [in Russian]
2.85 N.D. Sokolov: Probl. Kinet. Katal. 8, 141 (1955)
2.86 J.C. Mignolet: J. Chem. Phys. **21**, 1298 (1953)
2.87 P.M. Gundry, F.C. Tompkins: Trans. Faraday Soc. **56**, 847 (1960)
2.88 D.A. Dowden, D. Wells: in Proc. 2nd Int. Congr. Catalysis, Paris, 1960, p.1489
2.89 E.L. Nagayev: Zh. Fiz. Khim. **35**, 327 (1961) [in Russian]

2.90 E.L. Nagayev: Kinet. Katal. 3, 1907 (1962) [in Russian]
2.91 Ja. K. Syrkin: Probl. Kinet. Katal. **10**, 225 (1960) [in Russian]
2.92 S.Z. Roginsky: Zh. Fiz. Khim. **6**, 334 (1935) [in Russian]
2.93 O. Beeck: Discuss. Faraday Soc. **8**, 118 (1950)
2.94 J.B. Goodenough: *Magnetism and Chemical Bond* (Academic, New York 1963)
D.C. Mattis: *The Theory of Magnetism* 1, Springer Ser. Solid-State Sci., Vol.17 (Springer, Berlin, Heidelberg 1981)
2.95 W.A. Harrison: *Solid State Theory* (McGraw-Hill, New York 1970)
2.96 O.V. Krylov, V.F. Kiselev: *Adsorption and Catalysis on Transition Metals*, Springer Ser. Chem. Phys. (Springer, Berlin, Heidelberg 1985) to be published
2.97 N.D. Sokolov: Usp. Fiz. Nauk **57**, 205 (1955) [in Russian]
2.98 N.D. Sokolov, V.M. Chulanovsky: in *Vodorodnaya svyaz'* (The Hydrogen Band), ed. by V.M. Chulanovsky (Nauka, Moscow 1964)
2.99 J. Koutecky, P. Hochman, J. Paldus, R. Pollock, J. Cizek: in *Osnovnye problemy teorii fizicheskoi adsorbtsii* (Basic Problems in the Theory of Physical Adsorption) ed. by M.M. Dubinin (Nauka, Moscow 1970) p.36
2.100 J.W. Gibbs: *The Collected Works*, Vol.1, Thermodynamics (Longmans, New York 1928)
2.101 A.I. Rusanov: *Fazovye ravnovesiya i poverkhnostnye yavleniya* (Phase Equilibria and Surface Phenomena) (Khimiya, Leningrad 1967)
2.102 E.A. Flood: in *The Solid-Gas Interface*, Vol.1, ed. by E.A. Flood (Marcel Dekker, New York 1967) p.12
2.103 J.J. Bickerman: *Physical Surfaces* (Academic, New York 1970)
2.104 R.G. Linford: in *Solid State Surface Science*, 2, 1 (Marcel Dekker, New York 1973)
2.105 E. Guggenheim: Trans. Faraday. Soc. **36**, 397 (1940)
2.106 D.H. Everett: Trans. Faraday Soc. **46**, 453, 942, 957 (1950)
2.107 T. Hill: in *Advances in Catalysis*, Vol.4 (Academic, New York 1952) p.211
2.108 G.L. Kingston, J.C. Aston: J. Am. Chem. Soc. **73**, 1929 (1951)
2.109 W.D. Harkins, G. Jura: J. Am. Chem. Soc. **66**, 1362 (1944)
2.110 J.G. Dash: in *Progress in Surface Science*, 5, 119 (Pergamon, Oxford 1975)
2.111 B.P. Bering, M.M. Dubinin, V.V. Serpinsky: J. Colloid Interface Sci. **21**, 378 (1966)
V.V. Serpinski, T.S. Yakubov: in *Adsorbtsia v mikroporach*, ed. by M.M. Dubinin, V.V. Serpinski (Nauka, Moscow 1983) p.46 [in Russian]
2.112 M.M. Dubinin: Progr. Surf. Membrane Sci. **9**, 1 (1975)
2.113 B.P. Bering, A.L. Majers, V.V. Serpinsky: Dokl. Akad. Nauk SSSR **193**, 119 (1970) [in Russian]
2.114 B.P. Bering, V.V. Serpinsky: Izv. Akad. Nauk SSSR, Ser. Khim. 2679 (1973); 2427 (1974) [in Russian]
2.115 L. Yansen, E. Lombardi: Discuss. Faraday Soc. **40**, 78 (1965)
2.116 W.A. Steel: in *The Solid-Gas Interface*, Vol.1, ed. by E.A. Flood (Marcel Dekker, New York 1967) p.307
2.117 T.L. Hill: *Statistical Mechanics* (McGraw-Hill, New York 1956)
2.118 T.L. Hill: *Introduction to Statistical Thermodynamics* (University of Massachusetts, Amherst 1960)
2.119 S. Ono, S. Kondo: *Molecular Theory of Surface Tension in Liquids* (Springer, Berlin, Göttingen 1960)
2.120 J.H. de Boer: *The Dynamical Character of Adsorption* (Clarendon, Oxford 1953)
2.121 P. Broiter, A.A. Lopatkin: in *Osnovnye problemy teorii fizicheskoi adsorbtsii* (Basic Problems in the Theory of Physical Adsorption), ed. by M.M. Dubinin (Nauka, Moscow 1970) p.23
2.122 H. Stach, K. Fiedler: Zs. Phys. Chem. (Leipzig) **261**, 247 (1980)
2.123 T.L. Hill, N. Saito: J. Chem. Phys. **34**, 1543 (1961)
2.124 G.A. Martynov, A.L. Muler: in *Issledovaniya v oblasti molekularnykh*

sil (Investigations of Molecular Forces), ed. by B.V. Deryagin (Nauka, Moscow 1970) p.338
2.125 G.A. Martynov, A.L. Muler, I.B. Ivanov: in *Osnovnye problemy teorii fizicheskoi adsorbtsii* (Basic Problems in the Theory of Physical Adsorption), ed. by M.M. Dubinin (Nauka, Moscow 1970) p.338
2.126 A.N. Frumkin: Tr. Khim. Inst. im. L.Ya. Karpova, **4**, 56 (1925) [in Russian]
2.127 J.M. Honig: in *The Solid-Gas Interface*, Vol.1, ed. by E.A. Flood (Marcel Dekker, New York 1967) p. 371
2.128 R.H. Fowler, E.A. Guggenheim: *Statistical Thermodynamics* (Cambridge University, Cambridge 1939)
2.129 K. Huang: *Statistical Mechanics* (Wiley, New York 1963)
2.130 D.P. Woodruff, G.C. Wang, T.M. Lu: in *The Chemical Physics of Solid Surfaces and Heterogeneous Catalysis*, Vol.2, ed. by D.A. King, D.P. Woodruff (Elsevier, Amsterdam 1983) Chap.4
2.131 M.M. Dubinin, V.V. Serpinsky: Dokl. Akad. Nauk SSSR **99**, 1033 (1954) [in Russian]
2.132 A.A. Voskresensky: Dokl. Akad. Nauk SSSR **119**, 724 (1958) [in Russian]
2.133 A.R. Miller: *The Adsorption of Gases on Solids* (Cambridge University, Cambridge 1949)
2.134 B.P. Bering, V.V. Serpinsky: Dokl. Akad. Nauk SSR **58**, 1061 (1947) [in Russian]
2.135 B.P. Bering, V.V. Serpinsky: Dokl. Akad. Nauk SSR **205**, 611 (1972) [in Russian]
2.136 V.K. Fedyanin: Zh. Fiz. Khim, **44**, 495 (1970); **45** 2867 (1971); **46**, 119 (1972); **47**, 1287 (1973) [in Russian]
2.137 C. Kemball: in *Advances in Catalysis*, Vol.2 (Academic, New York 1948) p.233
2.138 S. Cerny: in *The Chemical Physics of Solid Surfaces and Heterogeneous Catalysis*, Vol.2, ed. by D.A. King, D.P. Woodruff (Elsevier, Amsterdam 1983) Chap.1
2.139 J. Fripiat, J. Chaussidon, A. Jelli: *Chimie-Physique des Phenomenes de Surface* (Masson, Paris 1971)
2.140 W. Orr: Trans. Faraday Soc. **35**, 1247 (1939); Proc. R. Soc. A **173**, 349 (1939)
2.141 T. Hill: J. Chem. Phys. **16**, 181 (1948); **17**, 520, 772, 775 (1949)
2.142 V.A. Bakayev, M.M. Dubinin: Izv. Akad. Nauk SSSR, Ser. Khim. 2156 (1967) [in Russian]
2.143 S.Z. Roginsky: *Adsorption und Katalyse an inhomogenen Oberflächen* (Akademie, Berlin 1958)
2.144 Ya.B. Zel'dovich: Acta Physicochim. URSS **1**, 449, 961 (1934)
2.145 G. Halsey: in *Advances in Catalysis*, **4**, 259 (Academic, New York 1952)
2.146 M.I. Temkin: Zh. Fiz. Khim. **15**, 296 (1941) [in Russian]
2.147 S.L. Kiperman, A.A. Balandin, I.R. Davydov: in *Kinetika i Kataliz* (Kinetics and Catalysis), ed. by S.Z. Roginsky (Akademiya Nauk SSSR, Moscow 1960) p.140
2.148 O.M. Todes, A.K. Bondareva: Zh. Prikl. Khim **21**, 693 (1948) [in Russian]
2.149 R. Sips: J. Chem. Phys. **18**, 1024 (1950)
2.150 J. Horiuti: J. Res. Inst. Catal. Hokkaido Univ. **9**, 108 (1961)
2.151 M.J. Sparnaay: Surf. Sci. **9**, 100 (1968)
2.152 D.J.C. Yates: in *Advances in Catalysis*, **12**, 265 (Academic, New York 1960)
2.153 M.M. Dubinin, A.I. Sakharov, V.F. Kononyuk: Dokl. Akad. Nauk SSSR **198**, 638 (1971); **206**, 901 (1972)
2.154 K.G. Krasil'nikov, N.N. Skoblinskaya: in *Svyazannaya voda v dispersnykh sistemakh* (Bound Water in Disperse Systems), ed. by V.F. Kiselev (Moscow University, Moscow 1972, 1974) Vol.2, p.66; Vol.3, p.33
2.155 H.L. McDermot, J.C. Arnell: Can. J. Chem. **33**, 913 (1955); **34**, 1114 (1956)
2.156 G.G. Fedorov, Yu.A. Zarifyants, V.F. Kiselev: Zh. Fiz. Khim. **37**, 2344 (1963) [in Russian]

2.157 J.H. de Boer, A.B.C. Van Doorn: Proc. K. Ned. Akad. Wet. B **61**, 242 (1958)
2.158 R. Tykody: Trans. Faraday Soc. **54**, 918 (1958)
2.159 R. Barrer: Trans. Faraday Soc. **40**, 347 (1944)
2.160 J.C.P. Mignolet: *Chemisorption* (Butterworths, London 1957)
2.161 E. Engel, R. Gomer: J. Chem. Phys. **52**, 5572 (1970)
2.162 I. Myuller: Izv. Akad. Nauk SSSR, Ser. Fiz. **38**, 345 (1974)
2.163 E.N. Figurovskaya, V.F. Kiselev: Dokl. Akad. Nauk SSSR **182**, 1365 (1968) [in Russian]
2.164 A. Rosenberg: J. Chem. Phys. **62**, 1112 (1958)
2.165 V. Kiselev: Z. Chem. **7**, 369 (1967)
2.166 M.I. Temkin: in *Problemy khimicheskoi kinetiki, kataliza i reaktsionnoi sposobnosti* (Problems of Chemical Kinetics, Catalysis and Reactivity), ed. by V.N. Kondratev, N.M. Emanuel (Akademiya Nauk SSSR, Moscow 1955) p.484
2.167 S.F. Timashev, V.K. Fedyanin: Dokl. Akad. Nauk SSSR **191**, 1333 (1970) [in Russian]
2.168 A.J. Bennet: Crit. Rev. Solid State Sci. **4**, 261 (1974)
2.169 N.D. Lang, A.R. Williams: Phys. Rev. B **18**, 616 (1978)
2.170 V.I. Osherov: Dokl. Akad. Nauk SSSR **130**, 117 (1960); **132**, 884 (1960) [in Russian]
2.171 F. Delanaye, M. Schmets, A.A. Lucas: J. Chem. Phys. **69**, 5126 (1978)
2.172 A.V. Kiselev: "Adsorbtsiya" (Adsorption) in *Fizicheskii entsiklopedicheskii slovar'*, T.1 (Encyclopedic Dictionary of Physics, Vol.1) (Sovet Entsiklopediya, Moscow 1960) [in Russian]
2.173 B.M.W. Trapnell: *Chemisorption* (Butterworths, London 1955)
2.174 V.A. Roiter, G.I. Golodets: *Vvedenie v teoriyu kinetiki i kataliza* (Introduction to the Theory of Kinetics and Catalysis) 2nd ed. (Naukova Dumka, Kiev 1971)
2.175 G.K. Boreskov: *Kataliz* 1 i 2 (Catalysis 1 and 2) (Nauka, Novosibirsk 1971) [in Russian]
2.176 J.A. Anderson, M. Boudart (eds.): *Catalysis, Science and Technology*, Vol.1, Vol.2 (Springer, Berlin, Heidelberg 1981)
D.E.W. Vaughan: *Heterogeneous Catalysis in Practice* (McGraw-Hill, New York 1980)
2.177 J. Baxendale: in *Advances in Catalysis*, **4**, 242 (Academic, New York 1952)
2.178 M. Boudart: in *Interactions on Metal Surfaces*, ed. by R. Gomer, Topics Appl. Phys., Vol.4 (Springer, Berlin, Heidelberg 1975) Chap.7
2.179 G.K. Boreskov: in *Geterogennyi kataliz v khimicheskoi promyshlennosti* (Heterogeneous Catalysis in the Chemical Industry), ed. by G.K. Boreskov (Goskhimizdat, Moscow 1955) p.5
2.180 I.I. Tretyakov, A.V. Sklyarov, B.R. Shub: Kinet. Katal. **12**, 996 (1971) [in Russian]
2.181 M. Boudart: in *Advances in Catalysis*, **20**, 153 (Academic, New York 1969)
M. Boudart, G. Djega-Mariadassou: *Cinetique des Reaction on Catalyse Heterogene* (Masson, Paris 1982)
2.182 O.V. Krylov: Kinet. Katal. **21**, 79 (1980) [in Russian]
2.183 V.I. Atroshchenko, S.I. Kargin: *Tekhnologiya azotnoi kisloty* (Nitric Acid Technology) (Goskhimizdat, Moscow 1949)
2.184 J. Gland, V.N. Korchak: J. Catal. **53**, 9 (1978)
2.185 K. Tamaru: *Dynamic Heterogeneous Catalysis* (Academic, London 1978)
2.186 M. Temkin: Acta Physicochim. URSS **8**, 141 (1938)
2.187 S.L. Kiperman: *Osnovy khimicheskoi kinetiki v geterogennom katalize* (The Fundamentals of Chemical Kinetics in Heterogeneous Catalysis) (Khimiya, Moscow 1979)
2.188 O.V. Krylov, M.U. Kislyuk, B.R. Shub, A.A. Gezalov, Yu.N. Rufov, N.D. Maksimova: Kinet. Katal. **13**, 598 (1972) [in Russian]
2.189 R.W. Maatman: J. Catal. **19**, 64 (1970); **43**, 1 (1976)
2.190 K. Miyahara, A. Kazusaka: J. Res. Inst. Catal. Hokkaido Univ. **24**, 65 (1976)

2.191 G.A. Somorjai: *Principles of Surface Chemistry* (Prentice Hall, Englewood Cliffs, NJ 1972)
G.A. Somorjai: *Chemistry in Two Dimensions: Surfaces* (Cornell U. Press, Ithaca 1981)
2.192 D.A. Frank-Kamenetskii: *Diffuziya i teploperedacha v khimicheskoi kinetike* (Diffusion and Heat Transport in Chemical Kinetics) 2nd ed. (Nauka, Moscow 1969)
2.193 M. Temkin, V. Pyzhev: Acta Physicochim. URSS **1**, 177 (1934); **2**, 473 (1935); Zh. Fiz. Chim. **13**, 851 (1939) [in Russian]
M. Temkin: in *Advances in Catalysis*, **28**, 173 (Academic, New York 1979)
2.194 L.D. Schmidt: *Physical Bases of Heterogeneous Catalysis* (New York 1975) p.491
2.195 C. Aharon, F.C. Tompkins: in *Advances in Catalysis*, **21**, 1 (Academic, New York 1970)
2.196 S. Roginsky, Ya. Zel'dovich: Acta Physicochim. URSS **1**, 554, 559 (1934) [in Russian]
2.197 S.Yu. Elovich, G.M. Zhabrova: Zh. Fiz. Khim. **13**, 1761 (1939) [in Russian]
2.198 S. Elovich, S. Roginsky: Acta Physicochim. URSS **7**, 295 (1937)
2.199 F. Characorin, S. Elovich: Acta Physicochim. URSS **5**, 325 (1936) [in Russian]
2.200 S.Z. Roginsky: *Teoreticheskie osnovy izotopnykh metodov izucheniya khimicheskikh reaktsii* (The Theoretical Fundamentals of Isotopic Methods of Studying Chemical Reaction) (Akademiya Nauk SSSR, Moscow 1956)
2.201 E.G. Christoffel: Catalysis Revs. Chem. Eng. **24**, 159 (1982)
2.202 A.D. Berman: Sov. Phys. Usp. Khim. **45**, 983 (1976)
2.203 J. Wei, C.D. Prater: in *Advances in Catalysis*, **13**, 203 (Academic, New York 1962)
2.204 S.Z. Roginsky: Probl. Kinet. Katal. **4**, 234 (1940) [in Russian]
2.205 N.N. Semenov, V.V. Voevodsky: in *Geterogennyi kataliz v khimicheskoi promyshlennosti* (Heterogeneous Catalysis in the Chemical Industry), ed. by G.K. Boreskov (Goskhimizdat, Moscow 1955) p.233
2.206 V.B. Kazansky, V.A. Shvets, M.Ya. Kon, V.V. Nikisha, B.N. Shelimov: in *Proceedings of the 5th International Congress on Catalysis, Miami Beach 1972*, Vol.2, ed. by J. Hightower (North-Holland, Amsterdam 1973) p.1423
2.207 O.V. Krylov: *Catalysis by Nonmetals. Rules for Catalyst Selection* (Academic, New York 1970)
D.A. King, D.P. Woodruff (eds.): *The Chemical Physics of Solid Surfaces and Heterogeneous Catalysis* (Elsevier, Amsterdam 1982)
2.208 S.Z. Roginsky: Dokl. Akad. Nauk SSSR **67**, 97 (1949) [in Russian]
2.209 O.V. Krylov: Kinet. Catal. (Engl. Transl.) **14**, 28 (1973)
2.210 D.A. Dowden: in *Proceedings of the 4th International Congress on Catalysis, Moscow 1968*, Vol.1, ed. by Ya.T. Eidus (Akademiai Kiado, Budapest 1971) p.163
2.211 O.V. Krylov, E.A. Fokina: Kinet. Katal. **5**, 284 (1964) [in Russian]
2.212 I.I. Ioffe, S.Z. Roginsky: Zh. Fiz. Khim. **31**, 612 (1957) [in Russian]
2.213 A.F. Shestakov, V.A. Matyshak, A.A. Kadushin, O.V. Krylov: Kinet. Katal. **20**, 189 (1979) [in Russian]
E.I. Odumah, J.C. Vickerman: J. Catalysis **62**, 95 (1980)
2.214 A.A. Balandin: J. Phys. Chem. B **2**, 289 (1928); B **3**, 167 (1929)
2.215 G.I. Golodets: *Heterogeneous Catalytic Reactions Involving Molecular Oxygen* (Elsevier, Amsterdam 1983)

Chapter 3

3.1 W.A.Weyl: in *Structure and Properties of Solid Surfaces*, ed. by T.N. Rhodin (Chicago Univ. Press, Chicago, 1953) p.147
3.2 J.J. Bickerman: *Physical Surfaces* (Academic, New York 1970)

3.3 R.G. Lindford: in *Solid State Surface Science* **2**, 5 (Marcel Dekker, New York 1973)
3.4 V.F. Kiselev: *Poverkhnostnye Yavleniya v poluprovodnikakh i dielektrikakh* (Surface Phenomena in Semiconductors and Dielectrics) (Nauka, Moscow 1970)
3.5 M. Lannoo, J. Bourgoin: *Points Defects in Semiconductors I,* Springer Ser. Solid-State Sci., Vol.22 (Springer, Berlin, Heidelberg 1981)
J. Bourgoin, M. Lannoo: *Point Defects in Semiconductors II,* Springer Ser. Solid-State Sci., Vol.35 (Springer, Berlin, Heidelberg 1983)
3.6 V.F. Kiselev: Probl. Kinet. Katal. **16**, 221 (1975)
C.D. Kohl: Appl. Phys. **A30**, 127 (1983)
3.7 A. Many, Y. Goldstein, N.B. Grover: *Semiconductor Surfaces,* 2nd ed. (North-Holland, Amsterdam 1971)
3.8 M.J. Sparnaay: *The Electrical Double Layer* (Pergamon, New York 1972)
3.9 F. Meyer, M.J. Sparnaay: in *Surface Physics of Phosphors and Semiconductors,* ed.by C.G. Scott, C.E. Reed (Academic, London 1975)
3.10 W.E. Spicer: CRC Crit. Rev. Solid State Sci. **6**, 317 (1976)
3.11 R.P.H. Gasser: Q. Rev. Chem. Soc. **25**, 283 (1971)
3.12 L. Eckertova: *Physics of Thin Films* (Plenum, New York 1977)
3.13 H.E. Farnsworth, R.E. Schlier, T.H. George, R.M. Burger: J. Appl. Phys. **29**, 1150, 1195 (1958)
3.14 G.K. Wehner: in *Methods of Surface Analysis,* Vol.1, ed. by A.W. Czanderna (Elsevier, New York 1975), Chap.1
F.D. Townsend, J.C. Kelly, N.E.W. Hartley: *Ion Implantation, Sputtering and Their Applications* (Academic, New York 1976)
3.15 H. Ryssel, H. Glawisching (eds.): *Ion Implantation Techniques,* Springer Ser. Electrophys., Vol.10 (Springer, Berlin, Heidelberg 1982)
3.16 E.W. Müller: Z. Phys. **106**, 541 (1937)
3.17 I.L. Sokoloskaya: in *Poverkhnostnaya diffuziya i rastekanie* (Surface Diffusion and Spreading), ed.by A.A. Lebedev (Nauka, Moscow 1969) p.108
3.17a A. Modinas: Surf. Sci. **70**, 52 (1978)
3.18 R.S. Hansen, N.C. Gardner: in *Experimental Methods in Catalytic Research,* ed. by R.B. Anderson, Vol.1 (Academic, New York 1968) p.169
3.19 B.E. Nicwenhuys, D.Th. Mejer, W.M. Sachtler: Surf. Sci. **40**, 125 (1973)
3.20 A.V. Creve: In *Surface Properties of Materials*, ed. by L.L. Levenson (North-Holland, Amsterdam 1975) p.152
3.21 E.W. Müller: in *Methods of Surface Analysis,* Vol.1, ed. by A.W. Czanderna (Elsevier, New York 1975) Chap.8
3.22 J.H. Block, A.W. Czanderna: in *Methods of Surface Analysis,* ed. by A.W. Czanderna, Vol.1 (Elsevier, New York 1975) Chap.9
3.23 A. Van Oostrom: CRC Crit. Rev. Solid State Sci. 4, 353 (1974)
T.T. Tsong: Surf. Sci. **70**, 211 (1978)
3.24 M.G. Lagally: Ultramicroscopy **11**, 103 (1983)
3.25 B.P. Lemke, D. Haneman: Phys. Rev. B**17a**, 1893 (1978)
3.26 G.A. Katrich: Prib. Tekh. Eksp. **6**, 213 (1965) (in Russian)
3.27 F. Prouix, P. Handler: Surf. Sci. **5**, 81 (1966)
3.28 G.S. Khodakov, P.A. Rebinder: Dokl. Akad. Nauk SSSR **127**, 1070 (1970) (in Russian)
3.29 G.S. Khodakov: *Fizika izmel'cheniya* (The Physics of Comminution) (Nauka, Moscow 1972) (in Russian)
3.30 V.F. Kiselev, K.G. Krasil'nikov, G.S. Khodakov: Dokl. Akad. Nauk SSSR **130**, 1273 (1960) (in Russian)
3.31 M.M. Egorov, V.F. Kiselev, K.G. Krasil'nikov: Zh. Fiz. Khim. **35**, 2234 (1961) (in Russian)
3.32 M.M. Egorov, V.F. Kiselev: Zh. Fiz. Khim. **40**, 1984 (1966) (in Russian)
3.33 G.I. Aleksandrova, V.F. Kiselev, M.P. Mayorova: Kolloidn, Zh. **24**, 3 (1962) (in Russian)
3.34 L.A. Feigin: Dokl. Akad. Nauk SSSR **127**, 313 (1959) (in Russian)

3.35 S.J. Gregg: Kolloid-Z. **169**, 5 (1960)
3.36 A.N. Streletsky, P.Yu. Butyagin: Kinet. Katal. **21**, 765, 770 (1980) (in Russian)
3.37 P.Yu. Butyagin: Usp. Khim. **40**, 1935 (1971) (in Russian); Zh. Vses. Khim. Ova. 18, 90 (1973) (in Russian)
3.38 G.B. Demidovich, R.B. Dzhanelidze, V.F. Kiselev: Fiz. Tekh. Poluprovodu **3**, 629 (1969) (in Russian)
3.39 V.M. Bermudez: Prog. Surf. Sci. **11**, 1 (1982)
R.Vanselow, R. Hove (eds.): *Chemistry and Physics of Solid Surface IV and V*, Springer Ser. Chem. Phys. Vols.20 and 36 (Springer, Berlin, Heidelberg 1982 and 1984)
3.40 K. Heinz, K. Müller, and T. Engel, K.H. Rieder: In *Structural Studies of Surfaces*, Springer Tracts Mod. Phys. Vol.91 (Springer, Berlin, Heidelberg, 1982) p.1 and 55
3.41 G.A. Somorjai: *Principles of Surface Chemistry* (Prentice-Hall, Englewood Cliffs, New Jersey 1972)
G.A. Somorjai: *Chemistry in Two Dimensions: Surfaces* (Cornell U. Press, Ithaca 1981)
P. Mark, S.C. Chang, W.F. Creighton, B.W. Lee: Crit. Rev. Solid State Sci. **5**, 189 (1975)
3.42 M.A. Van Hove, S.Y. Tong: *Surface Crystallography by LEED*, Springer Ser. Chem. Phys., Vol.2 (Springer, Berlin, Heidelberg 1979)
3.43 P.J. Estrup, E.G. McRae: Surf. Sci. **25**, 1 (1971)
F. Jona: Surf. Sci. **8**, 478 (1967)
F. Jona, A. Strozier Jr., W.S. Yang: Repts. Progr. Phys. **45**, 527 (1982)
3.44 S.G. Davison, J.D. Levine: *Surface States* (Academic, New York 1970)
3.45 M. Henzler: in *Festkörperprobleme*, **11**, 187 (Vieweg, Wiesbaden (1971)
3.46 H. Ibach (ed.): *Electron Spectroscopy for Surface Analysis*, Topics Current Phys., Vol.4 (Springer, Berlin, Heidelberg 1977)
3.47 E. Bauer: in *Techniques of Metals Research*, Vol.2, ed. by R.F. Bunshah (Wiley, New York 1969) Chap.15
3.48 G.W. Kammlott: Surf. Sci. **25**, 120 (1971)
3.49 E. Bauer: Surf. Sci. **7**, 351 (1967)
3.50 N. Cabrera: in *Structure et Properties des Surfaces des Solids*, ed. by N. Cabrera (CNRS, Paris 1970) p.270
3.51 C.B. Duke: Adv. Chem. Phys. **27**, 1 (1974)
3.52 C.B. Duke: CRC Crit. Rev. Solid State Sci. **4**, 371 (1974)
3.53 R.F. Willis (ed.): *Vibrational Spectroscopy of Adsorbates*, Springer Ser. Chem. Phys., Vol.15 (Springer, Berlin, Heidelberg 1980)
3.54 K. Hirobayashi: J. Phys. Soc. Jpn, **25**, 856 (1968)
3.55 D. Lichtman: in *Methods of Surface Analysis*, Vol.1, ed. by A.W. Czanderna (Elsevier, New York 1975) Chap.2
3.56 G.A. Somorjai, M.A. Van Hove: *Adsorbed Monolayers on Solid Surfaces*, Structure and Bonding, Vol.38 (Springer, Berlin, Heidelberg 1979)
3.57 M. Cardona, L. Ley (eds.): *Photoemission in Solids I*, Topics Appl. Phys., Vol.26 (Springer, Berlin, Heidelberg 1978)
3.58 B. Feuerbacher (ed.): *Photoemission and the Electronic Properties of Surfaces* (Wiley, New York 1978)
3.59 K. Siegbahn: *ESCA* (Almquist & Wiksells, Uppsala 1967)
3.60 T.N. Rhodin, J.W. Gadzuk: in *The Nature of Surface Chemical Bonds*, ed. by T.N. Rhodin, G. Ertl (North-Holland, Amsterdam 1979) p.113
3.61 W.M. Riggs, M.J. Parker: in *Methods of Surface Analysis*, Vol.1, ed. by A.W. Czanderna (Elsevier, New York 1975) Chap.4
3.62 Kh.M. Minachev, G.V. Antoshin, E.S. Shapiro: Probl. Kinet. Katal. **16**, 189 (1975) (in Russian)
3.63 W.N. Delgass, T.R. Hughes, C.S. Fadley: Catal. Rev. **4**, 149 (1970)
3.64 D.W. Dwight, W.M. Riggs: J. Colloid Interface Sci. **47**, 650 (1974)
3.65 P.E. Best: Phys. Rev. B**19**, 1054 (1979)

3.66 M. Pijolat, G. Hollinger: Surf. Sci. **105**, 114 (1982)
E.W. Plummer, W. Eberhardt: Adv. Chem. Phys. **49**, 533 (1982)
3.67 R.S. Bauer, J.C. McMenamin, H. Petersen, A. Bianconi: in *Proc. Intern. Conf. Physics of SiO_2 and Its Interfaces*, ed. by S.T. Pantelides (Plenum, New York 1978)p.225
R.S. Bauer, J.C. McMenamin: in *Intern, Conf. Physics of Semiconductors*, Edinburgh, 1978
3.68 A.P. Lukirsky, O.A. Ershov, T.M. Zimkina, E.P. Savinov: Sov. Phys.-Solid State 8, 1422 (1966)
3.69 C.J. Powell, R.J. Stien, P.B. Needham, T.J. Driscoll: Phys. Rev. B**16**, 1370 (1977)
3.70 A. Bianconi, R.Z. Bachrach, S.A. Flodstrom: Solid State Commun. **24**, 539 (1977)
3.71 A. Bianconi, L. Incoccia, S. Stipcich (eds.): *EXAFS and Near Edge Structures*, Springer Ser. Chem. Phys., Vol.27 (Springer, Berlin, Heidelberg 1983)
J.M. Thomas, R.M. Lambert (eds.): *Characterization of Catalysts* (Wiley, New York 1980)
T.L. Einstein: Appl. Surf. Sci. **11-12**, 42 (1982)
3.72 P. Eisenberger, B.M. Kincard: Science **200**, 1441 (1978)
3.73 C.C. Chang: in *Characterization of Solid Surfaces*, ed. by P.F. Kane, G.B. Larrabec (Plenum, New York 1974) p.509
A. Joshi, L.E. Davis, P.W. Palmberg: in *Methods of Surface Analysis*, Vol.1, ed. by A.W. Czanderna (Elsevier, New York 1975) Chap.5
J.T. Grant: Appl. Surf. Sci. **13**, 35 (1982)
3.74 W. Reuter: Surf. Sci. **25**, 80 (1971)
3.75 J.D. Place, M. Prutton: Surf. Sci. **82**, 315 (1979)
3.76 O. Meyer (ed.): *Ion Beam Surface Layer Analysis*, Vols.1,2 (Plenum, New York 1976)
3.77 G.A. Somorjai, S.B. Brumbach: Crit. Rev. Solid State Sci. **4**, 429 (1974)
G. Benedek, U. Valbusa (eds.): *Dynamics of Gas-Surface Interaction*, Springer Ser. Chem. Phys. Vol.21 (Springer, Berlin, Heidelberg 1982)
G. Benedek: Surf. Sci. **126**, 624 (1983)
3.78 A. Benninghoven, C.A. Ivans, Jr., R.A. Powell, R. Shimizu, H.A. Storms (eds.): *Secondary Ion Mass Spectrometry (SIMS II)*, Springer Ser. Chem. Phys., Vol.9 (Springer, Berlin, Heidelberg 1979)
A. Benninghoven, J. Giber, J. László, N. Riedel, H.W. Werner (eds.): *Secondary Ion Mass Spectrometry (SIMS III)*, Springer Ser. Chem. Phys., Vol.19 (Springer, Berlin, Heidelberg 1982)
A. Benninghoven, J. Okano, R. Shimizu, H.W. Werner (eds.): *Secondary Ion Mass Spectrometry (SIMS IV)*, Springer Ser. Chem. Phys., Vol.36 (Springer, Berlin, Heidelberg 1984)
3.79 A.W. Czanderna (ed.): *Methods of Surface Analysis*, Vol.1 (Elsevier, New York 1975)
3.80 P.E. Wierenga: *Reflectrometric Study of Surface States and Oxygen Adsorption on Clean Silicon Surfaces*, Ph.D. Thesis, Techniscke Hogeschool Twente, The Netherlands (1980)
3.81 D.E. Aspnes, A.A. Studna: Surf. Sci. **96**, 294 (1980)
3.82 J. Lunsford: Catal. Rev. **8**, 135 (1973)
J. Lunsford: in *Chemistry and Physics of Solid Surfaces*, I, ed. by R. Vanselow, S.Y. Tong (CRC Press, Boca Raton, FL 1980) p.325
3.83 G.C. Benson, K.S. Yun: in *The Solid-Gas Interface*, Vol.1, ed. by E.A. Flood (Marcel Dekker, New York 1967) p.203
3.84 G.C. Benson, P.T. Freeman, E. Dempsey: J. Chem. Phys. **39**, 302 (1963)
3.85 J.J. Schlier, H.E. Farnsworth: J. Chem. Phys. **30**, 917 (1959)
3.86 J.B. Pendry: *Low-Energy Electron Diffraction* (Academic, New York 1974)
3.87 G.A. Somorjai (ed.): *The Structure and Chemistry of Solid Surfaces* (Wiley, New York 1969)

3.88 C.B. Duke, R.J. Meyer, A. Paton, P. Mark: J. Vac. Sci. Technol. **16**, 847 (1979)
3.89 M. Henzler: J. Appl. Phys. **40**, 3758 (1969)
3.90 P.W. Palmberg, W.T. Peria: Surf. Sci. **6**, 57 (1967)
3.91 B.Z. Olshanetsky, A.A. Shklyaev: Surf. Sci. **67**, 581 (1977)
3.92 D.J. Chadi, R.S. Bauer: Phys. Rev. Lett. **44**, 799 (1980)
3.93 P.W. Palmberg: Surf. Sci. **11**, 153 (1968)
3.94 R.N. Thomas, M.H. Francombe: Surf. Sci. **25**, 357 (1971)
3.95 F. Jona, H.D. Shih, A. Ignatiev: J. Phys. C. **10**, 62 (1977)
3.96 J.T. Grant, T.W. Haas: Appl. Phys. Lett. **15**, 140 (1969)
3.97 T.M. French, G.A. Somorjai: J. Phys. Chem. **74**, 2489 (1970)
3.98 J. Markland, S. Anderson: Surf. Sci. **5**, 197 (1967)
3.99 E.G. McRae, C.W. Caldwell: Surf. Sci. **2**, 509 (1964)
3.100 L. Fiermans. E. Arijs, J. Vennik, W. Moemhout-Van der Vorst: Surf. Sci. **39**, 357 (1973)
3.101 C.B. Duke, R.J. Meyer, A. Paton, P. Mark: Phys. Rev. B**18**, 1223 (1978)
3.102 J.J. Lander, J. Morrison: J. Appl. Phys. **34**, 1403 (1963)
3.103 R. Seiwatz: Surf. Sci. **2**, 473 (1964)
3.104 D. Haneman: Phys. Rev. **70**, 705 (1968)
3.105 J.C. Phillips: Surf. Sci. **40**, 459 (1973)
3.106 K.A.R. Mitchell: Surf. Sci. **64**, 797 (1977)
3.107 W.A. Harrison: Surf. Sci. **55**, 1 (1976)
3.108 P. Mark, J.D. Levine, S.H. McFarlane: Phys. Rev. Lett. **38**, 1408 (1977)
3.109 J.D. Levine, S.H. McFarlane, P. Mark: Phys. Rev. B**16**, 5415 (1977)
3.110 M. Taubenblatt, E. So, P. Sih, A. Kahn, P. Mark: J. Vac. Sci. Technol. **15**, 1143 (1978)
3.111 A.R. Lubinsky, C.B. Duke, B.W. Lee, P. Mark: Phys. Rev. Lett. **36**, 1058 (1976)
3.112 J.D. Levine: Surf. Sci. **86**, 672 (1979)
3.113 D.J. Chadi: Phys. Rev. Lett. **43**, 43 (1979)
3.114 W. Mönch: Surf. Sci. **86**, 672 (1979)
3.115 T.B. Grimley, S.M. Walker: Surf. Sci. **14**, 395 (1969)
3.116 F. Jona, H.D. Shih: J. Vac. Sci. Technol. **16**, 1248 (1979)
3.117 A.A. Maradudin, D.L. Mills, S.Y. Tong: in *Structure and Chemistry of Solid Surfaces*, ed. by G. Somorjai (Wiley, New York 1969) Chap.16
3.118 I.M. Lifshitz, L.N. Rosenzweig: Zh. Eksp. Teor. Fiz. **18**, 1012 (1948) (in Russian)
3.119 R.E. Allen, G.P. Alldrige, F.W. De Wette: Phys. Rev. B**4**, 1048 (1971)
3.120 J.B. Theeten, L. Dobrzynski: Phys. Rev. B**5**, 1529 (1972)
3.121 P. Masri, L. Dobrzynski: Surf. Sci. **32**, 623 (1972)
3.122 V.I. Peresada, E.S. Syrkin: Sov. Phys.-Solid State **16**, 452 (1974)
3.123 V.I. Peresada, E.S. Syrkin: Surf. Sci. **54**, 293 (1976)
3.124 R.F. Wallis: in *Structure and Chemistry of Solid Surfaces*, ed. by G. Somorjai (Wiley, New York 1969) Chap.17
3.125 D. Tabor, J.M. Wilson, T.J. Bastov: Surf. Sci. **26**, 471 (1971)
3.126 V.F. Dvoryankin, A.Yu. Mityagin: Kristallografia **12**, 1112 (1967)
3.127 B.A. Nesterenko, A.D. Borodkin: Fiz. Tverd. Tela **12**, 2042 (1970) (in Russian)
3.128 B.A. Nesterenko, O.V. Snitko: *Fizicheskie svoistva chistoi poverchnosty poluprovodnikov* (Physical Properties of Atomic-Clean Semiconductor Surfaces) (Naukova Dumka, Kiev 1983)
3.129 H. Ibach: Phys. Rev. Lett. **24**, 1416 (1970); **27**, 253 (1971)
3.130 J.M. Poate (ed.): *Thin Films. Interdiffusion and Reactions* (Wiley, New York 1978)
3.131 J.A. Van Vechten: Phys. Rev. B**7**, 1479 (1973)
J. Ross, R.P. Andress: Surf. Sci. **106**, 11 (1982)
3.132 V.N. Zharkov, V.A. Kalinin: *Uravneniya sostoyaniya tverdykh tel pri vysokikh davleniyakh i temperaturakh* (The Equations of State for Solids) (Nauka, Moscow 1968)

3.133 J.J. Gilvarry: Phys. Rev. **102**, 308 (1955)
3.134 R.L. Gordon: J. Appl. Phys. **7**, 385 (1956)
3.135 M. Cignitti, A. Uva: Ric. Sci. **28**, 1915 (1958)
3.136 A. Liberti, G. Denitofrancesco: Chem. Ind. **51**, 1983 (1963)
3.137 P.B. Dempster, P.D. Ritchi: J. Appl. Chem. **3**, 182 (1953)
3.138 G.S. Khodakov, P.A. Rebinder: Dokl. Akad. Nauk SSSR **131**, 1316 (1960) (in Russian)
3.139 B.A. Nesterenko, B.I. Gorbachev: Sov. Phys.-Solid State **16**, 2284 (1974)
3.140 N.T. Gladkikh, V.I. Khodkevich: Ukr. Fiz. Zh. **16**, 1429 (1971) (in Russian)
3.141 V.I. Kvlividze, V.F. Kiselev, A.B. Kurzaev, L.A. Ushakova: Surf. Sci. **44**, 60 (1974)
3.142 S. Valeri, S. Mantovani: J. Chem. Phys. **69**, 5207 (1978)
3.143 N.H. Fletcher: in *International Symposium on the Physics and Chemistry of Ice*, Royal Society of Canada, Ottawa, 1972, p.20
3.144 M. Henzler, J. Clabes: Jpn. J. Appl. Phys. **2**, (Suppl.) 389 (1974)
3.145 M.F. Chung, D. Haneman: J. Appl. Phys. **37**, 1879 (1966)
3.146 D. Haneman, B.P. Lemke: Phys. Rev. Lett. **35**, 1379 (1975)
3.147 G.B. Demidovich, V.F. Kiselev, N.N. Leznev, O.V. Nikitina: J. Chim. Physique **65**, 1078 (1968)
3.148 G.B. Demidovich, V.F. Kiselev, O.V. Nikitina: Dokl. Akad. Nauk SSSR **205**, 383 (1972) (in Russian)
3.149 G.B. Demidovich: *Tezisy doklada na Vsesoyuznom simpoziume po elektronnym protsessam na poverkhnosti i v monokristallicheskikh plenkakh* (Minutes of the Report at the All-Union Symposium on the Electronic Processes on the Surfaces of Monocrystal Semiconductors)(Novosibirsk 1969) p.32
3.150 J. Higinbothom, D. Haneman: Surf. Sci. **34**, 450 (1973)
3.151 D.J. Miller, D. Haneman: Phys. Rev. B**3**, 2918 (1971)
3.152 D.J. Miller, D. Haneman: Surf. Sci. **19**, 45 (1970)
3.153 J. Higinbothom, D. Haneman: Surf. Sci. **32**, 466 (1972)
3.154 A. Akhmed-Zade, V.V. Votizanski, V.A. Zakrevsky, E.I. Tomashevsky: Sov. Phys.-Solid State **14**, 343 (1972)
3.155 G. Hochstrasser, J.E. Antonini: Surf. Sci. **32**, 644 (1972)
3.156 V.A. Radzig, A.V. Bystrikov: Kinet. Catal. (Engl.Transl.) **19**, 548 (1978)
3.157 M.F. Chung: J. Phys. Chem. Soc. **32**, 475 (1971)
3.158 G.B. Demidovich, V.F. Kiselev: Phys. Status Solidi B **50**, 33 (1972)
3.159 J. Higinbothom, D. Haneman: Surf. Sci. **19**, 39 (1970)
3.160 D.J. Miller, D. Haneman: Surf. Sci. **33**, 477 (1972)
3.161 V.F. Kiselev, O.V. Nikitina: Dokl. Akad. Nauk SSSR **171**, 374 (1966)
3.162 S. Mrozowski, J.F. Andrew: in *Proceedings of the 4th Conference on Carbon, Buffalo* (Pergamon, New York 1960) p.207
3.163 N.N. Gerasimenko, A.P. Dvurechensky, L.S. Smirnov: Fiz. Tekh. Poluprovodn. **5**, 1700 (1971) (in Russian)
3.164 W. Stöber: Kolloid-Z. **145**, 17 (1956)
3.165 S.P. Zhdanov, A.V. Kiselev: Zh. Fiz. Khim. **31**, 2213 (1957) (in Russian)
3.166 W.A. Weyl: Research **3**, 230 (1960)
3.167 V.A. Radzig: Kinet. Catal. **20**, 367, 375, 1022 (1979)
3.168 V.A. Radzig, V.A. Khalif: Kinet. Catal. **20**, 605 (1979)
3.169 K.G. Krasil'nikov, V.F. Kiselev: Zh. Fiz. Khim. **33**, 2621 (1959)
3.170 M.M. Egorov, V.F. Kiselev, K.G. Krasil'nikov: Zh. Fiz. Khim. **35**, 2032 (1961) (in Russian)
3.171 J.A. Hockey, B.A. Pethica: Trans. FAraday Soc. **57**, 2247 (1961)
3.172 D. Haneman, M.F. Chung, A. Talony: Phys. Rev. **170**, 719 (1968)
3.173 V.S. Kuznetsov, G.B. Demidovich, V.F. Kiselev: Vestn. Mosk. Univ. Fiz. Astronomiya **16**, 177 (1975) (in Russian)
G.B. Demidovich, S.N. Karjagin, V.F. Kiselev: Phys. Stat. Sol. (B) **114**, 705 (1982)
3.174 G.B. Demidovich, V.F. Kiselev: Vestn. Mosk. Univ. Fiz. Astronomiya **14**, 158 (1973) (in Russian)

3.175 P.A. Thomas, M.N. Brodsky, D. Kaplan, D. Lepine: Phys. Rev. B18, 3059 (1978)
3.176 H. Iwasaki, S. Nakamura: Surf. Sci. **81**, 627 (1979)
3.177 Yu.A. Zarifyants: Zh. Fiz. Khim. **38**, 2655 (1964) (in Russian)
3.178 C. Coulson: in *Proceedings of the 4th Conference on Carbon, Buffalo* (Pergamon, New York 1960) p.215
3.179 M. Nishida: J. Phys. C **11**, 1217 (1978)
3.180 C.T. White, K.L. Ngai: J. Vac. Sci. Technol. **16**, 1412 (1979)
3.181 P.Yu. Butyagin, A.N. Streletsky, Yu.N. Rufov: *Trudy 5-go Vsesoyuznogo simpoziuma po mekhanoemissii i mekhanokhimii* (Proceedins of the 5th All-Union Symposium on Mechanoemission and Mechanochemistry) (Tartu, U. Press, 1976) p.38
3.182 Yu.N. Jarym-Agaev, P.Yu. Butyagin: Dokl. Akad. Nauk SSSR **207**, 892 (1972) (in Russian)
3.183 J.C. Phillips: J. Non-Cryst. Solids **35**,**36**, 1157 (1980)
F. Yonezawa (ed.): *Fundamental Physics of Amorphous Semiconductors*, Springer Ser. Solid-State Sci., Vol.25 (Springer, Berlin, Heidelberg 1981)
Y. Nagaoka, H. Fukuyama (eds.): *Anderson Localization*, Springer Ser. Solid-State Sci., Vol.25 (Springer, Berlin, Heidelberg 1982)
3.184 R.K. Iler: *The Colloid Chemistry of Silica and the Silicates* (Cornell University Press, Ithaca, N.Y 1955)
3.185 J.H. De Boer, J.M. Vleskens: Proc. K. Ned. Akad. Wet., Ser. B**60**, 45, 54 (1957); B**61**, 2, 85 (1958)
3.186 S.I. Kol'tsov, V.B. Aleskovsky: *Silikagel', ego stronie i khimicheskie svoistva* (Silica Gel: Its Structure and Chemical Properties) (Goskhimizdat, Moscow 1963)
3.187 M.M. Egorov: Dokl. Akad. Nauk SSSR **140**, 401 (1961) (in Russian)
3.188 R.V. Prudnikov, V.F. Kiselev, M.M. Egorov: Dokl. Akad. Nauk SSSR **166**, 395 (1966)
3.189 H.P. Boehm, M. Hermann: Z. Anorg. Allg. Chem. **352**, 156 (1967)
3.190 M.M. Egorov, N.N. Dobrovol'sky, V.F. Kiselev, G. Furman, S.V. Khrustaleva: Zh. Fiz. Khim. **39**, 3070 (1965)
3.191 T. Tung, E. McInish: J. Catal. **3**, 229 (1964)
3.192 J.B. Peri: J. Phys. Chem. **69**, 211, 220 (1965)
3.193 J.B. Peri, A.L. Hensley: J. Phys. Chem. **72**, 2926 (1968)
3.194 V.I. Lygin, V.V. Smolikov, A.G. Chernov: Zh. Fiz. Khim. **50**, 498 (1976) (in Russian)
3.195 I.D. Mikheikin, I.A. Abronin: J. Mol. Catal. **1**, 326 (1978)
K. Takahashi: J. Coll. and Interface Sci. **88**, 286 (1982)
3.196 J.B. Peri: J. Phys. Chem. **79**, 1271 (1975)
3.197 G.G. Grigoryan, A.V. Mushegyan: Arm. Khim. Zh. **31**, 811 (1978) (in Russian)
3.198 A.V. Uvarov, T.V. Antipina, S.P. Tikhomirova: Zh. Fiz. Khim. **46**, 3059 (1967)
3.199 J.E. Dabrowski, J.B. Butt, H. Bliss: J. Catal. **18**, 297 (1970)
3.200 O. Glemser: Angew. Chem. **73**, 785 (1961)
3.201 U. Hofman, K. Endell, D. Wilm: Angew. Chem. **47**, 539 (1934)
3.202 E.K. Rideal: Trans. FAraday Soc. **32**, 4 (1936)
3.203 A.V. Kiselev: Kolloidn. ZH. **2**, 17 (1936) (in Russian)
3.204 P.C. Carman: Trans. Faraday Soc. **36**, 964 (1940)
3.205 V.I. Lyashenko, V.G. Litovchenko, I.I. Stepko, V.I. Strikha, L.V. Lyashenko: *Elektronnye yavleniya na poverkhnosti poluprovodnikov* (Electronic Phenomena on Semiconductor Surfaces) (Naukova Dumka, Kiev 1968)
3.206 V.A. Arslambekov: in *Mekhanizmy vzaimodeistviya metallov s gazami* (Mechanisms of Interaction Between Metals and Gases), ed. by B.V. Derjagin (Nauka, Moscow 1964) p.167
3.207 V.G. Litovchenko: *Osnovy fiziki poluprovodnikovykh sloistykh sistem* (Fundamentals of the Physics of Laminated Semiconductor Systems) (Naukova Dumka, Kiev 1980)

3.208 Yu. A. Zarif'yants, V.F. Kiselev, S.N. Kozlov, Yu.F. Novototsky-Vlasov: Vestn. Mosk. Univ. Fiz. Astronomiya **16**, 84 (1975) (in Russian)
3.209 H.R. Philipp: J. Phys. Chem. Sol. **32**, 1935 (1970)
3.210 D.B. York: J. Electrochem. Soc. **110**, 271 (1963)
3.211 E.K. Kazenas, D.M. Chizhikov, Yu.V. Tsvetkov, Yu.V. Vasyuta: Zh. Fiz. Khim. **43**, 695 (1973) (in Russian)
3.212 J.R. McCreary, R.J. Thorn, L.C. Wagner: J. Non-Cryst. Solids **23**, 293 (1977)
3.213 G.F. Romanova: Ukr. Fiz. Zh. **14**, 840 (1969) (in Russian)
3.214 S. Yoshioko, S. Takyanagi: J. Appl. Phys. Jpn. **4**, 829 (1965)
3.215 Ch. Kleint, B. Hartman, H. Meyer: Z. Phys. Chem. (DDR) **250**, 315 (1972)
3.216 S. Sinharoy, M. Henzler: Surf. Sci. **51**, 75 (1975)
3.217 H.G. Weiss, J.A. Knight, J.S. Shapiro: J. Am. Chem. Soc. **81**, 1823 (1959)
3.218 C. Naccache, B. Imelik: Bull. Soc. Chim. Fr. 553 (1961)
3.219 H.P. Boehm, M. Scheider: Z. Anorg. Allg. Chem. **301**, 326 (1959)
3.220 V.A. Tertykh, V.V. Pavlov, V.M. Mashchenko, A.A. Chuiko: Dokl. Akad. Nauk SSSR **203**, 1913 (1971) (in Russian)
3.221 J. Uytterhoeven, H. Navenau: Bull. Soc. Chim. Fr. 27 (1962)
3.222 K.D. Shcherbakova, K.I. Slovetskaya: Dokl. Akad. Nauk SSSR **111**, 855 (1956)
3.223 H.P. Boehm: Angew. Chem. **12**, 617 (1966)
3.224 J.J. Fripiat, A. Leonard, H. Barake: Bull. Soc. Chim. Fr. 122 (1962)
3.225 I.E. Neumark, B.M. Chertov: Dokl. Akad. Nauk SSSR **132**, 1356 (1960); **138**, 877 (1966)
3.226 R.G. Haldman, P. Emmett: J. Am. Chem. Soc. **78**, 2917 (1956)
3.227 L.T. Zhuravlev, A.V. Kiselev: Zh. Fiz. Chim. **39**, 453 (1965) (in Russian)
3.228 G.B. Bliznakov, I. Bakardjev, M. Maidraganova: J. Catal. **15**, 135 (1969)
3.229 I. Bakardjev, M. Maidraganova, G. Bliznakov: J. Non-Cryst. Solids **20**, 349 (1976)
3.230 V.B. Aleskovsky: *Stekhiometriya i sintez tverdykh soedineni* (Stoichiometry and the Synthesis of Solid Compounds) (Nauka, Leningrad 1976)
3.231 D.L. Kantro, S. Brunauer, L.E. Copeland: in *The Solid-Gas Interface*, Vol.1, ed. by. E.A. Flood (Marcel Dekker, New York 1967) p.414
3.232 J.J. Chessick, G.J. Joung, A.C. Zettlemoyer: Trans. Faraday Soc. **50**, 587 (1954)
3.233 G. Kraus: J. Phys. Chem. **59**, 343 (1955)
3.234 K.G. Krasil'nikov, V.F. Kiselev, N.V. Kapitonova, A.E. Sysoev: Zh. Fiz. Khim. **31**, 1448 (1957) (in Russian)
3.235 M.M. Dubinin, E.G. Zhukovskaya, E.D. Zaverina, I.E. Neumark, R.Yu. Sheinfain: Izv. Akad. Nauk SSSR, Ser. Khim. 588 (1960) (in Russian)
3.236 L.G. Ganichenko, M.M. Dubinin, E.D. Zaverina, V.F. Kiselev, K.G. Krasil'nikov: Izv. Akad. Nauk SSSR, Ser. Khim. 1535 (1960) (in Russian)
3.237 A.V. Kiselev (ed.): *Poverkhnostnye khimicheskie soedineniya i ikh rol' vyavleniyakh adsorbtsii* (Surface Chemical Compounds and Their Role in Adsorption Phenomena) (Moscow University, Moscow 1957) pp.90, 199
3.238 B.V. Il'in, V.F. Kiselev, K.G. Krasil'nikov: Vestn. Mosk. Univ. Fiz. Astronomiya **6**, 35 (1957) (in Russian)
3.239 M.M. Egorov, K.G. Krasil'nikov, V.F. Kiselev: Zh. Fiz. Khim. **32**, 2448 (1958) (in Russian)
3.240 J. Bastick: Bull. Soc. Chim. Fr. 437 (1953)
3.241 S.P. Zhdanov: in *Poverkhnostnye khimicheskie soedineniya i ikh rol' vyavleniyakh adsorbtsii* (Surface Chemical Compounds and Their Role in Adsorption Phenomena), ed. by A.V. Kiselev (Moscow University, Moscow 1957) p.129
3.242 M.M. Egorov, V.F. Kiselev: Zh. Fiz. Khim. **35**, 2031 (1961) (in Russian)
3.243 M.M. Egorov, V.F. Kiselev, V.I. Kvlividze, K.G. Krasil'nikov: Kolloid-Z. **212**, 126 (1966)
3.244 V.V. Morariu: Z. Phys. Chem. (DDR) **97**, 235 (1975)

3.245 M.M. Egorov, V.F. Kiselev, K.G. Krasil'nikov, V.V. Murina: Zh. Fiz. Khim. **33**, 65 (1959) (in Russian)
3.246 J.W. Whalen: Adv. Chem. **33**, 281 (1961)
3.247 J.A.G. Taylor, J.A. Hockey: Proc. Br. Ceram. Soc. **5**, 133 (1965)
3.248 L.G. Ganichenko, V.F. Kiselev, V.V. Murina: Kinet. Katal. **2**, 877 (1961) (in Russian)
3.249 R.V. Prudnikov: Vestn. Mosk. Univ. Fiz. Astonomiya **2**, 126 (1970) (in Russian)
3.250 A.E. Bashanova, V.F. Kiselev, S.T. Saveleva: Z. Chem. **9**, 52 (1969)
3.251 F. Gonzalez, G. Manuera: Rev. Chim. Miner. **7**, 1021 (1970)
3.252 M.J.D. Low, N. Madison, P. Rannamurthy: Surf. Sci. **13**, 238 (1969)
3.253 M.M. Egorov, V.F. Kiselev, K.G. Krasil'nikov, Yu.P. Simanov: Izv. Vyssh. Uchebn. Zaved., Khim. **2**, 360 (1953) (in Russian)
3.254 M.M. Egorov, Yu.A. Zarif'yants, V.F. Kiselev, K.G. Krasil'nikov, K.V. Topchieva: Zh. Fiz. Khim. **36**, 1882 (1962) (in Russian)
3.255 M.M. Egorov, Yu.A. Zarif'yants, V.F. Kiselev, K.G. Krasil'nikov: Dokl. Akad. Nauk SSSR **120**, 326 (1958)
3.256 J.H. de Boer, J.M.H. Fortuin, B.C. Lippens, W.H. Meijs: J. Catal. **2**, 1 (1963)
3.257 S.P. Zhdanov: Zh. Fiz. Khim. **36**, 2098 (1962) (in Russian)
3.258 A.N. Terenin, N.G. Yaroslavsky: Dokl. Akad. Nauk SSSR **66**, 885 (1949) (in Russian)
3.259 A.N. Terenin: in *Poverkhnostnye khimicheskie soedineniya i ikh rol' vyavleniyakh adsorbtsii* (Surface Chemical Compounds and Their Role in Adsorption Phenomena), ed. by A.V. Kiselev (Moscow University, Moscow 1957) p.206
3.260 A.N. Sidorov: Zh. Fiz. Khim. **30**, 995 (1956) (in Russian)
3.261 A.N. Sidorov: Opt. Spektrosk. **8**, 806 (1960) (in Russian)
3.262 L.H. Little: *Infrared Spectra of Adsorbed Species* (Academic, New York 1966)
3.263 M.L. Hair: *Infrared Spectroscopy in Surface Chemistry* (Marcel Dekker, New York 1967)
3.264 L.A. Ignat'eva, V.I. Kvlividze, V.F. Kiselev: in *Svyazannaya voda v dispersnykh sistemakh,* T.1 (Bound Water in Disperse Systems, Vol.1), ed. by V.F. Kiselev (Moscow University, Moscow 1970) p.56
3.265 A.A. Tsyganenko, V.N. Filimonov: in *Uspekhi fotoniki,* Vol.4, ed. by F.I. Vilesov (Leningrad University, Leningrad 1974) p.56 (in Russian)
3.266 A.V. Kiselev, V.I. Lygin: Infrared Spectra of Surface Compounds (Wiley, New York 1975)
3.267 C.R. Brundle, H. Morawitz (eds.): *Vibrations at Surfaces* (Elsevier, Amsterdam 1983)
R.F. Willis, A.A. Lucas, G.D. Mahana: in *The Chemical Physics of Solid Surfaces and Heterogenous Catalysis,* Vol.2, ed. by D.A. King, D.P. Woodruff (Elsevier, Amsterdam 1983) Chap.3
3.268 D.E.O. Reilly, H.P. Leftin, W.K. Hall: J. Chem. Phys. **29**, 970 (1958)
3.269 V.I. Kvlividze, N.M. Ievskaya, T.S. Egorova, V.F. Kiselev, N.D. Sokolov: Kinet. Katal. **3**, 91 (1962) (in Russian)
3.270 A.A. Chujko, V.A. Sobolev, V.A. Tertykh: Ukr. Khim. Zh. **38**, 773 (1972) (in Russian)
3.271 M. Baverez, J. Bastick: Bull. Soc. Chim. Fr. 4070 (1967)
3.272 R.S. McDonald: J. Phys. Chem. **62**, 1168 (1958)
3.273 T. Bernstein, H. Ernst et al.: Z. Phys. Chem. (Leipzig) **262**, 1123 (1981)
M. Holik, B. Matejkova: J. Chromatography **213**, 33 (1981)
D.W. Sindorf, G.E. Maclet: J. Phys. Chem. **86**, 5208 (1982)
3.274 S.P. Zhdanov: Zh. Fiz. Khim. **32**, 699 (1958) (in Russian)
3.275 A.A. Chuiko, V.M. Ogenko, V.A. Tertykh, V.A. Sobolev: in *Adsorbtsiya i adsorbenty,* T.3 (Adsorption and Adsorbents, Vol.3) (Naukova Dumka, Kiev 1975) p.69

3.276 H.A. Benesi, A.C. Jones: J. Phys. Chem. **63**, 179 (1959)
3.277 L.A. Ignat'eva, V.F. Kiselev, G.D. Chukin: Dokl. Akad. Nauk SSSR **181**, 914 (1968) (in Russian)
3.278 G.C. Cornier, M. Bavarez, J. Bastick: Bull. Soc. Chim. Fr. 2707 (1969)
3.279 I.A. Ignat'eva, G.D. Chukin, G.V. Yukhnevich: Zh. Prikl. Spectrosk. **12**, 318 (1970)
3.280 V.I. Kvlividze: Dokl. Akad. Nauk SSSR **157**, 158 (1964) (in Russian)
3.281 D. Michel: Z. Naturforsch. A **32**, 1751 (1966)
3.282 F. Reesing, P. Fink: Z. Chem. **12**, 1112 (1972)
3.283 V.M. Klimovich, A.A. Chuiko, V.A. Tertykh, I.E. Neumark: in *Adsorbenty* (Adsorbents), ed. by I.E. Neumark (Naukova Dumka, Kiev 1972) p.100
3.284 E. Colvin: *Silicon in Organic Synthesis* (Butterworth, London 1981) Chaps.2,3
3.285 H. Pfeifer: Phys. Lett. C **26**, 293 (1976)
T.M. Duncan, C. Dybovski: *Chemisorption and Surfaces Study by Nuclear Magnetic Resonance Spectroscopy* (North-Holland, Amsterdam 1981)
3.286 V.I. Kvlividze, V.F. Kiselev: Probl. Kinet. Katal. **12**, 302 (1968)
3.287 G.F. Golovanova, V.I. Kvlividze, V.F. Kiselev: in *Svyazannaya voda v dispersnykh sistemakh* (Bound Water in Disperse Systems, Vol.4), ed. by V.F. Kiselev (Moscow University, Moscow 1975) p.178
3.288 A.V. Rakov: *Spektroskopiya tonkoplenochnykh poluprovodnikovykh struktur* (Spectroscopy of Thin Semiconductor Films) (Sovetskoe Radio, Moscow 1975)
3.289 K.H. Beckman, N.J. Harrick: J. Electrochem. Soc. 118, 614 (1971)
3.290 B.E. Wagner, J.N. Helbert, E.H. Poindexter, R.D. Bates: Surf. Sci. **67**, 251 (1971)
3.291 Yu. A. Zarif'yants. V.F. Kiselev, S.V. Khrustaleva: in *Svyazannaya voda v dispersnykh sistemakh,* T.3 (Bound Water in Disperse Systems, Vol.3), ed. by V.F. Kiselev (Moscow University, Moscow 1974) p.74
3.292 A. Metcalfe, S.U. Shankar: Trans. Faraday Soc. **75**, 962 (1979)
3.293 J.B. Peri, R.B. Hannan: J. Phys. Chem. **64**, 1526 (1960)
3.294 L.A. Ignat'eva, G.D. Chukin, G.V. Bondarenko: Dokl. Akad. Nauk SSSR **181**, 393 (1968) (in Russian)
3.295 H. Knözinger, E. Ress: Z. Phys. Chem. (BRD) **59**, 49 (1968)
3.296 H.E. Hallam: J. Catal. **14**, 104 (1969)
3.297 G.F. Golovanova, V.I. Kvlividze, O.F. Kirina, T.V. Antipina: Zh. Fiz. Khim. **47**, 2549 (1973) (in Russian)
3.298 G.F. Golovanova, V.I. Kvlividze, V.F. Kiselev: Kinet. Catal. (Engl. Transl.) **16**, 652 (1975)
3.299 M.R. Basila: J. Phys. Chem. **66**, 2223 (1962)
3.300 D.E.O'Reilly: Adv. Catal. **12**, 31 (1960)
3.301 J. Fraissard, J. Elston, B. Imelik: J. Chem. Phys. **65**, 920 (1968)
3.302 V.M. Mastikhin, V.S. Levchuk, Yu. V. Chaikin: Kinet. Katal. **13**, 1337 (1962) (in Russian)
3.303 Yu. A. Zarif'yants, V.F. Kiselev, A.S. Petrov, R.V. Prudnikov, S. V. Khrustaleva, G.D. Chukin: kinet, Catal. (Engl.Transl.) **15**, 325 (1974)
3.304 J.M. Mays, G.W. Brady: J. Chem. Phys. **25**, 583 (1956)
M.A. Enriquez, C. Dorenieux-Morin, J. Fraissard: J. Sol. State Chem. **40**, 233 (1981)
3.305 D.J.C. Yates: J. Phys. Chem. **65**, 746 (1961)
3.306 K.E. Lewis, G.D. Parfit: Trans. Faraday Soc. **62**, 204 (1966)
3.307 A.V. Kiselev, A.V. Uvarov: Surf. Sci. **6**, 399 (1967)
3.308 M. Primet, P. Pichat, M.V. Mathieu: J. Phys. Chem. **75**, 1216, 1221 (1971)
3.309 P. Jones, J.A. Hockey: Trans. Faraday Soc. **67**, 2669, 2679 (1971)
3.310 D.M. Griffith, C.H. Rochester: Trans. Faraday Soc. **73**, 1471 (1977)
3.311 L.I. Burbulyavichus, Yu.A. Zarif'yants, V.F. Kiselev, S.V. Khrustaleva: Kinet. Katal. **12**, 922 (1971) (in Russian)
3.312 V.N. Filimonov: Opt. Spektrosk. **13**, 709 (1958) (in Russian)

3.313 F.G. Gartner, E. Mollwo: Phys. Status Solidi B **89**, 311 (1978); **90**, 33 (1978)
3.314 P.J. Anderson, R.F. Horlock, J.F. Oliver: Trans. Faraday Soc. **61**, 2754 (1965)
3.315 M. Faure: Bull. Soc. Chim. **69** (1970)
3.316 A.M. Deane, D.L. Griffiths, I.A. Lewis: Trans. Faraday Soc. **71**, 1005 (1975)
3.317 M.M. Egorov: in *Sovremennoe predstavlenie o svyazannoi vode v porodakh* (Current Views on Water Bound in Rocks), ed. by M.V. Popov (Akademiya Nauk SSSR, Moscow 1963) p.5
3.318 A.A. Tsyganenko, V.N. Filimonov: Dokl. Akad. Nauk SSSR **203**, 636 (1972) (in Russian)
3.319 A.A. Tsyganenko, V.N. Filimonov: J. Mol. Struct. **19**, 579 (1973)
3.320 L. Pauling: J. Phys. Chem. **56**, 361 (1952)
3.321 S.T. Pantelides, W.A. Harrison: Phys. Rev. B**13**, 2667 (1976)
3.322 A.G. Revesz: Phys. Status Solidi A **57**, 235 (1980)
3.323 K.A.R. Mitchel: Chem. Rev. **69**, 157 (1969)
3.324 A.G. Revesz: Phys. Rev. Lett. **27**, 1578 (1971)
3.325 D.J. Chadi: in *Proceedings of the International Conference, Physics of SiO_2 and Its Interfaces*, ed. by S.T. Pantelides (Plenum, New York 1978) p.55
3.326 R.N. Nucho, A. Madhukar: in *Proceedings of the International Conference, Physics of SiO_2 and Its Interfaces*, ed. by S.T. Pantelides (Plenum, New York 1978) p.60; Phys. Rev. B**21**, 1576 (1980)
3.327 F.J. Grunthamer, J. Meserjian: in *Proceedings of the International Conference, Physics of SiO_2 and Its Interfaces*, ed. by S.T. Pantelides Plenum, New York 1978) p.389
3.328 K.B. Yatsimirsky, V.V. Nemoshkalenko: in *Metallofizika*, (Metal Physics, Vol.60) (Naukova Dumka, Kiev 1975) p.60
3.329 V.F. Kiselev: Dokl. Akad. Nauk SSSR **170**, 124 (1967) (in Russian)
V.F. Kiselev, S.N. Kozlov: Poverchnost. Fizika. Khim. Mekhanika (Surface) **2**, 13 (1982)
3.330 F.H. Hambleton, J.A. Hockey, A.G. Taylor: Trans. Faraday Soc. **62**, 795, 801 (1965); Nature **208**, 138 (1965)
3.331 I.I. Novak, K.N. Kuksenko, V.P. Pukh: in *Tezisy 1-go Vsesoyuznogo simpoziuma po aktivoni poverkhnosti tverdykh tel* (Minutes of the 1st All-Union Symposium on Active Surfaces of Solids) (Tartu Univ. Press, Tartu 1975)
3.332 J.H. Moss, G.D. Parfitt, A. Wright: J. Colloid Polym. Sci. **256**, 1121 (1978)
3.333 D.E. Sayers, E.A. Stern, F.W. Lytle: Phys. Rev. Lett. 35, 584 (1975)
3.334 J. Karlo, J.H. Konnert: Phys. Rev. Lett. **14**, 823 (1976)
3.335 S.S. Batsanov: *Elektrootritsatel'nost' elementov i khimicheskaya svyaz'* (Electronegativity of Elements and the Chemical Bond) (Novosibirsk 1962)
3.336 K. Hauffe: *Reaktionen in und an festen Stoffen*, 2. Aufl. (Springer, Berlin, Heidelberg 1966)
3.337 F.A. Kröger: *The Chemistry of Imperfect Crystals* (North-Holland, Amsterdam 1964)
Z.M. Jarzebski: *Oxide Semiconductors* (Pergamon, Oxford 1973)
3.338 V.A. Vinetsky, G.A. Kholodar':*Radiatsionnaya fizika poluprovodnikov* (Radiative Physics of Semiconductors) (Naukova Dumka, Kiev 1979)
3.339 G.F. Neumark: J. Appl. Phys. **51**, 3383 (1980)
3.340 V.F. Kiselev, O.V. Krylov: *Electronic Phenomena in Adsorption and Catalysis* (Springer, Berlin, Heidelberg, New York) to be published
3.341 S.J. Teichner: Rev. Haut. Temp. **2**, 79 (1965)
3.342 M.N. Danchevskaya, E.N. Figurovskaya: Kinet. Katal. **10**, 930 (1969) (in Russian)
3.343 Yu.D. Tret'yakov: *Khimiya nestekhiometricheskikh okislov* (The Chemistry of Nonstoichiometric Oxides) (Moscow University, Moscow 1974)

3.344 B.S. Bokshtein, C.Z. Bokshtein, A.A. Zhukhovitsky: *Termodinamika, kinetika i diffuziya v tverdykh telakh* (Thermodynamics, Kinetics and Diffusion in Solids) (Metallurgiya, Moscow 1974)
3.345 H.P. Leftin, M.C. Hobson: Adv. Catal. **14**, 115 (1963)
3.346 A.N. Terenin: Probl. Kinet. Katal. **12**, 27 (1968) (in Russian)
3.347 V.Ya. Lodin, V.E. Kholmogorov, A.N. Terenin: Dokl. Akad. Nauk SSSR **159**, 187 (1964) (in Russian)
3.348 Yu.D. Pimenov, V.E. Kholmogorov, A.N. Terenin: Dokl. Akad. Nauk SSSR **163**, 935 (1965) (in Russian)
3.349 Yu.D. Pimenov: in *Molekulyarnaja fotonika* (Molecular Photonics), ed. by F.I. Vilesov (Nauka, Leningrad 1970) p.351
3.350 Yu.A. Zarif'yants, S.N. Karyagin, V.F. Kiselev, S.V. Khrustaleva, G.D. Chukin: Dokl. Akad. Nauk SSSR **202**, 109 (1972) (in Russian)
3.351 V.D. Kazaritsky, S.N. Kozlov, V.F. Kiselev, Yu.F. Novototsky-Vlasov: Dokl. Akad. Nauk SSSR **195**, 115 (1970) (in Russian)
3.352 Yu.D. Pimenov, V.E. Kholmogorov: Kinet. Katal. **10**, 169 (1969) (in Russian)
3.353 Yu.A. Zarif'yants, V.V. Kurylev: Sov. Phys.-Semicond. **7**, 414 (1973)
3.354 A.J. Tench, R.H. Nelson: Chem. Phys. Lett. **7**, 459 (1970)
3.355 C. Naccache: Chem. Phys. Lett. **9**, 33 (1974); **11**, 323 (1975)
3.356 M. Setaka, K.M. Sancier, T. Kwan: J. Catal. **16**, 44 (1970)
3.357 K. Hoffman, D. Hahn: Phys. Status Solidi A **24**, 637 (1974)
M. Watanabe: Jpn. J. Appl. Phys. **19**, 1853 (1980)
3.358 Y. Mizokawa, S. Nakamura: Jpn. J. Appl. Phys. **14**, 779 (1975)
3.359 V.E. Heinrich, H.J. Zeiger, G. Dresselhaus: *Electrocatalysis on Non-Metallic Surfaces*, No.455 (NBS, Washington, DC 1976) p.133
3.360 L.I. Burbulyavichus, Yu.A. Zarif'yants, S.N. Karyagin, V.F. Kiselev: Kinet. Katal. **14**, 1526 (1973) (in Russian)
3.361 B.P. Berkovsky, M.P. Votinov, A.I. Grachev, T.F. Ivanov, O.A. Yanzurzhinskaya: in *Fizika dielektrikov i perspetivy ee razvitiya* (The Physics of Dielectrics and Prospects for Its Development) Vol.1, ed. by M.P. Votinov (Nauka, Leningrad 1973) p.50
3.362 D. Tsur, M. Blau, M. Weger: J. Phys. Chem. Sol. **7**, 1257 (1974)
3.363 P. West: in *Positrons on Solids*, ed. by P. Hautojärvi, Topics Current Phys., Vol.12 (Springer, Berlin, Heidelberg 1979) Chap.3
V.I. Goldanski: Poverchnost. Fizika. Khim. Mekhanika **5**, 1 (1982)
3.364 E.I. Kotov, L.G. Baranova, A.A. Pankratov: in *Molekulyarnaya fotonika* (Molecular Photonics), ed. by F.I. Vilesov (Nauka, Leningrad 1970) p.334
3.365 V.I. Evreinov, V.B. Golubev, E.V. Lunina: Zh. Fiz. Khim. **47**, 215, 1018 (1973) (in Russian)
3.366 A.V. Bondarenko, V.F. Kiselev, K.G. Krasil'nikov: Kinet. Katal. **2**, 590 (1961) (in Russian)
3.367 F. Freund, H. Gentsch: Ber. Dtsch. Keram. Ges. **44**, 51 (1967)
3.368 F. Freund: J. Am. Ceram. Soc. **50**, 494 (1967)
3.369 V.A. Barachevsky, A.N. Terenin, V.E. Kholmogorov, L.G. Karakchiev: Probl. Kinet. Katal. **12**, 318 (1968) (in Russian)
3.370 V.N. Beznozdrev, V.B. Kazansky: Kinet. Katal. **14**, 1304 (1973) (in Russian)
3.371 Y. Nishi: Jpn. J. Appl. Phys. **10**, 52 (1971)
3.372 P.J. Caplan, J.N. Helbert, B. Wagner, E.H. Poindexter: Surf. Sci. **54**, 33 (1976)
3.373 M.M. Tagieva, V.F. Kiselev: Zh. Fiz. Khim. **35**, 1381 (1961) (in Russian,
3.374 V.B. Kazansky, G.B. Pariisky, V.V. Voevodsky: Disc. Faraday Soc. **31**, 203 (1961)
3.375 Z.L. Krylova, P.I. Dolin: Khim. Vys. Energ. 3, 152 (1969) (in Russian)
3.376 L. Abrams, A.O. Allen: J. Phys. Chem. **73**, 2741 (1969)
3.377 F.G. Feigl, W.B. Fowler, K.L. Yip: Solid State Commun. **14**, 225 (1974)
3.378 G. Kordos, B. Camara, H.J. Oel: J. Non-Crystall. Sol. **50**, 79 (1982)
R.A.B. Devine, A. Golanski: J. Appl. Phys. **54**, 3833 (1983)
3.379 S.N. Karyagin, P.K. Kashkarov, V.F. Kiselev, S.N. Kozlov: Phys. Status Solidi A **37**, K17 (1976)

3.380 G.F.J. Garlick, J.E. Nichols, A.M. Özer: J. Phys. C **4**, 2230 (1971)
3.381 S.S. Kas'yumova, V.A. Mashkov, D.M. Judin: Dokl. Akad. Nauk SSSR **207**, 1318 (1972) (in Russian)
3.382 C.L. Marquardt, G.H. Sigel: IEEE Trans. NS-**22**, 2234 (1975)
3.383 T. Izumi, T. Matsumori: Jpn. J. Appl. Phys. **14**, 1067 (1975)
3.384 R.A. Weeks, C.A. Nelson: J. Am. Chem. Soc. **43**,399 (1960)
3.385 H.J. Tiller, V. Busack, M. Kispert, F.W. Breitbarth: Krist. Tech. **12**, 743 (1977)
3.386 H.J. Tiller, C. Appel, G. Rudakoff: Wiss. Z. Friedrich-Schiller-Univ., Jena, Math. Naturwiss. Reihe **27**, 631 (1978)
3.387 E.J. Friebele, D.L. Griscom, M. Staperbrock: Phys. Rev. Lett. **42**, 1346 (1979)
3.388 H. Dunken, P. Fink: Acta Chim. Acad. Sci. Hung. **53**, 179 (1967)
3.389 H. Stolz, H. Knözinger: Kolloid-Z. Z. Polym. **243**, 71 (1971)
3.390 E.P. Parry: J. Catal. **2**, 371 (1963)
3.391 O.F. Kirina, T.V. Antipina, G.D. Chukin: Zh. Fiz. Khim. **47**, 441 (1973)
3.392 E.V. Lunina, M.N. Kurganov, V.B. Golubev: Zh. Fiz. Khim. **42**, 1539 (1968) (in Russian)
3.393 V.P. Galkin, V.B. Golubev, A.A. Kubasov, E.V. Lunina: Zh. Fiz. Khim. **46**, 1005 (1972) (in Russian)
3.394 J.J. Fripiat, A. Leonard, J.B. Uyetterhoeven: J. Phys. Chem. **69**, 3274 (1965)
3.395 J. Turkevich, F. Nozaki, D. Stamires: in *Proceedings of the 3rd International Congress Catalysis, 1964*, Vol.1 (Amsterdam 1965) p.586
3.396 D.H. Olson, E. Dempsey: J. Catal. **13**, 221 (1969)
3.397 J.B. Peri: J. Catal. **41**, 221 (1976)
3.398 G.D. Parfitt, J. Ramsbotham, C.H. Rochester: Trans. Faraday Soc. **67**, 841 (1971)
3.399 G.D. Chukin, S.V. Khrustaleva: Zh. Fiz. Khim. **47**, 2055 (1973) (in Russian)
3.400 Y. Kondratoff, C. Naccache, B. Imelik: J. Chem. Phys. **65**, 562 (1968)
3.401 A. Abraham, B. Bleaney: *Electron Paramagnetic Resonance of Transition Ions* (Clarendon, Oxford 1970)
3.402 O. Geschwind (ed.): *Electron Paramagnetic Resonance* (Plenum, New York 1972)
3.403 S. Al'tshuler, B. Kozyrev: *Elektronnyi paramagnitnyi rezonans soedinenii elementov promezhutochnykh grüpp* (EPR of Transition Elements), 2nd ed. (Nauka, Moscow 1972)
3.404 J.E. Wertz, J.B. Bolton: *Electron Spin Resonance* (McGraw-Hill, New York 1970)
3.405 E.H. Poindexter: J. Colloid Interface Sci. **38**, 412 (1972)
3.406 M.A. Ghantons, I.A. Clark, W.S. Moore: J. Phys. C. **9**, 1965 (1976)
3.407 J.A. Rabo (ed.): *Zeolite Chemistry and Catalysis*, ACS Monograph 171 (ACS, Washington, DC 1976)
3.408 B. Imelik, C. Naccache et al. (eds.): *Catalysis by Zeolites* (Elsevier, Amsterdam 1980)
3.409 I.E. Maxwell: *Advances in Catalysis* **31**, 231 (Academic, London 1982)
3.410 E. Dempsey: J. Phys. Chem. **73**, 3660 (1969)
3.411 F.S. Stone: Acta Cient. Venez. **24**, 32 (1973)
3.412 M.M. Dubinin: in *Osnovnye problemy teorii fizicheskoi adsorbtsii* (Basic Problems in the Theory of Physical Adsorption), ed. by M.M. Dubinin (Nauka, Moscow 1970) p.252
M.M. Dubinin: Izvestjy Akademii Nauk SSSR, Ser. Khim. **4**, 738 (1983)
3.413 L. Riekert: Adv. Catal. **21**, 281 (1970)
3.414 J.B. Uytterhoeven, R. Schonheudt, B.V. Liengme, W.K. Hall: J. Catal. **13**, 425 (1969)
3.415 M.R. Basila: Appl. Spectros. Rev. **1**, 289 (1968)
L.M. Kustov, V.Ju. Borovkov, V.B. Kazanski: Kinetika i Kataliz **25**, 147 (1984) (in Russian)

3.416 J.W. Ward: J. Phys. Chem. **72**, 12, 4211 (1968)
3.417 P.E. Eberly: J. Phys. Chem. **72**, 1042 (1968)
3.418 J.W. Ward: in *2nd International Conference on Molecular Sieve Zeolites* (Spec. Publ. Soc. Chem. Ind., Worcester 1970) p.682
3.419 V.I. Abramova, G.B. Vlasova, L.A. Ignat'eva, I.F. Moskovskaya, K.V. Topchieva: Vestn. Mosk. Univ., Khim. **4**, 446 (1970) (in Russian)
L.M. Kustov, V.Ju. Borovkov, V.B. Kazanski: J. Catalysis **72**, 149 (1981)
3.420 V.G. Gvakhariya, V.I. Kvlividze, V.F. Kiselev, M.B. Pylova, G.V. Tsitsishvili: Dokl. Akad. Nauk SSSR **188**, 379 (1969) (in Russian)
3.421 V.G. Gvakhariya, V.I. Kvlividze, G.V. Tsitsishvili: Dokl. Akad. Nauk SSSR **22**, 373 (1975) [Engl. transl.: Doklady Phys. Chem. **223**, 1005 (1975)]
3.422 E. Dempsey, D.H. Olson: J. Phys. Chem. **74**, 305 (1970)
3.423 P.E. Rickert, J.A. Rabo, E. Dempsey, V. Schomaker: in *Proceedings of the 3rd International Congress of Catalysis*, 1964, Vol.1 (Amsterdam 1965) p.714
3.424 A.L. Klyachko-Gurvich, A.T. Khudiev, A.N. Runstein: Dokl. Akad. Nauk SSSR **177**, 1379 (1967) (in Russian)
3.425 Ya.I. Isakov, A.L. Klyachko-Gurvich, A.T. Khudiev, Kh. M. Minachev, A.M. Rubinstein: in *Proceedings of the 4th International Congress on Catalysis*, Vol.2 (Akademiai Kiado, Budapest 1970) p.126
3.426 J.T. Richardson: J. Catal. **9**, 172, 182 (1967)
3.427 E. Dempsey: *Molecular Sieves* (Spec. Publ. Soc. Chem. Ind., London 1968) p.293
3.428 N.Y. Chen, W.E. Garwood: J. Catal. **52**, 453 (1978)
D.L. Kirg. J.A. Casumano, R.L. Garter: Catalysis Revs. **23**, 233 (1981)
3.429 P.B. Weisz: Plenary Lecture at 7th Intern. Congr. Catalysis, Tokyo, 1980, p.1
3.430 G.T. Kokotailo, S.L. Lawton, D.H. Olson, W.M. Meier: Nature **272**, 437 (1978)
D.H. Olson, W.O. Haag, R.M. Lago: J. Catalysis **61**, 390 (1980)
3.431 J. Vedrine: Lecture on Post-Congress Symposium on Catalysis, Japan, Kyushu July 1980
J.A. Rabo: Catalysis Revs. **23**, 420 (1981)
3.432 S. Brunauer, D.L. Kantro, C.H. Weise: Can. J. Chem. **34**, 1483 (1956)
3.433 M.M. Egorov, V.F. Kiselev: Zh. Fiz. Khim. **36**, 318 (1962) (in Russian)
3.434 V.F. Kiselev: Zh. Fiz. Khim. **34**, 698 (1960) (in Russian)
3.435 V.A. Sobolev, M.I. Khoma, V.I. Furman: Ukr. Khim. Zh. **42**, 142 (1976) (in Russian)
3.436 I. Bakardjew: Dokl. Akad. Nauk SSSR **168**, 618 (1966) (in Russian)
3.437 I. Bakardjew: Z. Phys. Chem. (DDR) **225**, 273 (1964)
3.438 I. Bakardjew, K.H. Schnabel, K. Wencke: Z. Phys. Chem. (DDR) **231**, 391 (1966); **239**, 65 (1968)
3.439 P.A. Rebinder, E.D. Tschukin: *Progress in Surface Science*, **3**, 251 (Pergamon, Oxford 1973)
3.440 B.V. Rozentuler, K.N. Spiridonov, O.V. Krylov: Dokl. Akad. Nauk SSSR **259**, 859 (1981) (in Russian)
3.441 E.G. Schlosser: Ber. Bunsenges. Phys. Chem. **73**, 358 (1969)
3.442 V.A. Bakaev, V.F. Kiselev, K.G. Krasil'nikov: Dokl. Akad. Nauk SSSR **125**, 831 (1959) (in Russian)
3.443 M.N. Plooster, S.N. Gitlin: J. Phys. Chem. **75**, 3322 (1971)
3.444 R.T. Pearson, W. Derbyshire: J. Colloid Interface Sci. **46**, 232 (1974)
3.445 A.B. Kurzaev, V,I. Kvlividze, V.F. Kiselev: in *Svyazannaya voda v dispersnykh sistemakh* (Bound Water in Disperse Systems), Vol.4, ed. by V.F. Kiselev (Moscow University, Moscow 1977) p.156
3.446 G.G. Litvan: Adv. Colloid Interface Sci. **9**, 253 (1978)
3.447 D. Armitage, F.P. Price: Chem. Phys. Lett. **44**, 305 (1976)

Chapter 4

4.1 Yu.A. Zarif'yants, V.F. Kiselev, G.G. Fedorov: Dokl. Akad. Nauk SSSR **144**, 151 (1962) (in Russian)
4.2 Yu.A. Zarif'yants, V.F. Kiselev, N.N. Lezhnev, O.V. Nikitina: Carbon **5**, 127 (1967)
4.3 G.B. Demidovich, V.F. Kiselev, N.N. Lezhnev, O.V. Nikitina: J. Chem. Phys. **65**, 1072 (1968)
4.4 N.N. Semenov: *Problems in Chemical Kinetics and Reactivity,* Vols. I and II (Princeton University Press, Princeton, NJ 1958)
4.5 Yu.A. Zarif'yants: Zh. Fiz. Khim. **38**, 2655 (1964) (in Russian)
4.6 G.B. Demidovich, V.F. Kiselev: Zh. Fiz. Khim. **41**, 684 (1967) (in Russian)
4.7 S. Mrozowski, J.F. Andrew: in *Proceedings of the 4th Conference on Carbon, Buffalo* (Pergamon, New York 1960) p.207
4.8 H. Hirabayashi: in *Proceedings of the 4th Conference on Carbon, Buffalo* (Pergamon, New York 1960) p.227
4.9 Yu.A. Zarif'yants, V.F. Kiselev, G.G. Fedorov: Zh. Fiz. Khim. **37**, 1619, 1846 (1963) (in Russian)
4.10 V.F. Kiselev, O.V. Nikitina: Dokl. Akad. Nauk SSSR **171**, 374 (1966)
4.11 J.M. Thomas, E.L. Evans, M. Barber, P. Swift: Trans. Faraday Soc. **67**, 1875 (1971)
4.12 M. Barber, E.L. Evans, J.M. Thomas: Chem. Phys. Lett. **18**, 423 (1973)
4.13 O.V. Nikitina, N.N. Lezhnev, V.F. Kiselev: Carbon **8**, 402 (1970)
4.14 O.V. Nikitina, R. Brants, V.F. Kiselev: Zh. Fiz. Khim. **42**, 943 (1968) (in Russian)
4.15 N.N. Dobrovol'sky, Yu.A. Zarif'yants, V.F. Kiselev, N.N. Lezhnev, G.G. Fedorov: Zh. Fiz. Khim. **38**, 506 (1964) (in Russian)
4.16 R. Beeb, M.H. Polley, W.R. Smith, C.B. Wendell: J. Am. Chem. Soc. **69**, 2214 (1947)
4.17 G.G. Fedorov, Yu.A. Zarif'yants, V.F. Kiselev: Zh. Fiz. Khim. **37**, 2344 (1963) (in Russian)
4.18 V.I. Kvlividze, Yu.A. Zarif'yants, V.F. Kiselev: Zh. Fiz. Khim, **39**, 461 (1965) (in Russian)
4.19 A.H. Boonstra: Philips Res. Rep. Suppl. **3**, 1 (1968)
4.20 H. Ibach, H.D. Bruchmann, H. Wagner: Appl. Phys. A **29**, 113 (1982)
G. Holliner, F.J. Himpsel: J. Vac. Sci. Technol. **A1**, 640 (1983)
P. Chen, DàBolmont, C.A. Sebenne: Solid State Commun. **46**, 689 (1983)
4.21 V.N. Ageev, N.I. Ionov: in *Progress in Surface Science*, **5**, 1 (Pergamon, New York 1975)
4.22 M.J. Sparnaay: Vac. Technol. **23**, 141 (1974)
4.23 Ch. Kleint: in *Physik der Halbleiteroberfläche,* **5**, 127 (Akademie der Wissenschaften, DDR 1974)
4.24 H. Ibach (ed.): *Electron Spectroscopy for Surface Analysis,* Topics Current Phys., Vol.4 (Springer, Berlin, Heidelberg, 1977)
4.25 G.A. Somorjai, M.A. van Hovè: *Adsorbed Monolayers on Solid Surfaces,* Structure and Bonding, Vol.38 (Springer, Berlin, Heidelberg 1979)
4.26 J. Hölzl, F.K. Schulte, H. Wagner: *Solid Surface Physics,* Springer Tracts in Modern Physics, Vol. 85 (Springer, Berlin, Heidelberg 1979)
4.27 C.M. Garner, I. Lindau, S.Y. Su, P. Pianetta, W.E. Spicer: Phys. Rev. B **19**, 3944 (1979)
4.28 A.A. Frantsuzov: in *Elementarnye fiziko-khimicheskie protsessy na poverkhnosti monokristallicheskikh poluprovodnikov* (Elementary Physicochemical Processes on the Surface of Monocrystalline Semiconductors) ed. by A.V. Rzanov (Nauka, Novosibirsk 1975) p.49
4.29 P. Mark, W.F. Creighton: Thin Solid Films **56**, 19 (1979) and references cited therein
C.B. Duke: J. Vac. Sci. Technol. B1, 732 (1983)

4.30 P. Mark: Catal. Rev. Sci. Eng. **12**, 71 (1975)
4.31 B.A. Joyce, J.H. Neave: Surf. Sci. **27**, 499 (1971)
4.32 T.T. Bykova, N.S. Tsypkina: Fiz. Tverd. Tela **5**, 262 (1971) (in Russian)
4.33 H. Ibach, K. Horn, R. Dorn, H. Lüth: Surf. Sci, **38**, 433 (1973)
4.34 Y. Margoninski, D. Segal, R.E. Kirby: Surf. Sci. **53**, 488 (1975)
4.35 Y. Margoninski: J. Appl. Phys. **47**, 3868 (1976)
4.36 J.H. Neave, B.A. Joyce: J. Phys. D. **10**, 243 (1977)
4.37 R.E. Kirby, J.W. Dieball: Surf. Sci. **41**, 467 (1974)
4.38 R.J. Madix, A.A. Susu: Surf. Sci. **20**, 377 (1970)
4.39 H. Ibach, J.E. Rowe: Phys. Rev. B**10**, 710 (1974)
4.40 M.J. Sparnaay, A.H. Boonstra, J. van Ruller: Surf. Sci. **2**, 56 (1964)
4.41 M. Henzler: Surf. Sci. **9**, 34 (1968)
4.42 D.R. Palmer, S.R. Morrison, C.E. Dauenlaugh: J. Phys. Chem. Solids **14**, 27 (1960)
4.43 R. Missman, P. Handler: J. Phys. Chem. Solids **8**, 109 (1959)
4.44 M. Green, A. Lieberman: J. Phys. Chem. Solids **23**, 1407 (1962)
4.45 P. Handler, W. Portnoy: Phys. Rev. **116**, 516 (1959)
4.46 B.A. Nesterenko, O.V. Snitko: Fiz. Tverd. Tela **6**, 2913 (1964) (in Russian)
4.47 L. Ernst: Phys. Status Solidi **19**, 98 (1967)
4.48 Yu.I. Belyakov, T.N. Kompaniets: Zh. Tekh. Fiz. **42**, 1278 (1972) (in Russian)
4.49 G.D. Demidovich, R.B. Dzhanelidze, V.F. Kiselev: Fiz. Tekh. Poluprovodn. **3**, 629 (1969) (in Russian)
4.50 G.B. Demidovich, V.F. Kiselev: Vestn. Mosk. Univ. Fiz. Astronomiya **14**, 158 (1973) (in Russian)
4.51 D. Brennan, B.O. Hayward, B.H.W. Trapnell: J. Phys. Chem. Solids **14**, 117 (1960)
4.52 G.V. Demidovich, A.A. Sklyankin: Vestn. Mosk. Univ. Fiz. **35**, 104 (1980)
4.53 R.F. Lever, H.R. Wendt: Surf. Sci. **19**, 435 (1970)
4.54 R.F. Lever: Surf. Sci. **9**, 370 (1968)
4.55 M. Green, M.J. Lee: in *Solid State Surface Science*, Vol.1, ed. by M. Green (Marcel Dekker, New York 1969) p.133
4.56 M. Chen, I.P. Batra, C.R. Brundle: J. Vac. Sci. Technol. **16**, 1216 (1979)
4.57 J. Stöhr, L.I. Johansson, I. Lindau, P. Pianetta: J. Vac. Sci. Technol. **16**, 1221 (1979)
4.58 B. Carriere, J.P. Deville: Surf. Sci. **80**, 278 (1979)
4.59 M. Nishijima, R. Fujiwara: Solid State Commun. **24**, 101 (1977)
4.60 J.E. Rowe, G. Magaritondo, H. Ibach, H. Frotzheim: Solid State Commun. **20**, 277 (1976)
G.L. Kellogg: Appl. Surf. Sci. **11/12**, 186 (1982)
S. Ciraci, S. Ellialtioğlu, S. Erkoc: Phys. Rev. B**26**, 5716 (1982)
4.61 C.M. Garner, I. Lindau, C.Y. Su, P. Pianetta: Phys. Rev. Lett. **40**, 403 (1978)
4.62 J. von Wienskowski, H. Frotzheim: Phys. Status Solidi B **94**, 429 (1979)
4.63 G.B. Demidovich, V.F. Kiselev: Phys. Status Solidi B **50**, K33 (1972)
4.64 R. Ludeko, A. Koma: Phys. Rev. Lett. **34**, 1170 (1975); Phys. Rev. B**13**, 739 (1976)
4.65 Y.W. Chung, W. Siekhaus, G.A. Somorjai: Surf. Sci. **58**, 341 (1976)
4.66 G.V. Gadiyak, A.A. Karpushin, Yu.N. Morokov: in *Problemy fizicheskoi khimii poverkhnosti poluprovodnikov* (Problems in the Physical Chemistry of Semiconductor Surfaces) ed. by A.V. Rzanov (Nauka, Novosibirsk 1978) p.72
4.67 G.B. Demidovich, V.F. Kiselev, O.V. Nikitina: Dokl. Akad. Nauk SSSR **205**, 383 (1972) (in Russian)
4.68 D. Haneman: Phys. Rev. **170**, 705 (1968)
4.69 M.F. Chung: J. Phys. Chem. Solids **32**, 475 (1971)
4.70 V.S. Kuznetsov, G.B. Demidovich, V.F. Kiselev: Vestn. Mosk. Univ. Fiz. Astronomiya **16**, 177 (1975) (in Russian)

4.71 G.B. Demidovich, S.N. Karyagin, V.F. Kiselev: Phys. Status Solidi **114**, 705 (1982)
4.72 Yu.I. Belyakov, N.I. Ionov, T.N. Kompaniets: Sov. Phys.-Solid State **14**, 2567 (1973)
4.73 H. Ibach, J.E. Rowe: Surf. Sci. **43**, 481 (1974)
4.74 K. Fujiwara: Phys. Rev. B **26**, 2036 (1982)
G. Schulze, M. Henzler: Surf. Sci. **124**, 336 (1983)
4.75 B.P. Lemke, D. Haneman: Phys. Rev. B **17**, 1893 (1978)
4.76 H. Iwasaki, S. Nakamura: Surf. Sci. **81**, 627 (1979)
4.77 R.J. Madix, J.A. Schwarz: Surf. Sci. **24**, 264 (1971)
4.78 P.E. Best: Phys. Rev. B **19**, 1054 (1979)
M. Seel, P.S. Bagus: Phys. Rev. B **28**, 2023 (1983)
4.79 M. Nishijima, T. Murotama: Surf. Sci. **39**, 441 (1973)
4.80 A.J. Van Bommel, F. Meyer: Surf. Sci. **6**, 391 (1967); 8, 381 (1967)
4.81 F. Meyer: Surf. Sci. **27**, 107 (1971)
4.82 A.J. Van Bommel, J. Crombeen: Surf. Sci. **36**, 773 (1973)
4.83 M. Henzler, J. Töpler: Surf. Sci. **40**, 388 (1973)
4.84 K. Fujiwara, H. Ogata, M. Nishijima: Solid State Commun. **21**, 895 (1977)
4.85 M. Fujiwara: Surf. Sci. **108**, 124 (1981)
4.86 S.T. Pantelides (ed.): *Proc. Int. Conf. Physics of* SiO_2 *and Its Interfaces* (Plenum, New York 1978)
4.87 M. Schulze, G. Pensl (eds.): *Insulating Films on Semiconductors*, Springer Ser. Electrophys. Vol.7 (Springer, Berlin, Heidelberg 1981)
4.88 S.P. Zhdanov: Dokl. Akad. Nauk SSSR **68**, 99 (1949); **100**, 1115 (1955) (in Russian)
4.89 M.M. Egorov, T.S. Egorova, V.F. Kiselev, K.G. Krasil'nikov: Dokl. Akad. Nauk SSSR **114**, 579 (1957); Zh. Fiz. Khim. **32**, 2624 (1958) (in Russian)
4.90 J.H. DeBoer, M.E. Herman, J.M. Vleskens: Proc. K. Ned. Akad. Wet. Ser. B **60**, 45, 54 (1957)
4.91 H.W. Kohlschütter, G. Kämpf: Z. Anorg. Allg. Chem. **292**, 289 (1957)
4.92 L.D. Belyakova, A.V. Kiselev: Zh. Fiz. Khim. **33**, 1534 (1959) (in Russian)
4.93 L.G. Ganichenko, M.M. Egorov, V.F. Kiselev, K.G. Krasil'nikov, G.S. Khodakov: Dokl. Akad. Nauk SSSR **131**, 597 (1960) (in Russian)
4.94 M.M. Egorov, V.F. Kiselev, K.G. Krasil'nikov: Zh. Fiz. Khim. **35**, 2234 (1961) (in Russian)
4.95 A.E. Bazhanova, V.F. Kiselev, S.I. Savel'yeva: Z. Chem. **9**, 52 (1969)
4.96 L.J. Slutsky, W.H. Wade: J. Chem. Phys. **36**, 2688 (1962)
4.97 L.L. Levenson: in *Applications of Piezoelectric Quarz Crystal Microbalance*, ed. by C. Lu, A.W. Czanderna (Elsevier, Amsterdam 1984) Chap.6
4.98 L.G. Ganichenko, V.F. Kiselev, V.V. Murina: Kinet. Katal. **2**, 877 (1961) (in Russian)
4.99 M.M. Egorov, N.N. Dobrovol'sky, V.F. Kiselev, G. Furman, S.V. Khrustaleva: Zh. Fiz. Khim. **39**, 3070 (1965) (in Russian)
4.100 R.V. Prudnikov, V.F. Kiselev, M.M. Egorov: Dokl. Akad. Nauk SSSR **166**, 395 (1966)
4.101 F. Gonzalez, G. Manuera: Rev. Chim. Miner. **7**, 1021 (1970)
4.102 M. Nagao, T. Morimoto, K. Yunoki, H. Muraishi: J. Phys. Chem. **73**, 3809 (1969); 82, 1032 (1978)
4.103 J.H. DeBoer, J.M.M. Fortuin, B.C. Lippens, W.H. Meijs: J. Catal. **2**, 1 (1963)
4.104 M.M. Egorov, Yu.A. Zarif'yants, V.F. Kiselev, K.G. Krasil'nikov: Dokl. Akad. Nauk SSSR **120**, 326 (1958) (in Russian)
4.105 M.M. Egorov, V.F. Kiselev: Zh. Fiz. Khim. **40**, 1984 (1966) (in Russian)
4.106 V.F. Kiselev: *Poverkhnostnye yavleniya v poluprovodnikakh i dielektrikakh* (Surface Phenomena in Semiconductors and Dielectrics) (Nauka, Moscow 1970)
4.107 M.M. Egorov, T.S. Egorova, V.F. Kiselev, K.G. Krasil'nikov: Vestn. Mosk. Univ. Fiz. Astronomiya **1**, 203 (1958)

4.108 L.G. Ganichenko, V.F. Kiselev, K.G. Krasil'nikov, V.V. Murina: Zh. Fiz. Khim. 35, 1718 (1961) (in Russian)
4.109 J.A. Hockey, B.A. Pethica: Trans. Faraday Soc. 58, 2017 (1962)
4.110 J. Bastick: Comptes Rendus **234**, 1279 (1952)
4.111 G. Bliznakov, R. Polikarova: J. Catal. **5**, 18 (1966)
4.112 A.V. Kiselev: in *Poverkhnostnye khimicheskie soedineniya i ikh rol' v yavleniyakh adsorbtsii* (Surface Chemical Compounds and Their Role in Adsorption Phenomena) ed. by A.V. Kiselev (Moscow State University, Moscow 1957) pp.91, 199
4.113 S.P. Zhdanov: Zh. Fiz. Khim. **32**, 699 (1958); 36, 2098 (1962) (in Russian)
4.114 A.V. Kiselev, G.G. Muttik: Kolloidn. Zh. 19, 562 (1967) (in Russian)
4.115 J. Hala's, G. Schay: Acta Chim. Acad. Sci. Hung. **14**, 315 (1958)
4.116 A.V. Kiselev, B.A. Frolov: Kinet. Katal. **3**, 774 (1962) (in Russian)
4.117 V.T. Bykov, O.E. Presnyakova: Kinet. Katal. 3, 774 (1962) (in Russian)
4.118 T.S. Egorova, V.F. Kiselev, K.G. Krasil'nikov: Dokl. Akad. Nauk SSSR 123, 1060 (1958) (in Russian)
4.119 T.S. Egorova, Yu.A. Zarif'yants, V.F. Kiselev, K.G. Krasil'nikov, V.V. Murina: Zh. Fiz. Khim. **36**, 1458 (1962) (in Russian)
4.120 G.J. Young, J.J. Chessick, F.H. Healey, A.C. Zettlemoyer: J. Phys. Chem. **58**, 313 (1954)
4.121 K. Miyata, T. Kawahara: J. Chem. Soc. Jpn. **81**, 353 (1960)
4.122 O.M. Dzhigit, A.V. Kiselev, G.G. Muttik: Kolloidn. Zh. **24**, 15 (1962) (in Russian)
4.123 D.C. McIver, H.H. Tobin, R.T. Barth: J. Catal. 2, 485 (1963)
4.124 H. Spannheimer, H. Knözinger: Ber. Bunsenges. Phys. Chem. **70**, 570 (1966)
4.125 B.A. Hendriksen, D.R. Pearse, R. Rudham: J. Catal. **24**, 82 (1972)
4.126 S.P. Zhdanov: in *Osnovnye problemy teorii fizicheskoi adsorbtsii* (Basic Problems in the Theory of Physical Adsorption), ed. by M.M. Dubinin (Nauka, Moscow 1970) p.184
4.127 M.M. Egorov, V.F. Kiselev; Zh. Fiz. Khim. **36**, 318 (1962) (in Russian)
4.128 A.V. Kiselev, V.I. Lygin: in *2nd International Congress on Surface Activity*, Vol.2 (Butterworths, London 1957) p.204
4.129 A.V. Kiselev, V.I. Lygin: *Infrared Spectra of Surface Compounds* (Wiley, New York 1975)
4.130 A.N. Sidorov: Zh. Fiz. Khim. **30**, 995 (1956) (in Russian)
4.131 A.N. Sidorov: Opt. Spektrosk. **8**, 806 (1960) (in Russian)
4.132 M. Folman, D.J.E. Yates: Trans. Faraday Soc. **54**, 1684 (1958)
4.133 L.A. Ignat'eva, V.F. Kiselev, G.D. Chukin: Dokl. Akad. Nauk. SSSR **181**, 914 (1968) (in Russian)
4.134 V.I. Kvlividze, N.N. Ievskaya, T.S. Egorova, V.F. Kiselev, N.D. Sokolov: Kinet. Katal. 3, 91 (1962) (in Russian)
4.135 V.I. Kvlividze: Dokl. Akad. Nauk SSSR **157**, 158 (1964) (in Russian)
4.136 G.F. Golovanova, V.I. Kvlividze, V.F. Kiselev: in *Svyazannaya voda v dispersnykh sistemakh*, T.4 (Bound Water in Disperse Systems, Vol.4), ed. by V.F. Kiselev (Moscow State University, Moscow 1976) p.178
4.137 D. Michel: Z. Naturforsch. A **22**, 1751 (1967)
4.138 W. Stöber: Kolloid-Z. **145**, 17 (1956)
4.139 A.V. Karyakin, G.A. Kriventsova: *Sostoyanie vody v organicheskikh i neorganicheskikh soedineniyakh* (The State of Water in Organic and Inorganic Compounds) (Moscow, Nauka 1973)
4.140 H. Noller, G. Magerböck, G. Zundel: Surf. Sci. **33**, 82 (1972)
4.141 K. West, L.L. Whatley, K.J. Laker: J. Am. Chem. Soc. **83**, 761 (1961)
4.142 V.F. Kiselev: Probl. Kinet. Katal. **13**, 249 (1968) (in Russian)
4.143 Ya.K. Syrkin: Usp. Khim. **31**, 397 (1962) (in Russian)
4.144 W.D. Treadwell, W. Weiland: Helv. Chim. Acta **13**, 842 (1930)
4.145 W.A. Weyl, E.A. Hauser: Kolloid-Z. **164**, 72 (1954)
4.146 A.G. Revesz, K.H. Zeininger: RCA Rev. **29**, 22 (1968)
4.147 A.G. Revesz: J. Non-Cryst. Solids **2**, 309 (1973)

4.148 S.M. Stishov, N.V. Belov: Dokl. Akad. Nauk SSSR **143**, 951 (1962) (in Russian)
4.149 B.M. Mitsyuk: Kolloidn. Zh. **36**, 800 (1974) (in Russian)
4.150 V.N. Pak, S.I. Koltsov, V.B. Aleskovsky: Kinet. Catal. (Engl. Transl.) **14**, 1390 (1973)
4.151 Yu.I. Gorlov, M.M. Konoplya: Zh. Eksp. Teor. Khim. **15**, 446 (1979) (in Russian)
4.152 M.M. Egorov, V.I. Kvlividze, V.F. Kiselev, K.G. Krasil)nikov: Kolloid-Z. **212**, 126 (1966)
4.153 F. Boccuzzi, S. Coluccla, G. Ghiott, O. Morterra, A. Zecchina: J. Phys. Chem. 62, 1298 (1978)
4.154 M. Hino, H. Sato: Bull. Chem. Soc. Jpn. **44**, 33 (1971)
4.155 Ya.I. Ryskin: Opt. Spektrosk. **4**, 532 (1958); **5**, 278 (1959) (in Russian)
4.156 V.A. Tertykh, V.M. Ogenko, E.F. Voronin, A. A. Chuiko: Zh. Prikl. Spektrosk. **23**, 464 (1975) (in Russian)
4.157 V.F. Kiselev, O.V. Krylov: *Electronic Phenomena in Adsorption and Catalysis*, Springer Ser. Chem. Phys. (Springer, Berlin, Heidelberg)
4.158 V.I. Davydov: *Germanii* (Metallurgizdat, Moscow 1964) (in Russian)
4.159 P. Fink, B. Camara: Z. Chem. **12**, 35 (1972)
4.160 W. Pohle, P. Fink: Z. Phys. Chem. Neue Folge **109**, 77 (1978)
4.161 J.D. Jorgensen: J. Appl. Phys. **49**, 5473 (1978)
4.162 Yu.A. Zarif'yants, V.F. Kiselev, A.S. Petrov, R.V. Prudnikov, S.V. Khrustaleva, G.D. Chukin: Kinet. Catal. (Engl. Transl.) **15**, 1091 (1974)
4.163 M. Primet, P. Pichat, M.V. Mattieu: J. Phys. Chem. **75**, 1216, 1221 (1971)
4.164 P. Jones, J.A. Hockey: Trans. Faraday Soc. **67**, 2669, 2679 (1971)
4.165 D.M. Griffith, C.H. Rochester: Trans. Faraday Soc. **73**, 1510 (1977)
4.166 V.S. Il'enko, A.V. Uvarov: Kolloidn. Zh. **37**, 1161 (1975) (in Russian)
4.167 L.A. Ignat'eva, G.D. Chukin, G.V. Bondarenko: Dokl. Akad. Nauk SSSR **181**, 393 (1968) (in Russian)
4.168 G.D. Chukin, T.V. Antipina, O.F. Kirina: Zh. Fiz. Khim. **47**, 846 (1973) (in Russian)
4.169 L.A. Ignat'eva, B.I. Kvlividze, V.F. Kiselev: in *Svyazannaya voda v dispersnykh sistemakh*, T.1 (Bound Water in Disperse Systems, Vol.1), ed. by V.F. Kiselev (Moscow State University, Moscow 1970) p.56
4.170 V.A. Bakaev, M.M. Dubinin: Izv. Akad. Nauk SSSR, Ser. Khim. **10**, 1869 (1966) (in Russian)
4.171 A. Knoppwast, W. Gunsser, H. Lechert: Z. Naturforsch. A **21**, 1200 (1966); A 23, 1343 (1968)
4.172 H. Lechert: in *3rd International Conference Molecular Sieve Zeolites* (Plenum, New York 1973) p.74
4.173 V.I. Kvlividze, V.F. Kiselev, V.V. Serpinsky: Dokl. Akad. Nauk SSSR **165**, 135 (1965) (in Russian)
4.174 V.G. Gvakhariya, V.I. Kvlividze, G.V. Tsitsishvili: Dokl. Phys. Chem. (Engl. Transl.) **223**, 27 (1975)
4.175 B.A. Morrow, I.A. Cody: J. Phys. Chem. **80**, 1998 (1976)
4.176 V.I. Kvlividze, R. Brants, V.F. Kiselev, G.M. Bliznakov: J. Catal. 13, 255 (1969)
4.177 M.D. Low, K. Matsushita: J. Phys. Chem. 73, 908 (1969)
4.178 S.N. Kozlov, V.F. Kiselev, Yu.F. Novototsky-Vlasov, A.A. Sklyankin, N.D. Morozova: Kinet. Katal. **12**, 789 (1971) (in Russian)
4.179 F.S. Stone, L. Whalley: J. Catal. **8**, 173 (1967)
4.180 Y. Amenomiya, J.H.B. Chenier, R.J. Cvetanovic: J. Phys. Chem. **68**, 52 (1964)
4.181 V.I. Yakerson, L.I. Lafer, V.Ya. Danyushevskaya, A.M. Rubinstein: Izv. Akad. Nauk SSSR, Ser. Khim. 2246 (1967)
4.182 V.N. Filimonov: in *Osnovnye problemy teorii fizicheskoi adsorbtsii* (Basic Problems in the Theory of Physical Adsorption), ed. by M.M. Dubinin (Nauka, Moscow 1970) p.116

4.183 J. Madema, J.J.G.M. Van Bokhoeven, A.E. Kuiper: J. Catal. **25**, 238 (1972)
4.184 V.F. Kiselev: Dokl. Akad. Nauk SSSR **176**, 124 (1967) (in Russian)
4.185 J.L. Bonardet, J.P. Fraissard: Chem. Phys. Lett. **22**, 75 (1973)
4.186 H. Dunken, P. Fink: Acta Chim. Acad. Sci. Hung. **53**, 179 (1967)
4.187 J.C. McManus, K. Matsushita, M.J.D. Low: Can. J. Chem. **47**, 1077 (1969)
4.188 J.C. McManus, M.J.D. Low: J. Phys. Chem. **72**, 2378 (1968)
4.189 J. Fraissard, S. Bielikoff, B. Imelik: in 4th Congress on Catalysis, Moscow 1968, Symposium "*the Mechanism and Kinetics of Complex Catalytic Reactions*", Preprint 1
4.190 K. Hirota, K. Fuckui, K. Shindo, Y. Nakai: Bull. Chem. Soc. Jpn. **32**, 1261 (1959)
4.191 B.F. Lewis, M. Mosesman, W.H. Weinberg: Surf. Sci. **41**, 142 (1974)
4.192 B.F. Lewis, W.M. Bowser, J.L. Horn, T. Lun, W.H. Weinberg: J. Vac. Sci. Technol. **11**, 262 (1974)
4.193 V.F. Kiselev, R.V. Prudnikov, G.G. Fedorov: Dokl. Akad. Nauk SSSR **208**, 387 (1973) (in Russian)
4.194 Yu. A. Zarif'yants, S.N. Karyagin, V.F. Kiselev, S.V. Khrustaleva: Kinet. Catal. (Engl. Transl.) **15**, 890 (1974)
4.195 A.A. Chuiko, K.I. Tkachenko, V.A. Tertykh, Yu.I. Gorlov: Koord. Khim. 2, 1609 (1976) (in Russian)
4.196 L.N. Kurbatov: in *Poverkhnostnye khimicheskie soedineniya i ikh rol' v yavleniyakh adsorbtsii* (Surface Chemical Compounds and Their Role in Adsorption Phenomena), ed. by A.V. Kiselev (Moscow State University, Moscow 1957) p.223
4.197 J.H. Anderson, K.A. Wickersheim: Surf. Sci. **2**, 202 (1964)
4.198 S. Brunauer: *The Adsorption of Gases and Vapors* (Princeton University Press, Princeton, N.J. 1945)
4.199 H.A. Resing: Adv. Mol. Relaxation Proc. **3**, 155 (1972)
4.200 T.M. Duncan, C. Dubovski: *Chemisorption and Surface Studies by Nuclear Magnetic Resonance Spectroscopy* (North-Holland, Amsterdam 1981)
4.201 V. Morariu, R. Mills: Z. Phys. Chem. (DDR) **79**, 1 (1972)
4.202 S. Olejnik, J.W. White: Nature **236**, 15 (1972)
4.203 P.A. Chollet, J. Messier, C. Rosilio: J. Chem. Phys. **64**, 1042 (1976)
4.204 P.A. Chollet: Thin Solid Films **52**, 343 (1978)
4.205 J. Klinovski, J.M. Thomas, G.A. Fyte, J.S. Hartman: J. Phys. Chem. **85**, 2590 (1981)
W.H. Dawson, S.W. Kaiser, P.D. Ellis, R.R. Inners: J. Am. Chem. Soc. **103**, 6780 (1981)
T. Bernstein, L. Kitaev, D. Michel, H. Pfeifer, P. Fink: J. Chem. Soc. Faraday Trans. **78**, 761 (1982)
4.206 R.E. Sempels, P.G. Rouxhet: Bull. Soc. Chim. Belg. **84**, 361 (1975)
4.207 K.G. Miesserov: Dokl. Akad. Nauk SSSR **87**, 809 (1952) (in Russian)
4.208 W. Sindorf, G.E. Maclet: J. Phys. Chem. **86**, 5208 (1982)
4.209 L.A. Ignat'eva: Dokl. Akad. Nauk SSSR **163**, 398 (1965) (in Russian)
4.210 H. Knözinger: Surf. Sci. **41**, 349 (1974)
4.211 A.N. Terenin: Probl. Kinet. Katal. **12**, 27 (1968) (in Russian)
F.S. Stone: in *Surface Properties and Catalysis of Non-Mettals*, Proc. NATO Adv. Study Inst. Lille (1982) (Reidel, Dodrecht 1983) p.237
4.212 L.A. Ignat'eva, G.D. Chukin: Dokl. Akad. Nauk SSSR **184**, 644 (1969) (in Russian)
4.213 S.R. Morrison: *The Chemical Physics of Surfaces* (Plenum, New York 1977)
4.214 P.W. Schindler, B. Fürst, R. Dick, P.U. Wolf: J. Colloid Interface Sci. **55**, 469 (1976)
4.215 N.E. Tret'yakov, V.N. Filimonov: Kinet. Katal. **13**, 815 (1972) (in Russian)
4.216 C.L. Tomas: Ind. Eng. Chem. **41**, 2564 (1949)
4.217 J.J. Fripiat, A. Leonard, J.B. Uytterhoeven: J. Phys. Chem. **69**, 3274 (1965)

4.218 M.R. Basila: J. Appl. Spectrosc. (Engl. Transl.) **1**, 289 (1968)
4.219 K.V. Topchieva: Probl. Kinet. Katal. **10**, 247 (1960) (in Russian)
4.220 A.V. Kiselev: Zh. Fiz. Khim. **41**, 2470 (1967) (in Russian)
4.221 M. Prettre: in *Proc. Colloquio Sobre quimica fisika an Superpieces* (Madrid University Press, Madrid 1965) p.217
4.222 K.G. Miesserov, O.E. Morozova, A.A. Petrov: Zh. Obshch. Khim. **25**, 2204 (1955)
4.223 K. Tanabe: *Solid Acids and Bases* (Academic, New York 1970)
4.224 J.A. Rabo, K.L. Angell, V. Schomaker: in *Proceedings of the 4th International Congress on Catalysis*, Vol.2 (Akademiai Kiado, Budapest 1970) p.310
4.225 K.V. Topchieva, B.V. Romanovsky: in *Sovremennye problemy fizicheskoi khimii* (Modern Problems in Physical Chemistry), ed. by Ja.I. Gerasimov (Moscow State University, Moscow 1970) p.407
4.226 J.B. Peri: J. Phys. Chem. **70**, 3168 (1966); **72**, 2917 (1968)
4.227 J.W. Ward: J. Catal. **9**, 225 (1967)
4.228 V.F. Kiselev, Yu.A. Zarif'yants: Probl. Kinet. Katal. **16**, 221 (1975) (in Russian)
4.229 V.N. Filimonov, Yu.N. Lopatin, D.A. Sukhov: Kinet. Katal. **10**, 2 (1969) (in Russian)
4.230 M. Herrman, H.P. Boehm: Z. Anorg. Allg. Chem. **368**, 73 (1969)
4.231 A.I. Buchneva, L.A. Ignat'eva, L.V. Levshin: Zh. Prikl. Spektrosk. **15**, 436 (1971) (in Russian)
4.232 V.P. Galkin, V.B. Golubev, A.A. Kubasov, E.V. Lunina: Zh. Fiz. Khim. **46**, 1005 (1972) (in Russian)
T. Wozniewski, E. Fedorunska, S.J. Malinovski: J. Côlloid and Interface Sci. **87**, 1 (1982)
4.233 L.A. Petermann: in *Progress in Surface Science*, **3**, 1 (Pergamon, New York 1973)
4.234 V.I. Yakerson, V.V. Rozanov: *Itogi nauki i tekhniki*, T.3 (Results in Science and Technology, Vol.3) (VINITI, Moscow 1974) p.47
4.235 A.V. Bondarenko, V.F. Kiselev, K.G. Krasil'nikov: Kinet. Katal. **2**, 590 (1961) (in Russian)
4.236 L.I. Burbulyavichus, Yu.A. Zarif'yants, V.F. Kiselev, R.V. Prudnikov: Kinet. Katal. **12**, 258 (1971) (in Russian)
4.237 E. Makoto, K. Shonachi, K. Shuichi, S. Tetsuro: Bull. Chem. Soc. Jpn. **51**, 3144 (1978)
4.238 V.G. Gvakhariya, V.I. Kvlividze, V.F. Kiselev, M.B. Pylova, G.V. Tsitsishvili: Dokl. Akad. Nauk SSSR **188**, 379 (1969) (in Russian)
4.239 S.P. Gabuda: Dokl. Akad. Nauk SSSR **146**, 840 (1963) (in Russian)
4.240 E.L. Saffar: J. Chem. Phys. **45**, 4643 (1966)
4.241 V.F. Chuvaev, R.A. Gazarov, V.I. Spitsyn: Dokl. Akad. Nauk SSSR **213**, 1134 (1973) (in Russian)
4.242 R.A. Gazarov, V.F. Chuvaev, V.I. Spitsyn: Zh. Neorg. Khim. **7**, 1902 (1974) (in Russian)
4.243 G. Feraris, D.W. Jones, J. Verkes: Acta Crystallogr. B **27**, 349 (1973)
4.244 M.M. Selim, S.P. Gabuda, A.A. Panich, Z.V. Gryaznova, N.K. Moroz: Dokl. Phys. Chem. (Engl. Transl.) **224**, 1078 (1975)
4.245 G.F. Golovanova, V.F. Kiselev, S.N. Kozlov, A.S. Petrov, E.A. Silaev: Phys. Status Sol. A **64**, 177 (1981)
4.246 I.E. Neumark: in *Osnovnye problemy teorii fizicheskoi adsorbtsii* (Basic Problems in the Theory of Physical Adsorption), ed. by M.M. Dubinin (Nauka, Moscow 1970) p.151
4.247 I.E. Neumark, R.Yu. Sheinfain: *Silikagel', ego poluchenie, svoistva i primenenie* (Silica Gel: Its Preparation, Properties and Use) (Naukova Dumka, Kiev 1973)
4.248 J.D. Chapman, M.L. Hair: J. Catal. **2**, 145 (1963)
4.249 N.W. Cant, L.H. Little: J. Catal. **12**, 134 (1968)

4.250 T.V. Antipina, O.V. Bulgakov, A.V. Uvarov: in *Proceedings of the 4th International Congress on Catalysis*, Vol.2 (Akademiai Kiado, Budapest 1971) p.376
4.251 G.F. Golovanova, I.I. Ivanova, V.I. Kvlividze, I.E. Neumark, S.V. Khrustaleva, G.D. Chukin, R.Yu. Sheinfain: Zh. Teor. Eksp. Khim. **9**, 383 (1973) (in Russian)
4.252 G.F. Golovanova, V.I. Kvlividze, O.F. Kirin, T.V. Antipina: Zh. Fiz. Khim. **47**, 2549 (1973)
4.253 G.F. Golovanova, V.I. Kvlividze, V.F. Kiselev: Kinet. Catal. (Engl. Transl.) **16**, 652 (1975)
4.254 A.N. Sidorov, I.E. Neumark: Zh. Fiz. Khim. **38**, 2784 (1964)
4.255 J. Makela, L. Niemela: Phys. Lett. A **56**, 307 (1976)
4.256 K.B. Yatsimirsky, V.V. Nemoshkalenko: in *Metallofizika*, T.60 (Metal Physics, Vol.60) (Naukova Dumka, Kiev 1975) p.60
4.257 G.F. Golovanova, V.F. Kiselev, S.N. Kozlov, A.S. Petrov, E.A. Silaev: Phys. Status Solidi A **64**, 177 (1981)
4.258 T.V. Antipina, O.F. Kirina: Dokl. Akad. Nauk SSSR **207**, 121 (1972)
4.259 G.D. Parfitt, J. Ramsbotham, C.H. Rochester: Trans. Faraday Soc. **67**, 841, 1500 (1971)
4.260 J.H. Moss, G.D. Parfitt, A. Wright: J. Colloid Polym. Sci. **256**, 1121 (1978)
4.261 S.S. Batsanov: *Elektrootritsatel'nost' elementov i khimicheskaya svyaz'* (The Electronegativity of Elements and the Chemical Bond) (Nauka, Novosibirsk 1962)
4.262 J. Reichardt, R.L. Johnson, L. Ley: Physica **117/118**, 877 (1983)
4.263 I.G. Ryss: *Khimiya ftora i ego neorganicheskikh soedinenii* (The Chemistry of Fluor and Its Inorganic Compounds) (Goskhimizdat, Moscow 1956)

Chapter 5

5.1 I. Langmuir: Trans. Faraday Soc. **17**, 642 (1922)
5.2 E.K. Rideal, H.S. Taylor: *Catalysis in Theory and Practice* (McMillan, London 1919)
5.3 Roginsky, Zel'dovich: Acta Physicochim. URSS **1**, 595 (1934)
5.4 N.N. Semenov, V.V. Voevodsky: in *Geterogennyi kataliz v khimicheskoi promyshlennosti* (Heterogeneous Catalysis in the Chemical Industry), ed. by S.Z. Roginsky (Goskhimizdat, Moscow 1955)
5.5 O.M. Poltorak: in *Geterogennyi kataliz v khimicheskoi promyshlennosti* (Heterogeneous Catalysis in the Chemical Industry), ed. by S.Z. Roginsky (Goskhimizdat, Moscow 1955)
5.6 Y. Scheve, J.F. Schultz: in *Proceedings of the 4th International Congress on Catalysis*, Vol.2 (Akademiai Kiado, Budapest 1971) p.483
R. Larsson: Chem. Scr. **12**, 78 (1979)
5.7 O.V. Krylov, M.U. Kislyuk, B.R. Shub, A.A. Gezalov, N.D. Maksimova, Yu.N. Rufov: Kinet. Katal. **13**, 598 (1972)
O.V. Krylov: in *Problemy kinetiki elementarnykh khimicheskikh reaktsii* (Problems in the Kinetics of Elementary Chemical Reactions), ed. by Ja.S. Lebedev (Nauka, Moscow 1973) p.115
O.V. Krylov: in *Vsesoyuznaya konferentsiya po mekhanizmu geterogenno-kataliticheskikh reaktsii, Moskva 1974* (All-Union Conference on the Mechanism of Heterogeneous Catalytic Reactions, Moscow 1974) (Nauka, Moscow 1977) p.25
5.8 J.R. Smith (ed.): *Theory of Chemisorption*, Topics Current Phys., Vol.19 (Springer, Berlin, Heidelberg 1980)
5.9 V.N. Kondratiev, E.E. Nikitin: *Gas-Phase Reactions. Kinetics and Mechanisms* (Springer, Berlin, Heidelberg 1980)

5.10 J.E. Lennard-Jones, A.F. Devonshire: Proc. R. Soc. **A156**, 6 (1936)
5.11 B. McCarrol, G. Ehrlich: J. Chem. Phys. **38**, 523 (1963)
5.12 V.V. Mazhuga, N.D. Sokolov: Probl. Kinet. Katal. **14**, 79 (1970) (in Russian)
5.13 V.P. Zhdanov, K.I. Zamarev: Catalysis Rev. **24**, 375 (1982)
5.14 A.A. Ovchinnikov: Zh. Eksp. Teor. Fiz. **57**, 263 (1969) (in Russian)
5.15 A. Maradudin: *Theory of Lattice Dynamics in Harmonic Approximation* (Academic, New York 1963)
5.16 R.L. Lundqvist, O. Gunnersson, H. Hjelmberg, J.K. Norskov: Surf. Sci. **89**, 196 (1979)
5.17 M.A. Kozhushner, B.R. Shub: Probl. Kinet. Katal. **17**, 11 (1978)
5.18 A. Nitzan, J. Jortner: Mol. Phys. **25**, 713 (1973); J. Chem. Phys. **60**, 3929 (1974)
5.19 M.A. Koznusher, V.G. Kustarev, B.R. Shub: Surface Sci. **81**, 261 (1979)
5.20 A. Förster: Ann. Phys. **2**, 55 (1948)
5.21 Th. Dexter: J. Chem. Phys. **21**, 836 (1953)
5.22 A.E. Morgan, G. Somorjai: J. Chem. Phys. **51**, 3309 (1969)
5.23 O.V. Krylov: Kinet. Katal. **21**, 79 (1980) (in Russian)
5.24 M. Kaminsky: *Atomic and Ionic Impact Phenomena on Metal Surfaces* (Springer, Berlin, Heidelberg, New York 1965)
R. Behrisch (ed.): *Sputtering by Particle Bombardment* I and II, Topics Appl. Phys. Vols.47 and 52 (Springer, Berlin, Heidelberg 1981 and 1983)
5.25 S.A. Kovalevsky: Probl. Kinet. Katal. **17**, 23 (1978) (in Russian)
5.26 S.V. Vincenc, V.F. Kiselev et al.: Dokladi. Akad. Nauk SSSR **268**, 373 (1982), **274**, 96 (1984)
5.27 J.E. Morgan, H.J. Schiff: Can. J. Chem. **41**, 903 (1963)
5.28 Yu.M. Gershenzon, V.I. Egorov, V.V. Rosenstein, S.Ya. Umansky: Chem. Phys. Lett. **20**, 77 (1973)
Yu.M. Gershenzon, V.B. Rosenshtein, S.Ya. Umansky: Probl. Kinet. Katal. **17**, 36 (1978) (in Russian)
5.29 Yu.M. Gershenzon, V.I. Egorov, V.B. Rosenshtein: Khim. Vys. Energ. **7**, 533 (1973)
5.30 S.A. Kovalevsky, B.R. Shub: Probl. Kinet. Katal. **17**, 29 (1978)
5.31 C. Black, H. Wise, S. Schechter, R.C. Sharpless: J. Chem. Phys. **60**, 3526 (1974)
5.32 R. Abonaf, F. Legay: J. Chem. Phys. **63**, 1393 (1966)
5.33 M. Kovacz, D.R. Rao, A. Lavan: J. Chem. Phys. **48**, 3339 (1968)
5.34 L. Doynette, M. Margottin-Maclon, H. Guegnen, A. Carion, L. Henry: J. Chem. Phys. **60**, 697 (1974)
5.35 O.V. Krylov: in *Problemy khimicheskoi kinetiki* (Problems in Chemical Kinetics), ed. by G.K. Boreskov (Nauka, Moscow 1979) p.245
5.36 M.U. Kislyuk, I.I. Tret'yakov, V.N. Korchak: Kinet. Catal. (Engl. Transl.) **17**, 838 (1976)
5.37 J.H. McCreery, G. Wolker: J. Chem. Phys. **63**, 4073 (1975)
5.38 G.A. Somorjai: *Chemistry in Two Dimensions. Surfaces* (Cornell U. Press, Ithaca 1981)
5.39 G.A. Somorjai: J. Colloid Surf. Sci. **58**, 150 (1973)
5.40 R.L. Palmer, J.N. Smith: Catal. Rev. **12**, 279 (1975)
5.41 R.J. Madix, J.A. Schwarz: Surf. Sci. **24**, 264 (1971)
R.J. Madix, R. Parks, A.A. Susu: Surf. Sci. **24**, 264 (1971)
5.42 G. Ertl: Surf. Sci. **89**, 525 (1979)
5.43 K.C. Jarda, J.E. Hurst et al.: Surf. Sci. **131**, 273 (1983)
5.44 R.J. Madix: J. Vac. Sci. Technol. **13**, 253 (1976)
5.45 S.L. Bernasek, G.A. Somorjai: J. Chem. Phys. **62**, 3149 (1975)
5.46 M.J. Temkin: Adv. Catal. **28**, 173 (1979)
5.47 R.L. Palmer, J.N. Smith: J. Chem. Phys. **60**, 1453 (1974)
5.48 B. Poelsma, R.L. Palmer, S.T. De Zwart, G. Gomsa: J. Vac. Sci. Technol. A **1**, 1078 (1983)

T. Matsushima: Surf. Sci. **123**, L669 (1982); **127**, 403 (1983)
D. Manteil, S.B. Ryalli, G.L. Haller, J.B. Fern: J. Chem. Phys. **78**, 4250 (1983)
5.49 G. Ertl: in *Proceedings of the 7th International Congress on Catalysis*, Tokyo 1980, Plenary Lecture P2
5.50 O.V. Krylov: Kinet. Katal. **22**, 13 (1981) (in Russian)
J. Harris, B. Kasemo: Surf. Sci. **105**, L281 (1981)
5.51 B. Prada-Silva, D. Loffler, B.C. Halpern, G.L. Haller, J.B. Fenn: Surf. Sci. **83**, 453 (1979)
5.52 M.E. Ryskin: Probl. Kinet. Katal. **17**, 18 (1978)
5.53 G. Ehrlich, C.N. Stuart: J. Chem. Phys. **62**, 4672 (1975)
5.54 H.R. Winters: J. Chem. Phys. **64**, 3495 (1976)
5.55 N.G. Basov, Z.M. Belenov, V.A. Isakov, Yu.S. Leonov, E.P. Markin, A.N. Oraevsky, V.I. Romanenko, N.B. Ferapontov: Zh. Eksp. Teor. Fiz. Pis'ma Red. **22**, 221 (1975) (in Russian)
5.56 A.V. Khmelev, V.V. Apollonov, V.T. Borman, V.I. Nikolaev, A.A. Sadykin, V.I. Troyan, K.N. Firsov, B.A. Frolov: Kvantovaya Elektron. **4**, 2271 (1971) (in Russian)
5.57 G.K. Boreskov: Kinet. Catal. (Engl. Transl.) **14**, 5 (1973)
5.58 R.R. Reeves, G.G. Mannella, P. Harteck: J. Chem. Phys. **32**, 946 (1960)
5.59 G.G. Mannella, R.R. Reeves. P. Harteck: J. Chem. Phys. **33**, 636 (1960)
5.60 P. Harteck, R.R. Reeves, G.G. Mannella: Can. J. Chem. **38**, 1648 (1960)
5.61 M.P. Weinraub, G.G. Mannella: J. Chem. Phys. **51**, 4973 (1969)
5.62 A.A. Tabachnik, B.R. Shub: Probl. Kinet. Katal. **17**, 59 (1978)
A.A. Tabachnik, B.R. Shub: Khimicheskaja Fizika (Chemical Phys. Russ.) **9**, 1242 (1983)
5.63 E.A. Ogryzlo, A.E. Pearson: J. Phys. Chem. **72**, 2913 (1968)
5.64 G.G. Mannella, P. Harteck: J. Chem. Phys. **34**, 2177 (1961)
5.65 A.M. Winter, K.D. Bayes: J. Phys. Chem. **70**, 302 (1966)
5.66 S.J. Arnold, M. Kubo, E.A. Ogryzlo: Adv. Chem. Ser. **77**, 138 (1968)
5.67 M.E. Ryskin, B.R. Shub: React. Kinet. Catal. Lett. **17**, 41 (1981)
M.E. Ryskin, B.R. Shub, A. Pavliczek, Z. Knor: Chem. Phys. Lett. **99**, 140 (1983)
5.68 K.N. Spiridonov, O.V. Krylov: Probl. Kinet. Katal. **16**, 7 (1975) (in Russian)
5.69 Yu.N. Rufov, A.A. Kadushin, S.Z. Roginsky: Dokl. Akad. Nauk SSSR **171**, 905 (1966) (in Russian)
5.70 Yu.N. Rufov: Probl. Kinet. Katal. **16**, 212 (1975) (in Russian)
V.T. Coor: Surf. Sci. **88**, L42 (1979)
5.71 Yu.N. Rufov, V.P. Sakun: Khimicheskaja Fizika (Chem. Phys. Russ.) **4**, 435 (1983)
5.72 B. McCarroll: J. Phys. Chem. **50**, 4758 (1969)
5.73 N.K. Saadeh, J. Olmsted: J. Phys. Chem. **79**, 1285 (1975)
5.74 V.P. Sakun, Yu.P. Ruffov, I.V. Alexandrov, V.I. Vladimirova, A.N. Il'ichev: Kinet. Catal. (Engl. Transl.) **20**, 340 (1979)
5.75 A.N. Il'ichev, Yu.N. Rufov: Kinet. Catal. (Engl. Transl.) **20**, 356 (1979)
5.76 V.V. Styrov, D.P. Popov: Zh. Prikl. Spektrosk. **4**, 749 (1973) (in Russian)
5.77 L.A. Kappers, R.L. Kroes, E.B. Hensley: Phys. Rev. B1, 4151 (1970)
5.78 M. Breusse, B. Claudel, M. Guenin, L. Faure: Chem. Phys. Lett. **30**, 149 (1975)
5.79 B.M. Antipenko, A.V. Dmitruk, V.R. Zubkova, G.O. Karapetyan, A.A. Mak: Izv. Akad. Nauk SSSR, Ser. Fiz. **37**, 466 (1973) (in Russian)
5.80 S.Z. Roginsky: *Elektronnye yavleniya v geterogennom katalize* (Electronic Phenomena in Heterogeneous Catalysis) (Nauka, Moscow 1975) p.138
5.81 V.A. Sokolov, V.D. Khorunzhy, V.V. Styrov: in *Spektroskopiya kristallov* (Spectroscopy of Crystals), ed. by N.S. Neporent (Nauka, Moscow 1975) p.295
5.82 F.F. Volkenstein, V.A. Sokolov, A.P. Gorban': *Luminestsentsiya i adsorbtsiya* (Luminescence and Adsorption) (Nauka, Moscow 1975)

5.83 A.A. Kovalevsky, B.R. Shub: in *Proceedings of the 3rd International Symposium, Heterogeneous Catalysis, Varna 1975* (BAS, Sofia 1978) p.138
5.84 V.I. Vladimirova, Yu.N. Rufov, O.V. Krylov: Kinet. Catal. (Engl. Transl.) **18**, 675 (1977)
5.85 I.V. Krylova: Sov.Phys.-Usp. Khim. **45**, 1101 (1976)
M.P. Cox, J.S. Foord, R.M. Lambert, R.H. Prince: Surf. Sci. **129**, 399 (1983)
5.86 J.P. Daugnet, J.P. Verhalgen, J. van Cokenbergen: Nature **223**, 825 (1969)
5.87 I.N. Bakulina, N.M. Blashenkov, G.Ya. Lavrent'ev, V.I. Paleev, B.N. Shustov: Zh. Tekh. Fiz. Pis'ma **1**, 170 (1975) (in Russian)
5.88 R.B. Woodward, R. Hoffmann: *The Conservation of Orbital Symmetry* (Academic, New York 1969)
5.89 F.D. Mango: Adv. Catal. **20**, 291 (1969)
5.90 O.V. Krylov, F.F. Kiselev: *Adsorption and Catalysis on Transition Metals and Their Oxides* (Springer, Berlin, Heidelberg) to be published
5.91 S.A. Surin, A.D. Shuklov, B.N. Shelimov, V.B. Kazansky: Probl. Kinet. Katal. **17**, 80 (1978) (in Russian)
5.92 R. Ugo: in *Proceedings of the 5th International Congress on Catalysis*, Vol.1 (North-Holland, Amsterdam 1973) p.19
5.93 S.Z. Roginsky, I.I. Tret'yakov, A.B. Shekhter: Zh. Fiz. Khim. **23**, 50 (1949) (in Russian)
5.94 J.W. May: Adv. Catal. **21**, 152 (1970)
5.95 Yu.K. Ustinov: Zh. Tekh. Fiz. **41**, 411 (1971) (in Russian)
5.96 Yu.K. Ustinov, R.Sh. Urazaev: Sov. Phys.-Solid State **20**, 597 (1978); Zh. Tekh. Fiz. **49**, 856 (1979) (in Russian)
5.97 E.Ya. Zandberg, E.V. Rut'kov, A.Ya. Tontegode: Zh. Tekh. Fiz. **46**, 210 (1976)
5.98 W. Göpel: Ber. Bunsenges. Phys. Chem. **80**, 481 (1976)
T. Takachi, A. Ebina: Appl. Surf. Sci: **11/12**, 268 (1982)
K. Jacoby: Surf. Sci. **132**, 1 (1983)
5.99 T. Engel, G. Ertl: Adv. Catal. **28**, 1 (1979)
5.100 T. Engel, G. Ertl: J. Chem. Phys. **69**, 1267 (1978)
5.101 B.V. Rozentuller, K.N. Spiridonov, O.V. Krylov: Dokl. Akad. Nauk SSSR **259**, 859 (1981) (in Russian)
5.102 A.Ya. Rozovsky: in *Khimiya Poverchnosti okisnykh katalizatorov* (Surface) Chemistry of Oxide Catalysts), ed. by A.Ya. Rozovsky (Nauka, Moscow 1979) p.128
5.103 A.V. Shestakov, V.A. Matyshak, A.A. Kadushin, O.V. Krylov: Kinet. Katal. **20**, 189 (1979) (in Russian)
5.104 E.L. Aptekar', O.V. Krylov: *Vsesoyuznaya konferentsiya po mekhanizmu geterogennogo kataliza* (All-Union Conference on the Mechanism of Heterogeneous Catalysis) (Nauka, Moscow 1979)
5.105 A.D. Berman, O.V. Krylov: Probl. Kinet. Katal. **17**, 102 (1978) (in Russian)
5.106 A.D. Berman, O.V. Krylov: Dokl. Phys. Chem. (Engl. Transl.) **227**, 201 (1976)
5.107 B. Delmon: *Introduction à la Cinetique Heterogene* (Technip, Paris 1969)
5.108 W.J. Dunning: in *Nucleation*, ed. by A.C. Zettlemoyer (Marcel Dekker, New York 1969) p.1
5.109 A. Nitzan, P. Ortoleva, J. Ross: J. Chem. Phys. **66**, 3650 (1977)
5.110 J. Heicklen, M. Luria: Int. J. Chem. Kinet. Symp. **1**, Proc. Symp. Chem. Kinet., Virginia 1975, p.567
5.111 J.B. Homer, A. Protheroi: Trans. Faraday Soc. 69, 673 (1973)
5.112 G.K. Boreskov: Kinet. Katal. **13**, 543 (1972) (in Russian)
5.113 S.V. Vasil'ev, S.I. Khudyakov, A.I. Vol'pert: Zh. Vychisl. Mat. Mat. Fiz. **3**, 683 (1973) (in Russian)
5.114 M. Kobayachi, Y. Maeda, N. Takahashi: J. Chem. Technol. Biotechnol. **33**A, 219 (1983)

5.115 B.L. Clark: J. Chem. Phys. **60**, 1481 (1974)
5.116 G. Nikolis, J. Portnoy: Chem. Phys. **73**, 365 (1973)
5.117 A.K. Jain, R.R. Hudgirs, P.L. Silveston: Can. J. Chem. Eng. **61**, 46 (1984)
5.118 O.V. Krylov: in *Proceedings of the 3rd International Symposium on Heterogeneous Catalysis, Varna 1975* (BAS, Sofia 1978) p.17
5.119 M.G. Slynko: Kinetics and Catalysis **24**, 803 (1983)

Subject Index

MIX
Papier aus verantwortungsvollen Quellen
Paper from responsible sources
FSC® C105338

If you have any concerns about our products,
you can contact us on
ProductSafety@springernature.com

In case Publisher is established outside the EU,
the EU authorized representative is:
Springer Nature Customer Service Center GmbH
Europaplatz 3, 69115 Heidelberg, Germany

Printed by Libri Plureos GmbH
in Hamburg, Germany